With This Ring

Divorce, Intimacy, and Cohabitation from a Multicultural Perspective

CONTEMPORARY STUDIES IN SOCIOLOGY, VOLUME 19

Dedication

S.L.B

To my parents, Willie and Bertha. Their 50-year marriage has been an inspiration to me.

R.R.M

To my family and my colleagues—in other words, to all those who had to put up with me.

With This Ring

Divorce, Intimacy, and Cohabitation from a Multicultural Perspective

Edited by R. ROBIN MILLER
Department of Behavioral Sciences
Drury University

and

SANDRA LEE BROWNING
University of Cincinnati

JAI PRESS INC.
Stamford, Connecticut

Library of Congress Cataloging-in-Publication Data

With this ring : divorce, intimacy, and cohabitation from a multicultural perspective / edited by R. Robin Miller and Sandra Lee Browning.

p. cm. -- (Contemporary studies in sociology ; v. 19)

Includes bibliographical references and index.

ISBN 0-7623-0569-X

1. Marriage–Cross-cultural studies. 2. Marriage law–Cross-cultural studies. 3. Divorce–Cross-cultural studies. 4. Unmarried couples–Cross-cultural studies. I. Miller, R. Robin. II. Browning, Sandra Lee. III. Series.

HQ518.W55 1999
306.81--dc21 99-088941
CIP

100 Prospect Street
Stamford, Connecticut 06901-1640

ISBN: 0-7623-0569-X

Library of Congress Catalog Number: 99-088941

Manufactured in the United States of America

CONTENTS

LIST OF CONTRIBUTORS

Suzanne M. Baker	Department of Sociology and Anthropology Creighton University
Sandra Lee Browning	Department of Criminal Justice University of Cincinnati
Wendy Y. Carter	Department of Sociology Arizona State University
Jui-shan Chang	Sociology Program, Department of Political Science University of Melbourne, Australia
Kate Davidson	Department of Sociology University of Surrey, UK
Susan De Vos	Department of Sociology University of Wisconsin-Madison
V. R. Dorjahn	Department of Anthropology University of Oregon
Aspasia Tsaoussis-Hatzis	Doctoral Candidate in Sociology of Law University of Chicago Currently teaching at American College of Thessaloniki, Greece
Lawrence K. Hong	Department of Sociology California State University, Los Angeles
Noriko Iwai	School of Commerce and Economics Osaka University of Commerce
Jae Kyung Lee	Department of Women's Studies Ewha Womans University, Seoul, Korea

Anna Mikheyeva — Institute of Economics and Industrial Engineering, at Siberian Branch of the Russian Academy of Sciences, Russia

R. Robin Miller — Department of Behavioral Sciences, Drury University

Helen E. Ullrich — A Cross-Cultural Psychiatrist in Private Practice, Tulane University Medical School

Barbara H. Vinick — Department of Social and Behavioral Sciences, Boston University School of Public Health

Paul Vlaardingerbroek — Tilburg University, the Netherlands; Honorary Judge, Court of Appeal in 's-Hertogenbosch

Deborah Woo — Department of Community Studies, University of California, Santa Cruz

ACKNOWLEDGMENTS

First and foremost we wish to thank our contributors for both their effort and their patience. We were continually surprised and delighted by the understanding extended by each—on a project that lasted just short of three years. They have our heartfelt respect and appreciation! Of course, this book would not have been possible without the moral support of our loved ones—special thanks to Rodney Coates, and Jo and Ren Miller. We also wish to thank a group of people without whom the project would have died: Louise Jacobs for her editorial work; Lindsay Claire for her production work; Linda Mercer for her work on the text; and Michael Helms for his invaluable work on so many stages of the project.

PREFACE

"With This Ring . . ." is a phrase that conveys an array of meanings. Vows accompanying the exchange of wedding rings can include: "With this ring I thee wed, with my body I thee worship, and with all my worldly goods, I thee endow" (Protestant); "Behold you are consecrated unto me with this ring, according to the law of Moses and Israel" (Jewish); and "I give this ring to you, my beloved, my friend, my wife/husband" (United Church of Christ).

While the symbolic nature of the exchange of rings is vested with widely accepted meanings, the origins of the use of them is little known. Both ancient Egyptian and Roman men reportedly gave rings as love tokens—a practice that evolved into the use of wedding rings worn, most commonly, by women (the "double-ring" ceremony is a relatively new convention). The specific use of diamonds set in wedding rings dates back to the time of Emperor Maximilian of Austria (AD 1477), and before whole veins of diamonds were found in Africa (approximately AD 1870) they were used almost exclusively by the aristocracy (Kunz 1973). Since that time, however, diamonds have become the most common choice of newlyweds, and the practice of wearing wedding rings is now normative in many cultures.

With This Ring is a collection of research and review papers that seeks to examine specific marital processes from the perspective of diverse cultures. In particular, this anthology will focus on law and divorce, intimacy, and cohabitation/alternative marital forms.

ISSUES OF DIVORCE

Given increasing rates of divorce in many nations in recent years, marital status has become a more important variable in describing the wellbeing of the individual and the nature of the social structure in a community (Anspach 1976; Ihinger-Tallman and Pasley 1986; Hurlbert and Acock 1990). As large numbers of individuals experience divorce, questions about how people can and will interact with those around them and what they can expect from their larger community become more significant.

In addition, divorce changes how society views the couple. Vaughan (1986) suggests that divorce is not a chaotic process but actually occurs in a uniform way whereby individuals revise their identities in order to separate from the couple union. While the divorce process changes the identities of divorcing individuals, it also changes the relationships these individuals maintain with others. With the breakup of that union comes changes in social networks (Milardo 1987). Since the marital union is reinforced by society, by extension, marital breakup affects the entire social fabric.

With an anthology of this type we can encourage the examination of changes in law around the globe that affect marriage. We can look at cultural and historical differences, and we can speculate on what appear to be patterns of change. We can look critically at common myths surrounding divorce: that divorce is uncommon in other cultures; that divorce is a new invention; and that divorce follows one set pattern regardless of where it is found.

ISSUES OF INTIMACY

Research suggests that the marital relationship creates a marital identity where the husband and wife together form a whole. Berger and Kellner (1975) suggest that each individual needs to feel validated by society and that this sense of affirmation emerges from his or her interactions with others. But the process by which individuals come together to construct the marital identity is complex. With the husband and wife, we have the integration of two different sets of conversations. We have the overlapping of two social circles, and the integration of at least two different networks of relationships. From this comes the construction of a joint biography which forms the marital identity (Berger and Kellner 1975).

Intimacy in marriage is formed (or not) in this difficult context. If culture shapes the institution of marriage in general, it is equally certain that culture shapes our intimate desires and behaviors. Communication patterns between partners, the degree of intimacy that is commonly expected, and cultural and structural roadblocks to intimacy are explored from the perspective of diverse cultures here.

ISSUES OF COHABITATION AND ALTERNATIVE MARITAL FORMS

While the marital union is supported and reinforced by society (Berger and Kellner 1975), the cohabiting couple is often stigmatized (Blumstein and Schwartz 1983). Largely ignored is the fact that cohabiting couples often marry (Macklin, 1983). Without an examination of the variations that are possible across cultures, the understanding that cohabitation may be a part of the marital process is never developed.

"Among the constantly recurring needs which each group must meet is to regulate the sexual behaviors of its members, provide for the safety and security of children, and promote emotional stability" (Henslin 1989, p. 4). Culture filters and shapes how these needs will be met—hence how the institutions of marriage and family will appear. Given different societal histories, marriage can look almost radically different from one culture to the next. For instance, cohabitation can be normative, and in some cultures, alternative unions (including cohabitation) have, in part, replaced marriage sanctioned by the state.

This text, and the research it contains, is useful because it empowers us to reject some of the flawed notions that many Americans take as common understanding: that cohabitation and other alternative marriage forms are new; that cohabitation occurs only in the United States; that these alternative forms are harmful to families; and that families that operate in these forms are necessarily dysfunctional.

ABOUT THIS BOOK

While there have been a multitude of texts on marriage, they have primarily focused on marriage in one specific culture. The "American Family," for instance, has been the topic of many recent books. However, it is a mistake to operate under the assumption that there exists one normative pattern of marriage from which other forms are to be regarded as mere deviations. Because the institution of marriage is given shape by the culture in which it is found, cultural variations and practices are vitally important. Works that pull together research from many cultures are clearly needed.

Exacerbating the nearly overwhelming difficulties presented by the dearth of cross-cultural research is the problem presented by the lack of real communication between disciplines. Many disciplines within the social sciences study marital processes, but from slightly different perspectives and using different terminologies. Respecting the socially constructed boundaries between closely related disciplines serves to cloak the fact that many of the same phenomena are being studied—and that improved communication is critical to better understanding.

Given the necessity of bridging the largely artificial gaps between the disciplines, we are delighted to be part of a project that includes works from sociologists, anthropologists, demographers, a gerontologist, a criminologist, and an honorary judge. In addition, other departments represented in this anthology include departments of economics, social policy, women's studies, law, and community studies.

This volume brings together original research and review articles on marital processes organized around a common set of issues and themes—law and divorce, intimacy, and cohabitation/alternative marital forms.

REFERENCES

Anspach, D. F. 1976. "Kinship and Divorce." *Journal of Marriage and the Family* 37: 323-330.

Berger, P. and H. Kellner. 1975. "Marriage and the Construction of Reality." Pp. 219-233 in *Life as Theatre,* edited by D. Brisett and C. Edgley. New York: Aldine De Gruyter.

Blumstein, P. and P. Schwartz. 1983. *American Couples: Money, Work, Sex.* New York: William Morrow.

Henslin, J., ed. 1989. *Marriage and Family in a Changing Society,* 3rd ed. New York: The Free Press.

Hurlbert, J. S. and A. C. Acock. 1990. "The Effects of Marital Status on the Form and Composition Of Social Networks." *Social Science Quarterly* 71: 163-174.

Ihinger-Tallman, M. and K. Pasley. 1986. "Remarriage and Integration Within the Community." *Journal of Marriage and the Family* 48: 395-405.

Kunz, G. F. 1973. *Rings For the Finger.* New York: Dover.

Macklin, E. D. 1983. "Nonmarital Heterosexual Cohabitation." Pp. 264-265 in *Family in Transition,* edited by A. S. Skolnick and J. H. Skolnick. Boston: Little, Brown.

Milardo, R. M. 1987. "Changes in Social Networks of Women and Men Following Divorce." *Journal of Family Issues* 8: 78-96.

Vaughan, D. 1986. *Uncoupling: Turning Points in Intimate Relationships.* New York: Oxford University Press.

PART I

LAW, DIVORCE, AND MARRIAGE

PART I

INTRODUCTION

In any given society there are mechanisms, both legal and cultural, for dealing with unsuccessful marriages. Individual rights, particularly women's rights under law, have experienced rapid change in recent times. Papers in this first part explore the intersections among law, marriage, and culture. Helen Ullrich discusses changes in law that have only begun to penetrate marriage custom in India in recent years. Her paper, entitled "A Thirty-year Perspective on Hindu Marriage and the Law: Cultural Confirmation of Change," examines the influence of increased female status on marital processes in conjunction with the decreased observation of menstrual taboos, an increased emphasis on education, and later ages at first marriage.

Jae Kyung Lee also explores recent changes in law in her paper, "Dual Ideological Structure of the Korean Family Law." The 1958 Family Law, which was added to Korea's Constitution, contained many provisions actually opposing equality between the sexes. Her research details these changes in marriage and family law, and suggests feminist alternatives.

In addressing the issue of unsuccessful marriages, V. R. Dorjahn examines "Divorce Among the Temne of Sierra Leone (West Africa): Alleged Causes, Effects of Polygyny, and Timing." His research investigates the degree to which polygyny (marriages where men have more than one wife) increases the probability of divorce and whether divorce is more likely to involve the first wife or subsequent wives. Also of interest in this paper is the peak number of years couples are married before divorce occurs.

Noriko Iwai also discusses factors increasing the probability of divorce throughout history in "Divorce in Japan: Historical Changes and Current Issues." This study finds that occupation, wives' employment, and presence of dependent children affect the incidence of divorce. In addition, changes in law have affected likelihood of divorce, child custody, visitation, financial provisions, and ease of remarriage.

The last paper in this part, Aspasia Tsaoussis-Hatzis's work, "Changes in Greek Marriage and Divorce Law: The Impact of the Family Law Reform of 1983," details the ways in which Greek marriages have diverged from common patterns found in industrial societies. Various impacts of the new Family Law in Greece are reviewed, finding that this law was progressive and responsive to the needs of women and children. Changes in law, including laws regarding marriage and divorce, are far-reaching and difficult to track—further investigations are indicated.

A THIRTY-YEAR PERSPECTIVE ON HINDU MARRIAGE AND THE LAW:

CULTURAL CONFIRMATION OF CHANGE

Helen E. Ullrich

INTRODUCTION

"Till Death Do Us Part" aptly described Havik Brahmin marriage in the South India village, Totagadde. Men could remarry less than two weeks after a wife's death. However, remarriage for a widow was culturally proscribed (Ullrich 1988). The children of those widows who remarried belonged to a special caste for illegitimate Brahmin children. On my first field trip to India, two women in Totagadde were treated as outcasts because they were children of widows. One was the daughter of a virgin widow whose second marriage was to a Havik Brahmin; the second, the illegitimate daughter of a widow who never remarried. In the second case, death ceremonies had been observed when the widow became pregnant. Although the death ceremony for the remarried widow occurred only after her death, Totagadde villagers regarded her daughter as illegitimate. Since the late 1960s, Havik Brahmin villagers reported that the law prevented outcasting. However they treated the daughter of a person whom they believed should have been outcast as an outcast. In this paper, I propose to discuss the impact of increased female status on the marriage system which is congruent with decreased observation of menstrual taboos, increased education, and changes in the legal system.

In 1969, the year I began field work in South India, Lévi-Strauss published *The Raw and the Cooked*. One South Indian informant's explanation of the origin of menstruation exemplifies Lévi-Strauss's dual characterization of nature versus culture (Lévi-Strauss 1969, p. 319):

> . . . initially there was no menstruation. Following an argument between the gods, Indra and Brahma, Indra cursed all humans with menstruation known as Brahma's curse *(brahmado:Sa*[1]). Men meditated and followed strict religious practices *(tapas)* and so escaped the curse. As women neither meditated nor followed strict religious practices, they experience four nights when they are in a ritually polluting state. Hence women remain cursed for their lack of discipline and ritual practices. (Ullrich 1992, p. 28)

Thus, menstrual taboos symbolized the failure of these Hindu women to transcend their nature. Accordingly, these Havik Brahmin women were too ritually impure to perform religious ceremonies or even to learn the mantras (sacred formulas) for religious ceremonies. In contrast, men achieved their status as religious, economic, and political leaders.

In the interval between 1965 and 1995, Havik Brahmin women found increased opportunities for achievement. Although the Havik Brahmins are the dominant caste ritually, economically, and politically, the women traditionally had little power. Indeed, in 1965 women described themselves as belonging to a lower caste *(ki:Lu ja:ti)*. Their fathers and male elders in the household held absolute decision-making power over all its members. By 1995, unmarried women could make their own decisions, a reflection of their increased autonomy and self-confidence. Moreover, they maintained this autonomy and self-confidence after marriage.

During this same interval, women also decreased their observation of and belief in menstrual ritual. Women who had been socialized to inferiority through menstrual taboos transcended traditional restrictions through education or urban residence, or both. Cultural emphasis for women went from the raw to the cooked, from the biological to the social, and from ascribed to achieved status. In their transition from ascribed to achieved status, Havik Brahmin women came to assume professional roles parallel to those of Havik Brahmin men. In this sense the society ceased to regard women as perpetually childlike, in need of protection by the traditional patriarchy. Along with the decrease in menstrual taboos have come increased educational and professional opportunities, which has led to increased autonomy. Thus, women have gone from deferring to their husbands as the sole decision makers to partnership in a nuclear family in which the husbands consult their wives. Moreover, women may be the prime decision makers for ritual observation, household management, and their own professional development. This paper examines changes noted in eight trips from 1964 to 1995.

RITUALS

In 1964, ceremonies from the first ritual haircut through the sacred thread ceremony *(upina:yana)* prepared men for leadership roles. Ritual impurity represented by menstrual taboos provided the stated rationale for the exclusion of women from the sacred. The ceremonies that included women focused on the family. The two rituals exclusively for women concerned reproductive capacity, namely menarche and the return of the wife to her maternal home during her first pregnancy. Other ceremonies that involved women also involved men. These included the rites of marriage and the welcoming of a newborn into the home.

Young women in the 1980s actually persuaded their male age mates to teach them the *ga:yatri,* the sacred formula that fathers teach sons at the *upina:yana.* The *upina:yana* is the ceremony in which a male achieves adult ritual status. At that ceremony his father and a priest teach the young man the ritual for worship *(pu:ja).* In 1994, some women even performed the daily worship *(pu:ja)* instead of their husbands or sons, while some men had ceased the daily worship *(pu:ja).* Data on women who perform worship *(pu:ja)* were difficult to elicit. However, a comment frequently heard in 1995 was that young men no longer learned the mantras for proper worship. They just "go through the motions."

Theoretically women are unable to acquire sufficient ritual purity to perform the worship and to learn the appropriate mantras. Nevertheless, I observed women as they sat at worship *pu:ja.* Only one woman allowed photographs while she performed *pu:ja.* In contrast, in 1964, no matter how late a man returned home, women did not serve the noon meal until he had performed worship *(pu:ja).* In 1987, some women performed the *pu:ja* when their husbands were late while other women preferred to do *pu:ja* and indicated that the men in the household were uninterested in performing it.

Thirty years ago, a family needed two sons to commemorate the ancestors. At that time some used the term "barren" to mean that a woman had no children or had only daughters. The term *ka:ge banje* (crow barren) referred to a woman with only one child—possibly only one of the two requisite sons. However, after couples chose to limit their families to one son, the term *ka:ge banje* (crow barren) dropped out of common usage. By 1994, couples with more than three daughters who planned to continue to try for a son had become a target of gossip and criticism. In one case, the woman's brothers reprimanded their brother-in-law for social and financial irresponsibility. In this case, the man's in-laws criticized his plan to strive for a son. My queries about the ritual importance of sons met with silence in 1995.

ASCRIBED STATUS VERSUS ACHIEVED STATUS

Although culture transforms everyone, it can be said that some are more transformed than others, or at the very least that cultural belief reflects a society in which some persons acquire greater status than others, despite similar efforts or accomplishments. Havik Brahmin men acquired the ritual purity requisite for worship that they performed for the benefit of the entire village as well as daily worship *(pu:ja)* for their families. Attainment of advanced education and employment in the professions changed ritual patterns. Young women in 1964 had a middle school education and no access to economic independence. By 1995, women focused on superior grades so that they could gain admittance to professional school. When education and professional accomplishments replaced ritual purity as a source of status, women transcended the limitations of ascribed status.

Ascribed status, in contrast to achieved status, reflects the innate imprint of a culture on its members. In contrast, achieved status may reflect the power of individual efforts over cultural limitations. In 1995, I noted a Havik Brahmin mother with her college student daughter. The mother observed menstrual ritual that her daughter had never observed and had no intention of observing. Other differences between this mother and daughter included educational level, assertiveness, aspirations, and resilience to adverse life events. The mother had a fifth-grade education, was hesitant to speak, had no professional options, and a history of depression and somatization (multiple physical complaints for which physicians discovered no etiology). Her daughter attended college, was outspoken, had plans to study medicine, and had alternate professional plans should she fail to qualify for medicine. The mother encouraged her daughter's assertiveness with the comment that all too soon her daughter would become passive *(sabya*[2]*)*. The daughter commented on her mother's intelligence and attributed her mother's reluctance to join conversations to self-consciousness about her lack of formal education. The mother herself exemplified the traditionally expected passivity *(sabya)*.

With the opportunity for educational and professional pursuits, Havik Brahmin women of Totagadde have become outspoken. The vocabulary describing women has also changed. The term, *jo:ru* (strong) has several meanings which paradoxically include assertive, outspoken, quarrelsome, and intelligent. The mother in the previously noted case would regard any person who labeled her as *jo:ru* as insulting her. In contrast, the daughter described herself as *jo:ru* in the sense of being intelligent, outspoken, and having the ability to take care of herself. For some *jo:ru* is a compliment; for others, an insult. The linguistic change to usage as a compliment reflects the cultural change to the expectation that women should protect themselves and cherish their own goals. Additionally incorporated into the meaning of this term is the ability to insist and to disagree with others in pursuit of an individual goal. With achieved status the rewards for accomplishment have replaced those for passivity *(sabya)*.

EDUCATION

> *"If one has no education, one is a cow." vidya vihimum pashu sa:ma:na.*
> *—A woman with a middle school education whose children and daughters-in-law all have college degrees*

Over the past 30 years, the culture in Totagadde has changed to accommodate women in educational and professional realms. Women now in their 50s and older had no such opportunities in their youth. However, they are the ones who currently verbalize their feelings of educational inferiority. In 1965, a husband criticized his wife when she gave a public lecture for the *mahila sama:j* (women's association). What was his criticism? Her poor command of literary Kannada, the appropriate Kannada dialect for public speeches, which showed her lack of education. The illiteracy of kinswomen, especially mothers, is a source of embarrassment. As such, older women have provided a strong motivating force for the education of their daughters (Ullrich 1994).

By 1984, a college-educated woman was the best public speaker in Totagadde and a source of village pride. As such, she spoke at many events. It was her mother who had provided me with the proverb. The mother has followed the menstrual taboos strictly, but her educated daughters-in-law of whom she is so proud have a rationale for ignoring the taboos—one works while the other lives in an urban area. Her daughter chose to marry a landholder with less formal education than she. The daughter is a hostess to her husband's prominent friends, a partner in the management of their land, and a mother. In the view of Totagadde Haviks, the daughter's education has prepared her for her many roles (Ullrich 1994).

There were some surprising positive associations with education. Increased education is associated with an increased percentage of the following: women with parents within one hour's travel, those with involvement in social activities, those reporting lasting relationships with college friends, those feeling useful, and those who think they know what is happening to friends and relatives. With increased education, a lower percentage of women reported that they receive monetary help or have people to run their errands. Moreover, a lower percentage of educated women reported loneliness. This is understandable in the context of their increased social and professional involvement.

Education has increased the geographical mobility of Totagadde Brahmins. In 1995, there were 22 adult Havik Brahmin men from 15 households living in urban areas. Eleven households had no sons residing in Totagadde. In five households, the sons had left Totagadde for permanent employment in urban areas. In the remaining six households there are only daughters, no children at all, or the mother has established her own household apart from her sons. Twenty-two daughters and 11 daughters-in-law reside in urban areas. Of these, 12 of the daughters and three of the daughters-in-law were employed. All of the

Table 1. Adult Havik Brahmin Population of Totagadde (1994)

	Adult Women	*Adult Men*	*Adult Men—Urban*
Total	63	48	22
> 65	11	8	0
40–69	25	25	3
20–39	24	13	16
College	3	2	3
Households	35	34*	15 Totagadde households

Note: * One household consists of a widow who partitioned her portion of the land from her son.

Table 2. Departures from Totagadde (1994)

	Women Married in Urban Areas	*Singe Adult Women-Urban*	*Professional Daughters-in-Law*	*Professional Daughters*
Total	22 daughters; 11 daughters-in-law	1 daughter	3 daughters-in law	12 daughters
> 65	0	0	0	0
40–69	0	1 daughter	0	1 daughter
20–39	22 daughters; 11 daughters-in-law	0	3 daughters-in-law	113 daughters
College	0	0	0	0
Households	12	1	3	7

daughters-in-law and 11 of the 12 daughters were in the age range between 20 and 39 years of age (Tables 1 and 2).

Despite the increased geographical mobility, the more highly educated the group of women the more likely they were to live within an hour's travel time of their parents. This finding suggests that family ties have remained stable or have even increased with women's education. Since education has enhanced a woman's marital prospects, she may have increased opportunities to find an appropriate husband close by. In contrast, one may need to look further away for grooms for uneducated women. The positive correlation of employment with satisfaction and the frequency with which one sees one's kin may reflect both decreased loneliness and efforts toward personal achievement.

Traditionally, the mother-daughter relationship has been a supportive, close relationship. From this central relationship a woman developed her model for other relationships. Daughters observed their mother's low status and heard mothers speak about the importance of education and economic independence. Some mothers with arranged marriages also openly expressed support for love marriages, perhaps as an antidote for their own misery when a husband or mother-in-law ignored them. However, in 1995, sons and daughters from even the most orthodox families had arranged their own love marriages. In 13 households, daughters or sons have arranged their own marriages. Fathers openly discussed the wisdom of accepting their child's marital choice rather than experience a period of friction and then reconciliation. Indeed, the other Havik Brahmins criticized the one father who refused to visit his daughter after her elopement. They visited her and supported her decision.

A girl's marriage is now arranged after she has completed her education. When a woman is more highly educated than her fiancé, he and members of his family attempt to ascertain whether the educational disparity bothers her. This is particularly the situation where the fiancé is a landowner. There are several cases of women given to Totagadde men as well as daughters of Totagadde who have chosen to marry less highly educated men. In eight of 35 households, a daughter or daughter-in-law is more highly educated than her husband. Mothers-in-law boasted about their daughters-in-law who have more education than their sons. In all of these cases the husband is a landowner rather than a professional.

Educated women are likely to have more choices for their mates. They are more likely to find a person who meets their ideals. These may refer to urban or rural living, a nuclear family, wealth, education, and choice of professional man or a landowner. They may have a greater choice at any age when they choose to marry. On the other hand, there is concern that both undereducation and overeducation will limit a woman's marital opportunities. Those who have too little education may be dull; or those with a Ph.D., for example, may wish a man with a similar level of education. Such a man may be hard to find.

Although married women are expected to obey their husbands, they do not necessarily follow this dictum. In the 1960s and 1970s, a husband might beat his wife for disobedience; now, a husband might ignore his wife's disobedience. In recent years, open disagreement between husband and wife has become common. I have heard mothers chastise their daughters for such outspokenness. When their daughters have complained to them, they have told the daughters to obey their husbands and discussed the rationale of the husband's requests. These same mothers also have expressed sympathy for their sons-in-law when their own daughters criticize or scold their husbands. However, these mothers also have supported their daughters-in-law's freedom of expression. These opinions are major indicators of how the patriarchy has changed over the past 30 years to allow for increased consultation with female members of the family.

As women became more educated, they began to question their inferior ritual position. Women from college students to widows now have more say over their own fates. Whereas in the past the fear of sexual harassment presented the patriarchy with a rationalization for keeping women at home, now college students report strategies to discourage unwelcome advances. A discretely used pin, for example, is one way to discourage inappropriate advances on public transportation.

Marriage now requires the woman's unequivocal consent. At the same time that an educated wife adds to her husband's status, she has increased autonomy related to her older age at marriage and her education and professional skills. When elders ignored the past emphasis on the ritual importance of a son, this further indicated the increase in female status. With their increase in economic and intellectual power, Havik Brahmin women have achieved the status to make their own decisions—whether in regard to ritual observation, professional accomplishment, or household management. Not surprisingly, women's autonomy has led to their questioning menstrual ritual and the ritual pollution associated with it.

MENSTRUAL OBSERVANCES

Menstrual ritual in 1964 stigmatized women and placed emphasis upon a woman's inferior ritual status. While menarche was a celebration that included gifts for the woman and a special dinner upon her return to ritual purity, it was also her initiation into ritual untouchability. Menstruating women observed extensive ritual taboos from social and physical exclusion to eating apart. They were the last to be served. The glance from a menstruating woman was considered to be so polluting that any food left after serving her had to be thrown away. At that time, if women attended school while in a ritually polluting state, they sat apart from their classmates, an indication that the ritual purity of classmates had priority over the woman's feelings of stigma and her desire for education. Accordingly, many women felt such shame that they stayed away from school for the duration of their ritual pollution. The period of menstrual pollution was the one time in a woman's life when she was isolated from others.

Ten years later, in 1974, female high school students no longer sat apart during their menses. However, upon returning home, they remained outside the house. These girls often spent their time at home crying about their ritual pollution. However, they avoided any mention of their state to classmates. Upon returning home, other schoolchildren removed their polluted clothing before entering the house. After all, there was no way to determine whether one's female classmates had transmitted pollution.

At that time menarche was strictly a family matter. I attended one menarche celebration by accident. I happened to make a morning visit and was included. However, until the invitation to stay for the ritual was extended, nobody had mentioned the event. Increased respect for the girl's feelings corresponded with her increased educational opportunities. High school became the minimum education formally acceptable for a good marriage match.

By 1984, male visitors sat on the same bench as a woman they knew to be menstruating. She could enter the house but not the area for worship. I doubt that young people even knew that the glance of a menstruating woman polluted food. Even if they had known, however, they were unlikely to have cared about the traditional ritual. I overheard one husband tell his wife that he did not believe in the taboos. Some older women openly discussed their change from belief to disbelief in menstrual taboos. Younger women ignored the ritual observation in college as did women living in the nearby town.

In 1994, some women indicated residents of another town no longer followed the taboos. However, 1995 was the first year some Totagadde women openly admitted that they ignored the ritual. Menarche was so silently noted that I was told in confidence about girls who had reached menarche. This suggests that menses has changed from a public ritual to private observation. However, even in 1995 an occasional orthodox father avoided visiting his daughter in town because he believed she had discontinued the observation of menstrual taboos. These fathers kept their own council and did not publicize reasons for avoiding a daughter's home in town.

The changes from 1994 to 1995 and the increased openness of discussion about ritual observation have provided an indicator of the ability of women to make their decisions in all arenas. Even the increased privacy of menarche is influenced by the increased openness. In the 1960s, women found out about menarche upon attaining it. Now they know before the event. Increased knowledge has contributed to decreased stigma and to the ending of menstrual stigma. Not only have women reconsidered this, but a highly respected psychiatrist, Ashok Pai, M.D., criticized the observance of menstrual taboos on his radio program. A respected Havik Brahmin headmaster and a female high school valedictorian have openly discussed the taboos. Open discussion has contributed to cultural endorsement for private observation, even the discontinuation of menstrual taboos.

MARRIAGE AND THE LAW: CULTURAL CONFIRMATION OF CHANGE

Law may reflect cultural change, but the connection between the legal system and behavior may be tenuous. Nevertheless, there have been dramatic changes in the law and in women's perception of the law since 1964. This extends from autonomy to age of marriage, remarriage, and land management. These are the four areas I will briefly discuss with regard to the law.

In 1964, *The Laws of Manu* reflected beliefs about the ideal behavior of women and their dependence upon their husbands and male relatives (Ullrich 1980, pp. 227-228):

> 146 Hear now the duties of women
>
> 147 By a girl, by a young woman, or even by an aged one, nothing must be done independently, even in her own house.
>
> 148 In childhood a female must be subject to her father, in youth to her husband, when her lord is dead to her son; a woman must never be independent.

149 She must not seek to separate herself from her father, husband, or sons; by leaving them she would make both [her own and her husband's families] contemptible.

151 Him to whom her father may give her, or her brother with the father's permission, she shall obey as long as she lives.

152 . . . The betrothal [by her father or guardian] is the cause [of the husband's] dominion [over his wife]. (Muller, 1967)

In 1964, women in Totagadde fled to the kitchen when guests came. Should the husband be absent, they might send a child to the door to indicate no one was home. When I asked a woman a question in the presence of her male relatives, the male relatives answered. One mother-in-law came to ask me why I bothered her daughter-in-law so. Should the daughter-in-law need to make a decision, the husband would make the decision. Husbands proudly indicated they were doing their wives a favor by allowing them to remain at home.

By 1975, a few women took college courses by correspondence, Ten years later, women were attending college. These young women traveled alone to the college in the nearby town. Since then others have gone even further away to college. By this time, *The Laws of Manu* had become obsolete—with one exception. That exception occurred when a mother wished to make the point that her daughter should be respectful and obey her husband. The mother regarded that as a marital obligation which the daughter ignored. The daughters placed respect into a category of something earned and, indeed, most had husbands whom they loved and respected. However, the wives were the ones who decided when and what to ask a husband. One example was a family who had many visitors at the same time. As her husband is a willing helper and I knew she had been up half the night washing dishes, I asked her why she did not tell her husband of her distress. Her response was that he was very busy in his work. The guests, his relatives, were on their first visit with the couple. She decided out of consideration for his needs to avoid more demands on his time.

By 1995, the elders proclaimed to me that men needed wives who could assist them in their work and in their entertaining. In short, they needed partners. While women gave up their belief in *The Laws of Manu* first, younger men followed. Finally, the entire Havik society has placed value on a wife who will be a partner rather than a subordinate.

The second topic is age of marriage. When I initially did field work, women had been married around puberty. However, marriages among Havik Brahmins at that time were occurring about age 18. This age for marriage is in accord with the Special Marriage Act, 1954 (ICSSR, 1975).

Ten years later, a wealthy suitor came to see a beautiful high school graduate. The woman told her parents she wanted to wait some time before her marriage. She hid in her attic while her family told him she had gone to visit relatives. He gave his horoscope and the girl's parents sent word that the horoscopes failed to tally. This incident created some difference of opinion among the village men.

Older men thought the girl's father was a fool to let such a wealthy suitor get away. But men with daughters who were recent graduates commented with wonder, "She is so young. She is too young to marry." This example serves to indicate how the 1954 law became custom. Yet not every group in Totagadde is ready to endorse marriages for women who are a minimum of 18 years old. As recently as 1985, a non-Brahmin bride in Totagadde proudly told me her age—14. Her worried mother-in-law quickly came out to tell me the bride was really 18. Meanwhile, the Havik Brahmin women regard as the ideal age of marriage the age when a woman has completed her education—anywhere from 18 to 25.

The third topic is remarriage. The three types of remarriage include remarriage for a widow, remarriage for a divorced woman, and polygamy. Although men have a long tradition of remarriage, women have only recently begun to remarry. In 1964, one widow had remarried after her husband had committed suicide. This was before I arrived in Totagadde, so I know little more than that the Havik women thought her remarriage appropriate, as she was both young and without children. In 1975, one of the beauties of the area married. Shortly thereafter her husband died from a cobra bite. She campaigned against marriage because she felt widowhood to be so punitive. At that time widows, especially those with children, did not remarry. They were set apart from society as a symbol of bad luck. By 1995, the general view was that women should have the option to remarry.

From my first trip, I noted that women recounted that one man in a nearby village kept marrying a second wife. After the older wife committed suicide, the man again remarried. These Havik women reported to me that bigamy is legal if the first wife gives her permission, but stated that they would commit suicide if their husbands took a second wife. This tale indicates the strength with which these Havik women opposed polygamy. Although they may have been aware that bigamy is illegal, they felt helpless in asserting their own wishes. Moreover, they would not have been in a position to bring a lawsuit.

Although the Hindu Marriage Act of 1955 prohibited bigamy (ICSSR 1975, pp. 41-42), these Havik Brahmin women believe bigamy is legal if they give their permission. In 1985, one Havik originally from a neighboring village had two wives. When the first wife was mute after the marriage, the father-in-law apologized for giving her in marriage and promptly offered his next daughter in marriage. There has been no divorce. The trio lived in a distant area of Karnataka State. Hence Totagadde people can ignore this situation.

Divorce has become more accepted than bigamy. If a woman has an abusive husband, people consider a divorce better. This is true only if the woman has an education and a way of supporting herself. A woman without an education may find herself stuck in an unhappy marriage. Divorce, however, is an option for dire circumstances. The only divorce in the area caused much distress for the new bride. Her husband declared he had married a man. Although the charge was absurd, this seemed rationalization for a divorce. He wanted his wife to serve alcohol and meat to his business associates in their urban home. She was a strict vegetarian and opposed to alcoholic beverages. Moreover, women who were approximately the bride's age thought that he had forced himself on her. She was such a gentle woman that after traumatization she would have avoided him. Totagadde men thought that the groom would choose a physician to examine her. With bribery, the physician would report that anatomically, she was a he. After the divorce, she became a banker and then found a husband with whom she shares many interests—including their children. This illustrates her ability to maintain her own belief system and the extent to which the husband felt he had to go for a divorce. Her successful career and excellent marriage have provided an example that a divorce is better than an abusive marriage.

The last topic, land management, shows a long evolution with a recent change in inheritance laws to give women an equal portion of the land with their brothers. Initially, one woman with an improvident husband sued for the management and the right to the land until her sons reached an age where they could manage the land. The next landmark was a professional man who had his wife manage the Totagadde land. The third situation was a woman who sued her brothers for her portion of the land. She won, but her brothers no longer visit her. Since then, some fathers have decided to earn extra money so they can give their daughters the same amount as their sons. This situation implies that the woman who sued her brothers was before her time, but the law reflected current changes in the culture.

The culture and the legal system may reflect different aspects of society. When people respond independently of the law to follow principle, then the law reflects culture. Currently the laws on age at marriage are a well-established part of the Havik Brahmin culture of Totagadde, as is land management. In both of these situations, legal statutes preceded cultural change. Although the laws on divorce date back to 1955, the occurrence of divorce is so recent and rare that one still needs extreme reasons to sue for divorce. Members of the culture, however, do value this as an option which they have only recently noticed as applicable to them. On the other hand, the change in inheritance laws, while still causing much controversy, seems to resonate with the current status of women in Havik culture.

CONCLUSION

Between 1965 and 1995, Havik Brahmin women went from limitation in their ascribed status to recognition for their achievements. The change in marriage patterns, as well as the observation of menstrual taboos, reflects the changes. Thirty years ago ritual observance had priority over educational opportunities. Whereas in the past, the fear of sexual harassment presented the patriarchy with a strong rationalization to keep women at home, now the assumption is that college girls are able to prevent unwelcome advances.

At the same time that an educated wife adds to her husband's status, she has increased autonomy associated with her older age at marriage, advanced education, and professional skills. Marriages now require the woman's consent. Couples may decide to limit their family to one daughter. When elders ignore the past emphasis on the ritual importance of a son, this further indicates the increase in female status. Women's autonomy has led to their questioning menstrual ritual and the associated pollution. With their increase in economic and intellectual power Havik Brahmin women have achieved the status to make their own decisions—whether they be in regard to ritual observation, professional accomplishment, or household management.

At times women challenged the legal system. When women sued for land management, they found a legal means to extend their rights. This also challenged cultural beliefs about female inferiority. Now women may inherit land. Widows may choose to keep their land intact for their daughters. At other times the legal system challenged the culture. Although the laws were in effect long before women married at older ages, the awareness of the law may have shaped cultural behavior. In the case of the woman who obviously married below the legal marriage age limit with her consent, her in-laws worried that someone would bring a case against them. Just ten years later, women in that social group marry at a more mature age.

This study indicates that there is mutual influence between culture and law. In 1964, *The Laws of Manu* represented the feelings of appropriate dependence of women on their husbands. Even when women had the same amount of education as their husbands, these Havik women felt ignorant compared to their husbands. Consequently, the dependence was a combination of respect and acknowledgment of a husband's greater experience in worldly matters. However, with the opportunity for further education for both men and women the relationship changed. This correlated with following the laws for a minimum marriage age of 18 for women. Land management as a form of contributing to the household became a possibility. Laws with regard to divorce and remarriage, as well as equal inheritance rights for men and women, further established the partnership ideal of marriage. According to law and cultural practice the marital relationship over the past 30 years has evolved toward accommodation to the needs of individuals.

ACKNOWLEDGMENTS

The research on which this paper is based was carried out with the support of grants from the American Institute of Indian Studies (1964-66, 1974-75, 1987, 1994, and 1995) and the American Philosophical Society (1985). The National Endowment for the Humanities provided additional support for the 1994 and 1995 trips. I gratefully acknowledge the comments of Barbara Lex and the two editors, Robin Miller and Sandra Browning. Carin Fernandez graciously assisted with editing the text.

NOTES

1. The following conventions are used for the transliteration of Kannada words: a capital represents retroflex phonemes; a colon, vowel length; a tilda, a nasalized vowel; and *sh,* a palatal sibilant.

2. This can also mean timid, fearful, apprehensive. I noted its application only to women. Since 1984, I observed this term which had initially reflected the appropriate passivity of young adulthood had acquired less positive connotations. Exuberant daughters whose mothers protected them against suggestions for more restricted behavior patterns retained the exuberance and even in college they showed no sign of tempering their high spirits.

REFERENCES

ICSSR (The Indian Council of Social Science Research). 1975. *Status of Women in India: A Synopsis of the Report of the National Committee on the Status of Women (1971-74).* New Delhi: Allied Publishers Private Limited.

Lévi-Strauss, C. 1969. *The Raw and the Cooked: Introduction to a Science of Mythology: I.* New York: Harper and Row.

Muller, M. 1967. *The Laws of Manu.* Delhi: Motilal Banarsidass.

Ullrich, H. E. 1980. "Clause Structure of Northern Havyaka Kannada (Dravidian): A Tagmemic Analysis." Dravidian Linguistics Association 32: 227-269.

______. 1988. "Widows in a South India Society: Depression as an Appropriate Response to Cultural Factors." *Sex Roles* 19: 169-188.

______. 1992. "Menstrual Taboos among Havik Brahmin Women: A Study of Ritual Change." *Sex Roles* 26: 19-40.

______. 1993. "Asset and Liability: The Role of Female Education in Changing Marriage Patterns Among Havik Brahmins." Pp. 187-212 in *Women, Education, & Family Structure in India,* edited by C. Chapnick Mukhopadhyay and S. Seymour. Boulder: Westview Press.

DUAL IDEOLOGICAL STRUCTURE OF THE KOREAN FAMILY LAW

Jae Kyung Lee

INTRODUCTION

The Korean Constitution guarantees equal rights, protection of the purity of marriage, and dignity as human beings to both men and women. According to the Constitution, all citizens have dignity and are valued as human beings (Article 8); all citizens are equal before the law, and there is to be no discrimination on the basis of sex in political, economic, social, or cultural life (Article 9, Paragraph 1); the purity of marriage is to receive the protection of the state (Article 31); and marriage and family life shall be entered into and sustained on the basis of individual dignity and equality of the sexes (Article 34, Paragraph 1). All basic rights acknowledged by the Constitution are automatically bestowed on women as well as men. Thus, there is no discrimination against women in their professional careers, political participation, educational opportunities, and marriage and family life.

However, in the current Family Law, which was enacted and put into effect under the Constitution, there are many provisions that oppose the equality of the sexes. These provisions have been criticized as patriarchal as well as unconstitutional. The Civil Code was adopted by the Parliament in 1958. From the time it was beginning to be formulated, the family law amendment movement was set in motion. After years of struggle, Korean women were successful in bringing about a series of modifications in the highly gender discriminatory family laws of 1962,

1979, and 1990. The Korean Family Law, after its third revision in 1990, has been recognized for its progressiveness. Although it includes provisions for equality, ideas about patriarchal ideology and traditional myths about the family still remain parts of the law. This essay aims to provide a critical analysis of the problems of the Korean Family Law from the feminist perspective.

The first part of this paper provides an overview of the Korean Family Law, beginning with the constitution of the Family Law, and then briefly describing the history of Family Law and Family Law amendments. Next, the discussion moves to the dual ideological structure of the law, namely patriarchal familism and the liberal principles of individualism. I argue that as long as the patriarchal ideology continues to prevail in the law, some equality provisions in the law cannot be realized in family life. In the concluding section, I propose judicial alternatives that might alleviate inequality in the family.

AN OVERVIEW OF KOREAN FAMILY LAW

Constitution of the Family Law

The Korean Civil Code follows the pandect system and is divided into five books. It regulates the two major domains of people's private lives, namely their private property and family lives. The first book of the Civil Code deals with general provisions. The second and third books take up the issue of property, including property rights and obligations. This portion of the Code is said to resemble the German Code. The fourth and fifth books treat the issue of family law, covering family matters and inheritance, which are based primarily on Korean customs.

The so-called Family Law is not a term that appears in the Civil Code. This phrase refers to the law prescribing marriage, family relationships, kinship relationships, and inheritance, which is referred to in the fourth and fifth books. The term "Family Law" was used at first by activists in family law amendment movements and now has become widely used by the Korean people. Book Four of the Code, entitled "Relatives," consists of eight chapters, and Book Five, entitled "Inheritance," of three, encompassing (1) inheritance, (2) wills, and (3) legally secured portions. Because the laws of inheritance do not deal with strictly family matters, they are beyond the scope of this essay. Therefore, I discuss only Book Four here. A description of the chapters and sections of Book Four is given in Table 1.

A Brief History of the Korean Family Law

The Korean Family Law was ratified and promulgated in 1958 and has been in effect since 1960. In principle, the Family Law was enacted on the basis of a

Table 1. Major Contents of Book Four "Relatives"

Chapter	Section	Detailed Subject Matter
I. General Provisions (Articles 767–777)		Definition and the scope of relatives by blood and by affinity
II. Head of Family and Family Members (Articles 778–796)		Definition of the head of family, scope of family members, and family register
III. Marriage (Articles 800–843)	1. Matrimonial Engagement 2. Formation of Marriage 3. Nullity and Annulment of Marriage 4. Effect of Marriage 5. Divorce	
IV. Parents and Children (Articles 844–927)	1. Children of Natural Father 2. Adoption 3. Parental Authority	
V. Guardianship (Articles 928–959)	1. Guardian 2. Functions of Guardians 3. Termination of Guardianship	
VI. Family Council (Articles 960–973)		Organization and the role of the family council, and eligibility of members
VII. Support (Articles 974–979)		The duty of supporting among relatives, and order of supporters.
VIII. Succession as Head of Family (Articles 980–995)	1. General Rules 2. Successor as Head of Family 3. Effect of Succession as Head of Family	

Source: KLRI (1992).

liberal democratic ideology, but traditional Confucian ideology and the idea of a patriarchal family system still survive in it. This dual character has been criticized a great deal not only in academic circles, but also in feminist circles. It has also been the nucleus of the controversy regarding the law over the years. Since the law became effective in 1960, it has been revised three times, in 1962, 1977, and 1989. The first amendment was made in December 1962 and became effective in March 1963. This amendment modified the law to a small extent but had little significance in changing the status of women.

Since the Family Law was put into effect, family law specialists and feminist groups have never tired of making propositions for an amendment they hope will reflect the Constitutional principle of equality. In 1973, 61 women's organizations founded the Pan-Women's Committee for the Expedition of the Amendment of the Family Law (PWC), which adopted ten propositions for a legal amendment. After three years of discussions and hearings, both pro and con, the revised bill was enacted in December 1977. The most problematic provisions of the bill were dropped, and the new law became effective in January 1979. In this process, the family law amendment movement had to cope with intense opposition from conservative Confucian groups. As a result, the revised law did not amend the most patriarchal provisions of the law, referring to the head of the family system and the ban on marriage between men and women from the same family of origin. It only mitigated sexual discrimination in marital property rights and inheritance rights. The PWC, which had worked for a long time for revisions, was disappointed. In fact, the second amendment of the Family Law contributed little to the improvement of women's status. But it seemed to appease, temporarily, the conflict between women's organizations and the opposing conservative groups.

In 1984, the Women's League for the Amendment of the Family Law was organized, with family law scholars and feminist activists continuing their campaign for a legal amendment. At this time, the patriarchal resistance from the conservertive Confucianist groups was more active and organized than before. They launched a "Million Signature-collecting Campaign" against the family law amendment, arguing that traditional norms and customs should be preserved. Conservative groups appeared to safeguard patriarchal prestige and privileges.

Despite strong patriarchal resistance, after years of struggle Korean women were successful in bringing about a drastic modification in the highly discriminatory family laws in 1989. The revised law became effective in January 1990. The passing of the revised bill demonstrated the triumph of feminist activism in the legislative process. The successful legal changes can be attributed to three factors: (1) the increasing power of women voters and the growth of feminist movements; (2) the ideal of democratization upheld by the recent consecutive regimes; and (3) the effort of the United Nations concerning women's rights and their status (Cho 1996). (The Korean government had joined the UN Convention for the Elimination of All Forms of Discrimination against Women in 1984.) These legal changes over two decades mark great progress in the history of women in Korea.

DUAL IDEOLOGICAL STRUCTURE OF THE FAMILY LAW

The Korean Family Law is characterized by an awkward mixture of traditional patriarchal ideology and modern principles of individualism and equality. The law reflects traditional values about the family, including values regarding patriarcal authority, paternal kinship, biological family, and gender divisions of labor. On the other hand, the provisions referring to marriage and conjugal relations are based on the individualistic principle, which establishes equal rights, duties, and

Table 2. Trends in Household Type Change

Year	Number of Generations in Household (%)				Single-person Household (%)	Average Size of Household Members (persons)
	One	Two	Three	Four		
1960	7.4	63.9	27.0	1.6	—	5.6
1970	6.8	70.0	22.1	1.1	0.0	5.2
1980	8.3	68.5	16.5	0.5	4.8	4.5
1990	10.7	66.3	12.2	0.3	0.3	3.7
1995	15.1	73.1	11.5	0.3	12.7	3.3

Source: KWDI, 1986, 1994; KBS 1996.

reponsibilities between a husband and a wife. This dual ideological structure of the law hinders the process of democratization of the family and the struggle for equality among the sexes.

In the following pages, I examine the dual ideological characters embedded in the law. I also show that as long as the dual ideological structure persists and the law internalizes patriarchal ideology, gender equality, although formally guaranteed in the Constitution, cannot be realized in family life.

Patriarchal Familism: Formation and Succession of the Paternal Family

Patriarchal familism had dominated family life in traditional Korean society. The ideology of patriarchal familism is characterized by constructs of the primogeniture stem family, authority of the patriarch (father, husband, eldest son), patrilineal succession, paternal kinship, subordination of women, and divisions of labor by gender.

In recent years, traditional family structures and values have been rapidly destroyed through the interaction of social changes and the way in which the family members actually conduct their lives. Decrease in family size, predomination of the nuclear family, and changes in family functions are attributed to rapid industrialization and urbanization. As shown in Table 2, during the past few decades the occurrence of traditional types of households with three or four generations living together decreased from 28.6 percent in 1960 to 11.8 percent in 1995. In addition, the average number of household members decreased from 5.6 persons in 1960 to 3.3 persons in 1995.

Many observers note that this seemingly nuclearization of the family structure involves complicated dynamics between tradition and modernity, between the ideology of equality and the entrenched patriarchal system. The traditions of patrilineal succession and paternal kinship continue to work effectively through ancestor worship, the system for succession as head of the family, and an obligation to support aged parents (Lee 1994). The mixture of tradition and modernity also appears in the legal system. The Civil Code still retains the family headship system, which serves no other purpose than to assure the continuation of the male lineage. Still, marriage between those from the same clan is prohibited.

Head of the Family (Hojoo) System

The Civil Code deals with the system for succession as head of the family in detail in Articles 778 to 799. This was the most controversial issue of the Family Law amendments in 1977 and 1989. It has also been seen as the "must-be-abolished provision" by feminists. The term "family" in the context of the Family Law is defined as a group composed of a head of the family and the family members, and is legally bound by rights and obligations existing among its constituent members. This does not mean the nuclear family, and it is not identical with the family in the same residence or household. Rather, it refers to the large paternal family whose names are entered in the same family register. According to the law, the family is composed, in principle, of the head of the family, his spouse, his relatives by blood, and the latter's spouse (Article 779). The head of family is defined as a person who has succeeded to the family lineage or has set up a branch family (Article 778).

The functions of the system for the head of the family in the modern context are twofold. The first centers on the legal function of the family register. The legal domicile of a family is called the family register *(Hojok)*, which is arbitrarily decided by the head of the family and may not, therefore, be necessarily identical with the actual residence of common life of the family. The family register *(Hojok)* is an official record in which the names, domiciles, dates of birth, and relations of the head of the family are recorded. It should have one head of the family *(Hojoo),* who is its representative. The family register is the basis for public certification of one's personal status. In Korea, most official documents require the name of the family head and the family register certificate for verifying one's identification. Regardless of the legal function of the family head and family register, however, the existence of system for the succession as head of the family cannot be justified. Furthermore, there are two kinds of family register systems in Korea; the family register and the resident register. The resident register contains information about household members (usually family members) at the same address. In fact, the resident register could replace the family register; thus, the Korean people may no longer need to rely on *Hojok*.

The second function centers on the system for succession as head of the family, which has symbolic and ideological significance for the patrilineal traditions. Some argue that, under the current law, the system for succession as head of the family has become loosened and exists only in name, having little practical validity (KWDI 1991; Kim 1993). This occurred because the legal regulations of this system were drastically changed in the third amended bill. First, the succession of the head of the family changed from obligatory to non-obligatory. The old law banned a waiver of succession rights, whereas the new law stipulates the annulment of this ban. Second, the rights and obligations of the head of the family were greatly curtailed. Most stipulations on the rights of the family head, including privileges related to inheritance, were deleted in the third revision in 1989.

Although the system for succession as head of the family seems to be a nominal rule, its ideological attributes have had a tremendous effect on the Korean family. Feminists and most family law specialists agree that the existing system must be abolished because it is no more than an ideological framework for maintaining male supremacy. Only when this law is repealed can a democratic family law that is consistent with the idea of equality between the sexes be created.

The Prohibition of Marriage

Most civilized states prohibit certain marriages as incestuous. The prohibited relationship may arise from consanguinity (i.e., blood relationship) or from affinity (i.e., relationship by marriage). The present Korean Civil Code stipulates the ban of such marriages in Articles 809 and 815. Under the title of "Nullity of Marriage" in Article 815, the law forbids intermarriages within the eighth degrees of the relationship by consanguinity or affinity, as in many societies. Article 809 prohibits marriage between women and men from the same place of family origin. This provision, along with the system for succession as head of the family, is a telling example characterizing the patriarchal family in the law.

Marriage is not allowed between two people who are or were related lineally by blood as indicated by identical surname and origin, or who are or were collateral relatives by blood within the eighth degree of relationship, or when one of the parties is or was a spouse to a relative of such consanguinity. However, there are many instances of individuals who have identical surnames and same family origins, but for whom no actual kinship exists. For instance, the total number of person who are *Kim* with *Kimhae* origin, which is the biggest surname in Korea, is estimated at about four million. Among them, the number of young people who reach marriageable age is estimated at about one million (Han 1996). By law, these one million men and women cannot marry each other.

This is an absurd regulation founded on an illogical and unbiological assumption. This regulation reinforces paternal kinship and can be viewed as discriminating against women. It also hinders the freedom to marry whom one chooses for unfounded reasons. It has been proposed that the ban on marriage between persons with identical surnames and with the same origin be revoked, leaving only a prohibition against marriage between immediate relatives. Family law specialists (Pak 1981; Shin 1987; Kim 1993; Han 1996) agree that the prohibition of consanguineous marriage (Article 815) is sufficient to regulate incestuous marriages.

Article 809 has been a social and feminist issue over the decades. This ban injures thousands of couples and their children. The number of couples who have been maintaining de facto marriages is estimated at about 20 thousand now, approximately 2 percent of the total number of marriages in Korea (*Chosun Daily News* 1997). Revisions have not been made, though, because of strong opposition from Confucianist groups.

These couples suffered from several disadvantages. Because their marriages are considered unlawful, the wives cannot enter their names into the family register. This means they cannot receive certified marriages. Furthermore, their children are regarded as illegitimate children. As a result, these couples live together and conduct normal family lives, but they cannot get legal protection whenever they need it, and they do not enjoy various tax privileges and benefits from social security services. This prohibition is disadvantageous to women, in particular. If the husband leaves the family, the wife becomes an unwed mother with illegitimate children and cannot be protected by the law.

Over the decades these problems have become serious, and strong opposition against the ban has prevailed. Temporary enactments for a "Special Law for Marriages" in 1977, 1986, and 1995 were made as a compromise, and these laws were effective as of the following years—1978, 1988, and 1996 (Han 1996). The temporary enactments have transformed many illegal marriages into lawful ones. Those couples who had been maintaining de facto marriages had their relationships legalized by reporting their "marriages" to the authorities. The number of de facto marriages legalized under this special law was estimated at about 4,500 couples in 1978, 12,400 couples in 1988, and 27,805 couples in 1996. These temporary actions of the government substantiate that Article 809 is unfeasible and has little validity. The only function of this ban is as a means for reinforcing the ideology of patriarchal familism.

Individualism in Conjugal Rights and Duties

The Family Law Amendment of 1989 has been recognized for its progressiveness in the laws of marriage and divorce. Conservatives have criticized it as being too progressive in reflecting reality and as threatening the traditional family norms and customs. Feminist groups praise the new law as making a distinct improvement in gender equality, while they worry that this legal equality may be no more than formal egalitarianism (Shin 1987; Cho 1996). With respect to this concern, I argue that as long as patriarchal provisions continue to prevail in the law, some equality provisions in conjugal rights and duties cannot be realized in family life. The so-called equality provisions in the new law refer to conjugal rights and duties and matrimonial property rights.

Conjugal Rights and Duties

The Civil Code stipulates duties of the husband and wife (Article 826), including duties concering cohabitation, matrimonial residence, and the family register. This Code states that it is the duty of the spouses to live together as far as their circumstances permit, but differences may arise between them as to where the matromonial home is located. The present law ordains that "the husband and wife should live together" (Article 826, Paragraph 2) and "the place where husband and wife live together shall be determined by an agreement between them" (Article 826, Paragraph 3).

The present Civil Code also prescribes that "a husband and wife shall support and cooperate with each other" (Article 826, Paragraph 1). In terms of joint living expenses, the Code stipulates that "living costs necessary for conjugal cohabitation shall be borne jointly by them, unless a special stipulation has been made between them" (Article 833). The present Civil Code is seen as weakening the ideology of the division of labor based on gender and as improving gender equality in the family.

Before the legal revision of 1989, the old law incorporated the ideology of gender division of labor, which gave authority in the family to the men. It also assumed that men were sole breadwinners and considered women as economically dependent protegees of their husbands. The old law required a wife to have her name entered in her husband's family register; the place of residence was defined as the husband's address; joint living expenses were the responsibility of the husband; and the priority of paternal power was given to the father.

Despite the provisions for equality in the new law, the patriarchal provision still exists in marriages. In Korea a wife enters her husband's family as a rule and becomes a member of his family. The wife shall have her name entered in the husband's family register (Article 826, Paragraph 3). This has not changed since the third amendment in 1989. Thus, the current regulation of family registration perpetuates patrilineality and reinforces the hierachical, patriarchal relationship between a husband and a wife.

Although a wife enters her name in her husband's family register, she maintains the surname of her parental lineage as a rule. Thus, women may appear to be equal to men at first glance; however, the appearance of equality here must be viewed with a eye toward the Confucian connotation, which treats a woman as a second-class member of her husband's family. In fact, most Korean wives regret that they are the sole people in their families who have different names. In this way, they remain permanent outsiders in the family.

Matrimonial Property System

The Civil Code stipulates that any personal property possessed by either half of a couple before marriage or acquired in his or her name during marriage is regarded as that person's personal property (Article 830, Paragraph 1) and may be used, managed, or disposed of as that person wishes (Article 831). On the other hand, any of the couple's property whose ownership is uncertain is assumed to be the joint property of the couple (Article 830, Paragraph 2). Before the partial revision of the Civil Code in 1977, such property was assumed to be the personal property of the husband.

A divorced spouse's right to claim a division of property was established in the revised Civil Code of 1989, and this is considered an epoch-making provision from the feminist viewpoint. The present law postulates that the couple's equal right to share in the property, which the couple are assumed to have made through joint effort, be recognized.

The old law held that in a divorce proceeding, the spouse found at fault is the one responsible for divorce, and allowed the injured spouse (usually the wife) to demand compensation for financial losses and mental suffering. However, the amount of the compensation given to a divorced wife usually was minimal. As there was no provision insuring the divorced spouse's right to claim for division of property, there was no compensation for the wife's household work and little practical protection provided by the law. This meant that a woman who had made a life-long contribution toward building the family fortune could very well be thrown out of the house with no share of the property or responsibility for raising the children. Because a priority of parental authority was given to the father, according to the old law, the wife lost her maternal rights upon divorce.

The new law recognizes the economic contributions of housewives, who take care of the household. In spite of the legal recognition of the homemaker's contribution to the marriage, "equitable division" of the property has been difficult to achieve for the following reasons. First, the Korean Civil Code adopted the principle of a separate property system. There are no problems with the stipulation that property owned individually by either the husband or the wife prior to marriage remains the personal property of the party concerned. There are problems, however, with the stipulation that property acquired in the name of the husband or wife during marriage remains that party's personal property. Under the separate property system, it is hard to recognize the contribution of full-time housewives to their family wealth (Choi 1996; Lee 1996).

Second, under the property law it is hard for a full-time housewife to hold real estate, such as a house or land. The law, in principle, prohibits a person who does not have a job or any other income source from owning real estate. If a wife does not have any income, the law recognizes her share of co-ownership as a gift from her husband. So even if women are allowed to have property in their names, their claims to ownership are challenged by threats of a heavy gift tax. If the property is cooperatively bought by the husband and wife, for instance, and the wife takes care of the household while the husband is out working, or the wife works as an unpaid family worker in the family business, there is no legal proof of the wife's earnings. Furthermore, family property is under the husband's name as a rule in Korea. Even if full-time working wives have legal proof of their earnings, it is a common practice for the husbands to hold family property in their own names.

Third, under the current law there is no way of compensating for new property. Recognition of the homemaker's contribution to the marriage is increasingly

Table 3. Summary of the Revised Stipulations in the Korean Civil Code

	Before the Third Revision	*After the Third Revision*
Family and Kinship Relations		
Head of the family system	Obligatory Succession Ban on waiver of succession right	Non-obligatory succession Waiver of succession right
The scope of relatives	Patrilineal Eighth degree of paternal side Fourth degree of maternal side	Bilineal Eighth degree of both paternal and maternal sides
Marriage		
Formation of marriage	Ban on marriage between parties with identical surname and origin	No revision
Conjugal rights and duties	Husband's reponsibility to provide living expenses Decision on the matrimonial residence by the husband Conjugal domicile bears the husband's family register	Joint responsibility Decision on the matrimonial residence by agreement No revision
Parental authority	Priority given to the father	Joint authority
Property rights	Separate property system	No revision Establishment of divorced spouse's right to claim for division of property

becoming part of the criteria for the distribution of marital property. The contribution may include helping put a spouse through schooling or a profession, such as law or medicine, helping to create goodwill for a business, or contributing inherited money to the investment of family (Choi 1996).

Owing to the aforementioned reasons and the sexist bias of the law makers, it is often difficult to obtain a fair division of property between husband and wife. In this way, the present system does not fully protect women's benefits. The legal provisions discussed in this section are summarized in Table 3. Clearly it is too early as yet "to speak of men and women in equal terms," and thus, the constitutional principle cannot be tenable.

CONCLUSION: FEMINIST ALTERNATIVES

As the preceding discussion has demonstrated, the dual ideological structure of the law hinders the development of gender equality and the democratization of the Korean family. The family law reforms have exihibited contradictory tendencies. They legitimate the patriarchal family while also promoting women's independence and gender equality. The state has been unable to mediate successfully in the conflicting premises of liberal individualism and patriarchal familism. In concluding, I offer some suggestions for judicial alternatives that might substantially alleviate inequality in the family.

First, I propose an end to the "Family Law" as we know it, with suggestions that family and kinship be abolished as a legal category, and that inheritance law be separated from discussions about family matters. Book Four of the Civil Code, which is basically designed for extended paternal kinship, mainly regulate marriage, parent-child relationships, and kinship relations. The term "family" (*ka*) in the present Civil Code presupposes the paternal stem family, which is ensured by the system of succession as the head of the family. In fact, except for laws of marriage and parent-child relationships, such kinship relations are no longer the object of legal regulations in most developed states. As shown in Table 1, Book Four is composed of many chapters and sections. Thus, it may be desirable to deconstruct Book Four, and to amend the law based on a relational structure in the nuclear family, one that redefines conjugal rights and duties, and parent-child relations

Second, I propose that the legal reform be carried out in accordance with feminist principles, rather than formal egalitarianism. The legal amendments in past decades have placed emphasis on the revision of individual articles rather than on the structure of the family law itself. Feminists also have made an issue of patriarchal stipulations in the Civil Code. It is beyond doubt that the the amendments of patriarchal provisions help to achieve gender equality in the legal system. By focusing on the separate provisions rather than the legal structure for the amendment, however, inconsistencies among provisions often result, which sometimes contradicts feminist principles. For instance, in the revision of 1990, the scope of kinship was exended to the eighth degree of relatives on both the paternal and maternal side. Some feminists have praised this revision for an attainment of sexual equality in the scope of relatives, namely a change from patrilineality to bilineality, but this kind of revision may reinforce or even extend traditional conceptions of kinship, which contradict modern principles of equality and individualism.

REFERENCES

Cho, H. 1996. "Gender Equality in Law and Sex/Gender Politics." Pp. 261-296 in *Gender Equality in the Korean Legal System,* edited by H. Cho. Seoul: Ewha Womans University Press.

Chosun Daily News. 1997. "Current Situations in de facto Marriages." August 17. Seoul: Chosun Daily News Co.

Choi, K. -S. 1996. "Gender Equality in the Property Law." Pp. 210-260 in *Gender Equality in the Korean Legal System,* edited by H. Cho. Seoul: Ewha Womans University Press.

Han, B. H. 1996. "Review of Prohibition of Marriage between Persons with Identical Surnames and with the Same Origin." *Journal of Korean Family Law Studies* 10: 37-50.

Kim, C. 1987. *Korean Law Study Guide*. San Diego: Cross-Cultural Associates.

Kim, C. S. 1993. *Korean Family Law and Its Task.* Seoul: SamYoung Sa.

Korea Bureau of Statistics (KBS). 1996. *Social Indicators in Korea*. Seoul: Korea Bureau of Statistics.

Korea Legislation Research Institute (KLRI). 1992. *Current Laws of the Republic of Korea*. Seoul: Korea Legislation Research Institute.

Korean Women's Development Institute (KWDI). 1986. *Social Statistics and Indicators on Women*. Seoul: Korean Women's Development Institute.

______. 1991. *A Study on Family Law Reform and Family Law Amendment Movement*. Seoul: Korean Women's Development Institute.

______. 1994. *Social Statistics and Indicators on Women*. Seoul: Korean Women's Development Institute.

Lee, E. Y. 1980. "Women and Legal System." *Korea Journal*, 20(7): 8-14.

______. 1996. "The Legal Status of Women in Work." Pp. 110-134 in *Gender Equality in the Korean Legal System*, edited by H. Cho. Seoul: Ewha Womans University Press.

Lee, J. K. 1994. "Anti-social Family in Modern Korea." Pp. 21-32 in *Philosophy and Reality*. Seoul: Institute for Philosophy and Culture.

Pak, P. H. 1981. "Family Law." Pp. 1-18 in *Modernization and Its Impact upon Korean Law*, edited by P. Pak, C. Kim, and K. Chong. Berkeley, CA: Center for Korean Studies, Institute of East Asian Studies.

Shin, I. R. 1987. "The Inequality of Women in the Korean Legal System." *Korea Journal* 27: 21-33.

DIVORCE AMONG THE TEMNE OF SIERRA LEONE (WEST AFRICA): ALLEGED CAUSES, EFFECTS OF POLYGYNY, AND TIMING

V. R. Dorjahn

INTRODUCTION

In this chapter, I am concerned first with the alleged causes of divorce offered by rural and urban Temne[1] of both sexes and by secondary school students of both sexes in Magburaka. Gender and rural-urban differences are apparent as are modest changes over time. There can be little doubt that whether a particular divorce occurs or not can involve a variety of factors, mostly personal-psychological, social, and economic. These factors, in combination, give rise to a societal or population divorce rate, but any particular divorce occurs when it does as a result of any one or a number of "precipitating incidents."

Next, the oft-made assumption that divorce is more frequent in polygynous than monogamous marriages is examined in the Temne context. The question of which polygynous marriages are more likely to end in divorce, those involving the first wife or subsequently married wives, is also addressed.

Finally, I examine the question of how long after marriage a divorce occurs. The peaking in the five- to nine-year bracket suggests a Temne "seven-year itch." A concluding section provides a summary of the arguments made.

The data utilized below were collected by the author in rural Kolifa Mayoso chiefdom in 1955, 1963, and 1976, and in urban Magburaka, the administrative headquarter's town of Tonkolili District, in 1963 and 1976.[2] My wife assisted me in collecting the 1955 and 1963 data; the 1976 Magburaka data were collected for me by Dr. Antony H. Smith. The rural Kolifa Mayoso data are derived from 100 percent enumerations; the urban Magburaka data, from one in five block samples. In addition, I use data on the alleged or stated "causes" of divorce taken from supplemental "opinion questionnaires" administered in 1963 and 1976 in rural Kolifa Mayoso, in 1976 in urban Magburaka, and in 1963 and 1976 from secondary school students in Magburaka.

The Temne system of kinship and marriage differs from that with which American readers are most familiar in several ways. Among the more important are the following:

1. The Temne kinship system is unilineally based on agnatic links; that is, it is a patrilineal system as opposed to the American bilateral system, which includes both agnatic and uterine links on both father's and mother's sides. In the Temne system, agnatically related families are grouped into extended kin groups, namely patrilineages, and agnatically related lineages into patriclans. Lineages and clans are both discrete groups; that is, there is no overlapping membership—every person is a member of one and only one lineage and one clan and nobody can belong to two or more. The Temne kinship system sets up a sharp, unambiguous distinction between "kin," those within the patrilineal group, and "others," those who are not. A double standard of morality exists; one treats kinsmen differently than outsiders.
2. The Temne system not only permits, but encourages polygyny (Dorjahn 1988). For rural Temne the incidence of polygyny has declined since I began fieldwork in 1955; for urban Temne, the decline has been even greater. The importance of polygyny is shown by the percentages of children living at my census times who were born to mothers polygynously married at the time of birth. For rural Temne the figures were 75.1 percent (1955), 77.0 percent (1963), and 78.8 percent (1976); for the urban series, the figures were 74.8 percent (1963) and 79.8 percent (1976). Growing up, at least for a time, in a polygynous situation is the norm for three out of every four children. The implications of this regarding attitude formation are significant.
3. Most rural households are based around a family, monogamous, polygynous, or even extended, while often containing additional individuals, related or not, to a member of the core family (Dorjahn 1977). This was traditionally the basic farming unit, the basis of a household or domestic mode of production. The more laborers/people in a household, the larger the farm(s) that can be made and the better able a household is to cover/compensate for ill or absent labor. Land for farming and for building a house in

the village is held by the extended family or patrilineage and allocated to lineage men. The "big family" or "big household" ideal is real among Temne, and polygyny is a key element in achieving it. The basic motive for polygyny is economic, not sexual.

4. Temne marriage (Dorjahn 1973) is based on the transfer of wealth from the groom and his family to that of the bride. The bride-wealth is a significant, not a negligible, sum for the majority of men. The amount of bride-wealth and other minor exchanges is agreed upon by the older men in the bride's and groom's families or lineages. A dowry or counter-payment moves from the bride's family to the newly married couple. While not of great monetary value, the pans, kitchen knives, clay pots, and so on are welcome necessities.
5. Within traditional families there was an "upward flow of wealth" that greatly benefitted older men (Caldwell 1976, 1978, 1981). Women and younger men did most of the agricultural labor but received only that small part of the product of their labors which the family head chose to redistribute (Dorjahn 1989). This contributed to dissatisfaction, intergenerational and intergender tension, outmigration, and the reduction or breakup of big families and big households (Dorjahn 1975).
6. The authority of older men, manifest in the "upward flow of wealth," was also expressed in the past in the arrangement of marriages for their children. Older informants were agreed that in the past most marriages were arranged, but differed in assessing the roles, if any, of the two principals. By the time of my field research, I found consultation to occur between each principal and their parents and senior kinsmen. In the event of disagreement, neither party consistently prevailed. Overall, I think the authority of elders to dictate or veto a particular potential marital partner is reduced from what it was in the more distant past.
7. The major tensions within the Temne system of kinship and marriage are evident when the husband is accused by one or more members of his sibling group (brothers and sisters) or his wife (wives) and children, of favoring one over the other. Quarrels can arise between a husband and his co-wives or between jealous co-wives. Outside mediators in a domestic dispute can, on occasion, increase such tensions, especially when their own agenda is better served by exacerbating rather than mitigating a dispute. Fruitful analysis of domestic quarrels begins with a look at the tactics and strategies employed by the principals in these "marital games." While some quarrels can be overcome, reasonably harmonious family relations restored, and the marriage saved, other quarrels end in divorces, which ends marital games (Dorjahn 1990).

Concerns about divorce often become considerations in the marital game, and sometimes the marital game becomes a veritable divorce game. The goal of reconciliation and the continuance of a marriage can easily change to the goal of divorce and the ending of the marital game. Certain aspects of divorce in the Temne system must be mentioned before turning to the focal issues of this paper (Dorjahn 1990).

1. In divorce settlements the major bones of contention are bride-wealth repayment (in whole or in part) and the custody and support of any children born to the union. The opinions of men and women differ markedly on both of these.
2. Divorce has become more frequent over time among both rural and urban Temne. Divorce is also more frequent in urban than rural locations. Although no systematic comparison can be made, the few data sets available for agrarian West African populations suggest the Temne are medial in terms of divorce frequency.
3. A variety of developments during the Colonial and post-Colonial periods seem to have contributed to the increasing divorce frequency over time and the rural-urban differences among the Temne. First, the expanded primary school system after World War II meant that more children were exposed by teachers and teaching materials to western ideas of romantic, companionate, monogamous marriage between more or less equals. The polygynous alternative paled for some by comparison. Furthermore, schooling enabled some boys and girls to acquire skills enabling them to make it, at least in urban areas, when unmarried or even as a single parent with children. Divorce became a viable alternative for such. Second, in rural areas, a worsening man-land ratio and increased rural-urban migration led to smaller rural households that were less likely to contain older persons to mediate marital disputes. Third, while the incidence of polygynous marriages decreased a bit, this was offset by a lowered acceptance of polygyny by educated girls, and hence to higher divorce rates. Fourth, increased wealth and wealth differentials led to a "marrying up" pattern for both men and women, and this also entailed higher divorce rates. Finally, Temne opinions regarding the traditional "ground rules" for marriage and divorce began to change. This was especially true with regard to partial return of bride-wealth in some cases, and in joint or split custody of children in other divorce settlements.

CAUSES OF DIVORCE

In his "Life" of Aemilius Paulus, Plutarch notes that history provided no explanation or cause for Aemilius's divorcing his seemingly admirable wife of many years, Papiria, but he cites a similar account of another, unnamed Roman's divorce that he thought instructive:

> This person being highly blamed by his friends, who demanded, was she not chaste? was she not fair? was she not fruitful? holding out his shoe, asked them, Whether it was not new? and

> well made? Yet, added he, none of you can tell where it pinches me. Certain it is, that great and open faults have often led to no separation; while mere petty repeated annoyances, arising from unpleasantness or incongruity of character, have been the occasion of such estrangement as to make it impossible for man and wife to live together with any content (Clough 1911, vol. 2, p. 173).

Concern with why a divorce occurs in a particular instance and why certain behaviors lead to a divorce in one instance but not in another have puzzled many since Plutarch's day. Cross-cultural comparison subsequently has shown that the frequency or rate of divorce can and does vary widely from one society to another and seemingly for different reasons (Dorjahn 1990).

In the preceding quotation, Plutarch noted two important points about the reasons or causes leading to divorce: (1) only a principal in any particular divorce can know the "real reasons" and not always then, and (2) what is sufficient cause for one principal, is insufficient for another. When we move from attempting to explain why a specific divorce occurred, to what cause or causes might explain most divorces in a given society at a given time, Plutarch's points are among the major difficulties. There is a high probability that some individuals who were divorced do not know precisely why the divorce occurred. One reason they do not know is that many times there are multiple causes for a divorce. Other individuals may know the reason or reasons, but conceal it because it puts them in a bad light, is embarrassing, and so on. A divorce is one possible result in the ongoing process characterized elsewhere as the marital game. There are many elements in that game and different people assess them differently (Dorjahn 1990).

Mitchell (1963, pp. 255-256) has suggested that there are three kinds of factors involved in marital stability:

1. Idiosyncratic personality factors that lead to conjugal incompatibility,
2. Factors stemming from the membership of spouses in groups with different beliefs or values,
3. The social system, which provides the framework ". . . within which marriages take place and are dissolved."

This is based on the assumption that personal factors determine the incidence, but not the rate, of divorce. In Mitchell's words: "These personal factors . . . will act adventitiously and not systematically to give rise to random variation around the rate, but the rate itself is likely to be determined by features of the social system."

In another account, Mitchell (1961, p. 314) notes: "The multiplicity of factors, therefore, operate adventitiously and not systematically, in the same way that random factors affect the trajectory of a missile to give rise to a cone of fire and not to a constant error." This view is analogous to that often employed in the analysis of migration, where it is believed that the rate is determined primarily by perception of economic conditions, but the incidence or cause for a particular person migrating at a particular time (the "precipitating incident") could be ridicule, an arrest, the death of a child, or any in a wide range of personal idiosyncratic events (Lee 1966). It is analogous also to Durkheim's (1897) explanation of the incidence and rate of suicide. In what follows it is apparent that some respondents tried to explain (a given) instance while others were concerned with a (societal) rate; thus, idiosyncratic, communal, and structural reasons are provided. It is because of this indeterminantness that I speak of "alleged causes."

Table 1. Alleged Causes of Divorce, Male Respondents

	Rural		*Urban*
Reason	*1963 Number*	*1976 Number*	*1976 Number*
Wife to blame:	32 (61%)	59 (61%)	70 (77%)
Wife unfaithful	26	6	27
Wife disobedient/not well trained	5	18	29
Wife wants wealthier husband	2	0	4
Wife wastes money	1	0	0
Wife dislikes farm work, prefers town	0	25	0
Wife visits parents too often	0	4	0
Wife lazy	0	1	1
Wife childless/sterile	0	3	2
Wife bewitches children	0	0	1
Co-wife jealousy	0	0	6
Husband to blame:	5 (10%)	16 (16%)	11 (12%)
Husband mistreats wife	4	7	1
Husband poor	0	0	3
Husband fails to support wife	0	0	4
Husband unfaithful	0	0	2
Husband spoils wife	0	0	7
Husband jealous	0	1	0
Husband does not satisfy wife sexually	1	0	0
Husband ill/drunk	0	1	1
Indeterminate/both to blame:	15 (29%)	22 (23%)	10 (11%)
Arranged marriage system	1	0	0
Bridewealth system	0	9	0
Incompatibility/jealousy	12	10	8
Will of God	2	0	2
Poverty	0	3	0
Totals	52	97	91

I thus assume that the rate of divorce may differ between rural and urban Temne and change over time for either sector as a result of altered group composition (e.g., increased interethnic or interreligious marriage) or in social structure (e.g., increased wealth differences or age differences between spouses, differences in the size and kin-composition of households), or both. Mitchell (1961, p. 318), following Gluckman (1950), believed that ". . . the major determinants of marriage stability in tribal societies lie in the kinship system" and probably this was once true of the Temne. In the time period under consideration here, however, changing life circumstances have emphasized other factors as well; for example, the economic stress stemming from the inability to meet rising expectations, the spatial separation from family arising from rural-urban migration, and so on.

Table 2. Alleged Causes of Divorce, Female Respondents

	Rural		*Urban*
Reason	*1963 Number*	*1976 Number*	*1976 Number*
Wife to blame:	5 (11%)	14 (14%)	36 (36%)
Wife disobedient/not well trained	5	4	10
Wife unfaithful	0	0	9
Wife lazy	0	1	4
Wife childless/sterile	0	0	3
Wife wastes money	0	0	3
Wife wants wealthier husband	0	9	4
Husband to blame:	33 (75%)	72 (73%)	53 (54%)
Husband mistreats wife	14	60	9
Husband fails to support wife	9	0	14
Husband unfaithful	3	7	15
Husband disregards wife's kin	5	1	2
Husband sterile	1	2	0
Husband does not love wife	1	0	2
Husband does not respect wife's contribution	0	0	5
Husband wants additional wife	0	0	6
Husband treats co-wives unequally	0	2	0
Husband drinks to excess	0	0	0
Indeterminate/both to blame:	6 (13%)	12 (12%)	10 (10%)
Incompatibility/jealousy	0	7	9
Arranged marriage system	1	5	1
Polygynous marriage system	5	0	0
Totals	44	98	99

Respondents were asked to indicate what they believed to be the major causes of divorce in Sierra Leone, and first responses are tabulated in Table 1 for males and Table 2 for females. It is not surprising that males tended to blame females and female actions and shortcomings while females more frequently indicated that males were at fault. These tendencies seem a bit stronger among rural than urban respondents, suggesting that gender hostility may be higher in the more traditional areas.

Those men who blamed women stressed the unfaithfulness, real or suspected, of wives and the "fact" that wives were disobedient because they were badly raised or trained. The only major deviance is in the 1976 rural series where 25 men indicated that wives disliked farming and rural life, and preferred town living. Interview data corroborated that men believed this.

That minority of men who felt husbands and their actions caused most divorces felt mistreatment of and failure to support a wife (or just being poor) were the main reasons. Again, the men in the 1976 rural series were exceptional, nearly half feeling men caused divorces by spoiling wives, by being overindulgent, to keep them from the rural-urban migration stream.

The majority of those men who felt blame for divorce was most often indeterminate or shared, cited incompatibility or jealousy most often and second, blamed the traditional system of marriage, one that featured arrangement by parents or kin, bride-wealth, and polygyny.

Female respondents who blamed husbands stressed the mistreatment of wives, the unfaithfulness of men, and the failure to support a wife. A wide variety of other causes was cited. Those who blamed wives felt the causes were disobedience or lack of training, the wife's unfaithfulness, and economic reasons—namely the wife wasting the husband's money or the wife wanting a wealthier husband. The economic reasons, limited to the 1976 series, parallel the rural male awareness that wives dislike farm work and prefer urban life. Respondents who favored indeterminate blame indicated incompatibility or jealousy and aspects of the traditional marriage system. Some responses clearly indicated a superficial and impersonal commitment in marriage.

Males in the school series more frequently attributed divorce to indeterminate or both-to-blame causes, and less frequently to the women involved than either rural or urban adult males. Incompatibility or jealousy was the cause most often cited in all male series within the indeterminate or both-to-blame category. There is a deviation within the husband-to-blame category in that the 1976 school series overwhelmingly cited "husband poor" as a cause. Economic hard times in Sierra Leone in 1976, and the expectations of male secondary school students regarding an elite lifestyle probably heightened their awareness of the "wages of poverty," poverty being synonymous with failure.

Females in the school series also more frequently attributed divorce to indeterminate or both-to-blame causes; instead of singling out incompatibility or jealousy, however, they emphasized aspects of the traditional marriage system, chiefly

Table 3. Alleged Causes of Divorce, Secondary School Series

Reason	Males 1963 Number	Males 1976 Number	Females 1963 Number	Females 1976 Number
Wife to blame:	26 (43%)	31 (38%)	5 (9%)	3 (7%)
Wife unfaithful	14	9	1	1
Wife disobedient/not well trained	6	18	4	2
Wife does not respect husband's parents	0	2	0	0
Wife wastes money	0	1	0	0
Wife wants wealthier husband	4	1	0	0
Wife lazy	1	0	0	0
Wife childless/sterile	1	0	0	0
Husband to blame:	5 (8%)	20 (25%)	26 (49%)	18 (42%)
Husband poor	0	17	0	4
Husband unfaithful	0	3	15	9
Husband mistreats wife	3	0	4	4
Husband does not respect wife's parents	1	0	0	1
Husband does not satisfy wife sexually	1	0	0	0
Husband does not support wife	0	0	3	0
Husband drinks to excess	0	0	4	0
Indeterminate/both to blame:	29 (48%)	30 (37%)	22 (42%)	22(51%)
Arranged marriage system	5	1	8	2
Polygynous marriage system	3	4	5	18
Incompatibility/jealousy	17	24	9	1
Will of God	0	1	0	0
Different religions	0	0	0	1
Poverty	4	0	0	0
Totals	60	81	53	43

arranged marriage and polygyny. Both of these were and remain anathma to the Western-modernist-feminist view, which prevailed among secondary school teachers, a significant minority being American Peace Corps and British VSO, both in 1963 and 1976. The female school series in both years were also less likely to blame the women involved in divorce than either rural or urban adult females.

POLYGYNY AND DIVORCE

Numerous writers over the years have felt that polygynous marriage increases the chances of divorce (e.g., reference citations in Dorjahn 1958, pp. 849-850), but few quantitative data have been offered in support of opinions. Olusanya (1971, p.

Table 4. Polygyny and Divorce

	Mayoso			*Magburaka*	
	1955	*1963*	*1976*	*1963*	*1976*
Number of men married at time of inquiry	279	374	414	179	173
Number of men married polygynously	165	148	158	61	47
Percent married polygynously	59	53	38	34	27
Number of women married at time of inquiry	414	584	634	272	218
Number of women married polygynously	330	357	378	154	92
Percent married polygynously	74	61	60	57	42
Number of marriages terminated in divorce	64	97	161	48	106
Number which were polygynous	49	80	133	41	95
Percent which were polygynous	76	82	83	83	90

174), for example, simply states: "A high frequency of divorce is, however, an important feature of polygynous marriages." Hagen (1983, p. 192), writing on the Effutu of Winneba, Ghana, notes: "Several people noted that while acting as a salve to some of the domestic crises, polygyny also tends to lead to the dissolution of marriages." Ademola (1982, p. 16), however, reports on his Yoruba data that: "Of the 110 women divorcées only ten, about 9.3 percent, had no co-wife when they were divorced"; that is, 91 percent of all divorces took place in polygynous as opposed to monogamous unions. Cohen's (1971, Table 11:7) Kanuri data show that divorce was more frequent for polygynous men and women than for their monogamous counterparts. He accounts for this by noting that one divorce does not change a polygynist's marital state, he is still a married man, but that a divorce changes a monogamist's status to unmarried, and reduces that man's status significantly. Second, he remarks on the higher possibility for conflicts in polygynous unions, leading to higher divorce rates.

The Temne data are summarized in Table 4. The bottom row shows that between 76 percent and 90 percent of all reported divorces took place in polygynous unions. At no time of inquiry did the percent of polygynously married females, let alone males, approach the percentage figures in the bottom row, except for rural Mayoso in 1955. It appears, then, that divorce is more likely to occur in polygynous than monogamous unions, and that the tendency may be increasing over the time spans of both rural and urban data. Over time, in both rural and urban contexts, fewer (percentage-wise) married Temne men and women are polygynous, but they have an increasing proportion of divorces.

If, then, polygynous marriages are more likely to end in divorce than monogamous marriages, it is reasonable to assume that polygynously married men are more likely to be recidivists, that is, have more than one divorce. The data on men having had multiple divorces are summarized in Table 5. It is conceivable that a

Table 5. Men Having Multiple Divorces

	Mayoso			*Magburaka*	
	1955	*1963*	*1976*	*1963*	*1976*
Number of Divorces					
1 divorce	44	51	73	24	35
2 divorces	6	10	16	5	16
3 divorces	0	6	7	2	7
4 divorces	2	2	6	2	2
5 divorces	0	1	0	0	2
6 divorces	0	1	0	0	0
Totals, divorced men	52	69	104	33	62
Percent of all divorced men having had more than 1 divorce	15	26	30	27	43
Totals, divorced women	64	97	161	48	106
Percent of all divorced women having been divorced from a man who had at least one other divorce	31	47	55	50	67

man could have had three or four divorces without ever having been polygynously married—it happens in American society—but in the Temne series presented here, all were polygynous at the time of any particular divorce. The data show that an increasing percentage of divorced men have had two or more divorces and that an increasing percentage of divorced women are divorced from a man who has been divorced previously. It is interesting that Aguirre and Kirwan (1986) note that in the United States, the number of times spouses have been married leads to higher divorce rates.

Having recognized that divorce is more frequent in polygynous than monogamous unions, one must then address the question of which polygynous marriages are more likely to end in divorce. Is it the first wife who is usually the head wife, or is it later, usually junior co-wives? The data in Table 6 indicate that for all three years of investigation in rural Mayoso and for all three divorce ratios, the polygynous A wives had lower ratios than the polygynous C+ higher order wives.[3] This is a contrast between first, as opposed to third and higher order marriages. Among rural Temne the first wife married seems to stick it out and prevail over higher order, usually junior, co-wives. Several factors appear to contribute to this: first wives are usually older, are usually head wives, and in the time span covered, were less likely to have been to school and thus lacked skills and opportunities to support themselves. Over time the A-to-C+ gap widens with respect to ratios A and C. This reflects several factors: the younger C+ wives are progressively more likely

Table 6. Divorce Ratio for First, Second, Third, and Higher Order Wives' Polygynous Marriages, and for Monogamous Marriages

	(1) Number Wives Living and Married	*(2) Number Wives Divorced*	*(3) Number Wives Deceased*	*(4) Number Wives Ever Married*	*Divorce Rations**		
					A (2 ÷ 4)	*B (2 ÷ 2 + 3)*	*C (2 ÷ 1)*
Mayoso, 1955							
Monogamous†	111	17	—	128	13	—	15
Polygynous A	111	13	27	151	9	32	12
Polygynous B	124	15	12	151	10	56	12
Polygynous C+	98	19	15	132	14	56	19
Totals	444	64	54	562	11	54	15
Mayoso, 1963							
Monogamous†	161	17	—	161	10	—	10
Polygynous A	118	23	55	213	11	29	19
Polygynous B	180	20	13	213	96	1	11
Polygynous C+	138	37	26	201	18	59	27
Totals	597	97	94	788	12	51	16
Mayoso, 1976							
Monogamous†	183	28	—	183	15	—	15
Polygynous A	135	23	42	228	10	35	17
Polygynous B	170	35	23	228	15	60	20
Polygynous C+	146	75	29	250	30	72	51
Totals	634	161	94	889	18	63	25
Magburaka, 1963							
Monogamous†	87	7	—	87	8	—	8
Polygynous A	63	19	4	93	20	83	30
Polygynous B	79	11	3	93	12	78	14
Polygynous C+	43	11	10	64	17	52	26
Totals	272	48	17	337	14	74	18
Magburaka, 1976							
Monogamous†	43	11	—	43	26	—	26
Polygynous A	38	33	26	108	31	56	87
Polygynous B	71	26	11	108	24	70	37
Polygynous C+	66	36	9	111	32	80	54
Totals	218	106	46	370	29	70	49

Notes: * The divorce ratios used are those defined by Barnes (1949, p. 44 and 1967, p. 61): Ratio A relates the number of marriages ended in divorce as a percent of all marriages contracted; ratio B relates the number of marriages ended in divorce as a percent of all completed marriages; and ratio C relates the number of marriages ended in divorce as a percent of all marriages except those ended by death.

† Owing to the method of data collection, marital histories of males married at the time of inquiry, it was unlikely that cases of monogamous marriages ended by the death of the wife would have been recorded. In fact, none were; hence no ratio B was computed for the monogamous category of wives.

to have been to school, to have visited Magburaka and other towns, and, probably, to have developed attitudes less favorable to polygynous marriage and more receptive to the possibility of divorce. Younger women who are single, divorced, or widowed are more likely to migrate to cities than older women. Cohen (1971, pp. 142-144) has demonstrated that for the Kanuri, a three-wife (and presumably also more than three) family increases the probability of divorce for the senior wife. Among rural Temne it does not, for the probability of divorce is higher for C+ than A wives. Cohen's data on Kanuri also show that in a two-wife family, the probability of divorce increases for the second, usually younger wife. A comparison of divorce rates of polygynous A to polygynous B wives for rural Mayoso shows essential equality in 1955, slightly higher rates for polygynous A in 1963, and slightly lower rates for polygynous A in 1976. No trend in the probability of divorce is apparent.

For urban Temne in Magburaka, the contrast between polygynous A and C+ wives in 1963 shows (all three ratios) that polygynous A wives have the higher probability of divorce. This is contrary to the rural Temne situation and comparable to the Kanuri. For 1976 the situation is less clear: for polygynous A, the divorce ratio A is about equal, ratio B is lower, and ratio C is higher. Why, in urban Magburaka, should polygynous A wives be less likely to prevail over younger, higher order (C+) co-wives than in rural areas? I think the primary reason is that the polygynous A wives have lived longer in town, know of ways to support themselves, and thus see less point to sticking it out. In Magburaka in 1976, polygynous males had been practicing a kind of overlapping serial monogamy: with each additional marriage, a previous wife was subsequently divorced. The polygynous A to polygynous B comparison shows that polygynous A wives have higher divorce ratios, all three ratios in 1963 and ratios A and C in 1976, a situation contrary to that among the Kanuri. Gage-Brandon's (1992) study suggests that two-wife unions are the most stable and that unions with three or more wives have higher divorce rates.

TIMING OF DIVORCE

It is also of interest to see how long after marriage a divorce occurs. The relevant data are in Table 7 and suggest, first of all (except for Mayoso, 1955), a peaking in the fifth to ninth years of marriage—a Temne "seven-year itch." Second, it appears that divorce comes earlier, after fewer married years, in the urban context. Looking at the cumulative percent entries, more than 50 percent of all divorces occur before the tenth year in Mayoso (excluding 1955) and more than 60 percent in Magburaka.[4] Third, it is striking that in rural Mayoso more than 10 percent of all divorces occur after 20 years of marriage; in 1955, 21.9 percent occurred at this late time.

Table 7. Timing of Divorce in Terms of Number of Years Married

	Number of Years Married at Time of Divorce					
Number of Divorces	*0–4*	*5–9*	*10–14*	*15–19*	*20+*	*Totals*
Mayoso, 1955						
Number of divorces	9	14	13	14	14	64
Percent	14.0	21.9	20.3	21.9	21.9	
Cumulative percent	14.0	35.9	56.2	78.1	100.0	
Mayoso, 1963						
Number of divorces	17	39	24	7	10	97
Percent	17.5	40.2	24.7	7.2	10.3	
Cumulative percent	17.5	57.7	82.4	89.6	99.9	
Mayoso, 1976						
Number of divorces	32	50	36	25	18	161
Percent	19.9	31.1	22.4	15.5	11.2	
Cumulative percent	19.9	51.0	73.4	88.9	100.1	
Magburaka, 1963						
Number of divorces	5	25	10	5	3	48
Percent	10.4	52.1	20.8	10.4	6.2	
Cumulative percent	10.4	62.5	83.3	93.7	99.9	
Magburaka, 1976						
Number of divorces	32	41	19	8	6	106
Percent	30.2	38.7	17.9	7.5	5.7	
Cumulative percent	30.2	68.9	86.8	94.3	100.0	

With respect to the "seven-year itch," I would argue that data from male marital histories show it is usually five to nine years after his first marriage, in the rural context, that a man is ready and financially able to take a second wife. The "seven-year itch" thus reflects the problems attendant upon making a family polygynous. The five- to nine-year peak includes first wives who object and leave because another wife is brought into the house. Some leave before the husband's second marriage is fact, others leave afterwards. When a second or third wife is driven out, that process often takes time, sometimes five years or more.

Earlier divorce in Magburaka goes along with the greater frequency of divorce in urban areas. It reflects the smaller urban household—often without resident mediators, the greater economic and marital opportunities for women, probably the greater temptations in towns, and the reduced influence of kin, most of whom reside elsewhere. The strains of attempting to keep up with the "revolution of rising expectations" were apparent in interviews.

Finally, there is the question of the relatively high proportions of divorces in rural Mayoso which take place after 20 or more years of marriage. Three situations predominate: first, after 20 or more years of marriage a woman has no surviving children; second, a markedly younger co-wife or co-wives arrive; and third, a wife's older children have married and are economically independent, thus giving her the option of divorce and moving in (usually) with a married son. In the first instance there was traditionally the tendency to blame co-wives or the husband's kin of utilizing supernatural means to prevent conception, to abort a pregnancy, or to kill an infant or child. The second instance includes cases of a man marrying a second or third wife the age of his daughters by the first wife (the Temne "trophy-wife"), and cases of second and often higher order wives banding together successfully to drive out the first wife. The third instance reflects the fact that until recently, rural women had few options and tended to remain married even when the circumstances were unpleasant, as when the husband favored a younger co-wife. Such a woman who has one or more children who are adult and reasonably well off may divorce her husband and move in with the next generation.

CONCLUSION

The assigning of a cause (or causes) for a divorce is as difficult as determining a cause of death. There may be one or more possible primary causes, various complicating secondary causes, and probably a precipitating incident or two, easily elevated to primary causal status by observers or analysts. With divorce, the problem or explanation is one of multiple causes, further complicated by the failure of a principal to know the cause(s) and often, whether consciously or unconsciously, to prevaricate when the presumed cause is embarrassing or puts the principal in a bad light. The aforementioned survey questions were phrased generally to refer to the societal situation, so as to remove the respondent from any particular instance of divorce known to them. It was apparent that while some respondents answered in this vein, others were explaining a particular divorce or two of which they had knowledge. To cover this variance I have used the phrase "alleged causes."

On the societal level, I have assumed with Mitchell that divorce frequency in general is best accounted for by a complex of social structural variables (polygynous marriage, rising expectations, urban residence, and so on). Any one divorce results from a particular combination of idiosyncratic events, experiences, and so on. This explanatory scheme is similar to that used by demographers to account for the frequency of migration (Lee 1966).

Analysis of survey results shows that both rural and urban Temne men most frequently blame wifely failings or shortcomings, while both rural and urban females most frequently blamed males. Men stressed "unfaithfulness," real or imagined, and wifely "disobedience" (which led to quarreling and stemmed from poor training or upbringing). Women cited husband's mistreating wives (physically, by withholding money or attention, and so on), the "unfaithfulness" of husbands, and the lack of economic provision or support (ranging from insufficient food to failure to provide new clothes, gold jewelry, and so on). Both these tendencies to blame the opposite sex were slightly more pronounced in the rural than the urban series. Minorities of both male and female respondents indicated the same sex or indeterminant or joint blame. The secondary school students from Magburaka more frequently attributed divorce to indeterminant or both-to-blame reasons (e.g. incompatibility, mutual jealousy, the traditional marriage system, and so on). As one result of formal schooling at this level, some students come to accept the modernist-Western view of marriage as a monogamous, companionate partnership of more or less equals. Such individuals view marriage, and hence the causes of divorce, in a way that is quite different from more traditional Temne.

The Temne data show divorce to be more frequent in polygynous than in monogamous situations, the discrepancy increasing over time. In rural Mayoso, A wives had lower divorce ratios than C and higher order wives. In urban Magburaka, A wives had higher divorce ratios than C+ wives. In each case the difference increases over time. For the time period under consideration, the major reasons seem to be three. First, in the rural area, older first wives found it difficult to make it on their own while younger C+ wives were in a better position to marry again or emigrate to an urban area. Second, in Magburaka it is the older, first wives who have learned how to make it on their own, who have the supplier contacts and customer networks for successful independent retailing while the younger, higher order wives with shorter urban experience lack these. Third, more women have acquired modest skills and knowledge through increased schooling over time and have become less tolerant and less patient in difficult marital situations: male-female roles in marriage are viewed differently, less traditionally, than in the past, especially by women.

Finally, the question of the timing of divorce is considered: after how many years of marriage does divorce occur? In both rural and urban series there is a peak in the five-to-nine years of marriage category, a "seven-year itch." This corresponds to the amount of time it takes a man to acquire a second wife and for subsequent problems to develop between co-wives or between the husband and one of his two wives. The cumulative percents in Table 7 show a tendency for divorce to occur after fewer years of marriage in the urban than in the rural situation. This reflects the greater economic prospects for women in Magburaka and the increased strains from trying to keep up with the "revolution of rising expectations." The 1955 rural data reflect a situation two or three generations ago that manifested only a rudimentary cash economy, the existence of few schools, and the presence of an upward flow of wealth, and hence the greater authority of older males. At that time, more people stuck out and coped with difficult marital situations as few had viable alternatives. Finally, the 1955 rural data show a high proportion of divorces coming after 20 or more years of married life. It is suggested that three factors are involved. First, childless women, those with few children, and those whose last born children have died, come to view their situations as intolerable when concern about old age security from their children grows. Second, this reflects the situation where an older polygynist marries a young girl, a situation Temne women frequently joke about as "an old goat thinks he's a young buck syndrome," which is embarrassing to some older wives. Third, a woman's eldest son had a chance at 25 years of age not only of marrying but also of having his own farm and thus independence so that his mother could divorce her husband and move in with him and attain old-age security (Dorjahn 1989).

High divorce rates and, indeed, the institution of divorce are seen as significant social problems by many in our own society. The same has been and is true of the last two generations of Temne in Sierra Leone. Neither society has come up with a reasonable solution to the problems. Although the problem is essentially the same in both American and Temne societies, it results from different factors. Assessment of these different factors is an essential first step if we, or the Temne, are to come up with anything approximating solutions.

NOTES

1. The Temne comprise about one-third of Sierra Leone's estimated population of 4.5 million (Population Reference Bureau 1995). They are concentrated in the Northern Province. Cultural summaries are found in Dorjahn (1978, 1995). The classic early accounts of the Temne are those of Laing (1825) and Thomas (1916).

2. I am grateful to the National Science Foundation (Washington, D.C.) for funding my 1955 and 1963 research and to the National Institutes of Child Health and Human Development, Center for Population Research (Bethesda, Maryland) for supporting my 1976 research.

3. I follow convention in indicating the marital order of a man's wives using letters; thus a man's first wife is his A wife, his second is his B wife, and so on. The divorce ratios utilized (Barnes 1949, 1967) differ from the form of data used in Western countries: annual rates showing the number of divorces per year per thousand population or various components thereof, or the number of divorces per year, either (1) per number of new marriages, or (2) per number of extant marriages, both during the same time period. See notes to Table 6 for ratios and Solivetti (1994, Tables 5 and 6) for rates.

4. Amoateng and Heaton (1989) present information on the timing of divorce in Ghana using data from the 1979/1980 World Fertility Survey. They state that within ten years of marriage 23 percent have dissolved, and that for couples married in the 1970s the figure will rise to 32 percent.

REFERENCES

Ademola, A. 1982. "Changes in the Patterns of Marriage and Divorce in a Yoruba Town." *Rural Africana* 14: 1-24.

Aguirre, B. E. and P. Kirwan. 1986. "Marriage Order and the Quality and Stability of Marital Relationships: A Test of Lewis and Spanier's Theory." *Journal of Comparative Family Studies* 17: 247-276.

Amoateng, A. Y. and T. B. Heaton. 1989. "The Sociodemographic Correlates of the Timing of Divorce in Ghana." *Journal of Comparative Family Studies* 20: 79-96.

Barnes, J. A. 1949. "Measures of Divorce Frequency in Simple Societies." *Journal of the Royal Anthropological Institute* 79: 37-62.

______. 1967. "The Frequency of Divorce." Pp. 47-99 in *The Craft of Social Anthropology*, edited by A. L. Epstein. London: Tavistock.

Caldwell, J. C. 1976. "Toward a Restatement of Demographic Transition Theory." *Population and Development Review* 2: 321-366.

______. 1978. "A Theory of Fertility: From High Plateau to Destabilization." *Population and Development Review* 4: 553-577.

______. 1981. "The Mechanisms of Demographic Change in Historical Perspective." *Population Studies* 35: 5-27.

Clough, A. H. 1911. *Plutarch's Lives, the Translation Called Dryden's*. New York: Bigelow Brown.

Cohen, R. 1971. *Dominance and Defiance, a Study of Marital Instability in an Islamic African Society.* Washington, DC: American Anthropological Association.

Dorjahn, V. R. 1958. "The Factor of Polygyny in African Demography." Pp. 87-112 in *Continuity and Change in African Cultures*, edited by W. R. Bascom and M. J. Herskovits. Chicago: University of Chicago Press.

______. 1973. "Some Rural-Urban Marriage Differentials: The Temne of Magburaka Town and Its Environs." *Urban Anthropology* 2: 161-181.

______. 1975. "Migration in Central Sierra Leone: The Temne Chiefdom of Kolifa Mayoso." *Africa* 45: 28-47.

______. 1977. "Temne Household Size and Composition: Rural Changes over Time and Rural-Urban Differences." *Ethnology* 16: 105-127.

______. 1978. "The Temne of Sierra Leone (West Africa)." Pp. 774-779 in *Ethnographic Survey of The Muslim World*, edited by R. V. Weekes. Westport, CT: Greenwood.

______. 1988. "Changes in Temne Polygyny." *Ethnology* 26: 367-390.

______. 1989. "Where Do the Old Folks Live? The Residence of the Elderly among the Temne of Sierra Leone." *Journal of Cross Cultural Gerontology* 4: 257-278.

______. 1990. "The Marital Game, Divorce, and Divorce Frequency among the Temne of Sierra Leone." *Anthropological Quarterly* 63: 169-182.

______. 1995. "Temne." Pp. 341-346 in *Encyclopedia of World Culture,* vol. 9, edited by J. Middleton and A. Rassam. Boston: Hall.

Durkheim, É. 1897. *Le Suicide: Étude de sociologie*. Paris: F. Alcan.

Gage-Brandon, A. J. 1992. "The Polygyny-Divorce Relationship: A Case Study of Nigeria." *Journal of Marriage and the Family* 54: 285-292.

Gluckman, M. 1950. "Kinship and Marriage among the Lozi of Northern Rhodesia and the Zulu of Natal." Pp. 166-206 in *African Systems of Kinship and Marriage*, edited by A. R. Radcliffe-Brown and D. Forde. London: Oxford.

Hagen, G. P. 1983. "Marriage, Divorce and Polygyny in Winneba." Pp. 192-205 in *Female and Male in West Africa*, edited by C. Oppong. London: Allen and Unwin.

Laing, A. G. 1825. *Travels in the Timanee, Kooranko and Sulima Countries of Western Africa*. London: John Murray.

Lee, E. S. 1966. "A Theory of Migration." *Demography* 3: 47-57.

Mitchell, J. C. 1961. "Social Change and the Stability of African Marriage in Northern Rhodesia." Pp. 316-329 in *Social Change in Modern Africa*, edited by A. Southall. London: Oxford.

______. 1963. "Marriage Stability and Social Structure in Bantu Africa." Pp. 255-263 in *Proceedings of the International Union for the Study of Population*, vol 2. New York: United Nations.

Nabaitu, J., C. Bachengana, and J. Seeley. 1994. "Marital Instability in a Rural Population in South-West Uganda: Implications for the Spread of HIV-1 Infection." *Africa* 64: 243-251.

Olusanya, P. O. 1971. "The Problem of Multiple Causation in Population Analysis, with Particular Reference to the Polygamy-fertility Hypothesis." *Sociological Review* 19: 165-178.

Population Reference Bureau. 1995. *World Population Data Sheet.* Washington, DC: Population Reference Bureau.

Solivetti, L. M. 1994. "Family, Marriage and Divorce in a Hausa Community: A Sociological Model." *Africa* 64: 252-271.
Thomas, N. W. 1916. *Anthropological Report on Sierra Leone,* vol. 1. London: Harrison.

DIVORCE IN JAPAN
HISTORICAL CHANGES AND CURRENT ISSUES

Noriko Iwai

INTRODUCTION

Japan is now in the midst of a long trend of increasing divorce rates which started in the mid-1960s. Although the level of the divorce rate is still lower than in most other industrialized countries, people in Japan now wonder whether Japanese society is becoming a high-divorce society, and some see this as a breakdown of the family system in Japan.

In this chapter, I examine divorce trends in Japan and related current issues from comparative perspectives, both historical and cross-societal, especially with respect to the United States. Contrary to the rather conservative image of present-day Japanese society, Japan had experienced higher divorce rates earlier in the modern era. Therefore, we can examine the differences between the circumstances of divorce in earlier years and in the present, and whether there was a crisis in the family system then because of such high divorce rates. From a cross-societal perspective, the increase in divorce occurring in Japan is clearly in line with social changes that have occurred in other Western countries. Not only divorce, but also people's ideas about individuality, family, and marriage have been changing in accordance with changes in industrial systems. By comparing Japanese divorce trends with those in other countries, one can obtain insights that may help explain family changes in Japan, as well as those in industrialized countries in general.

As a consequence of increasing divorce rates, more people and governments are facing new problems, among them: how to decrease the feminization of poverty, how to guarantee the best interests of children psychologically and economically, how to guarantee the parental right of the non-custodial parent, and how to revise the laws on marriage and divorce to fit the current situation, but not facilitate it. Japanese society has just begun to address these problems, while more experience and knowledge have been accumulated in the United States.

In the first part of this chapter, I examine the historical trends in Japanese divorce rates and those of several other industrialized countries. Next, I focus on institutional and attitudinal changes regarding divorce from the early modern age to present-day Japan. In particular, I focus on types of divorce, procedures, and grounds for divorce specified by law, timing of divorce, reasons for divorce, and social norms for divorce. Based on my findings, I examine the relationship between an increase in divorce and an increase in Japanese women's employment. In the second part of this chapter, I discuss the effects of divorce on children, and aspects of life after divorce, such as single-mother families, single-father families, and remarriage. Based on the preceding discussion, I return to the question of whether or not increasing divorce rates will lead to a breakdown of the family system. In the final part of this paper, I discuss the revision of the civil law which is under investigation by the Japanese government.

HISTORICAL CHANGES IN DIVORCE IN JAPAN

Divorce Rates in Japan

Figure 1 shows the number of divorces and their annual crude rates from 1883 to 1996. The divorce rate in 1883 was the first recorded and also the highest recorded in Japanese history, at 3.38 per 1,000 people. The actual divorce rate was even higher, because there were many unrecorded marriages. People did not feel much obligation to report their marriages to a public office, and there was a variable time lag between the marriage date and that of reporting the marriage. The divorce rate continued to be high until 1897 (2.93), then it decreased radically after July of 1898, when the civil law took effect, falling to 1.53 in 1899. Thereafter, it slowly decreased until 1938, when it hit the lowest recorded rate of 0.63.

After World War II, a period of confusion, the divorce rate stayed relatively high. It began to decrease again in 1951 and continued to decrease slowly so that by 1963, the divorce rate was 0.73, the lowest rate recorded since World War II. The rate then began to increase and continued to increase annually for 20 years until 1983 (1.51).

This peak is referred to as "the first divorce boom." After this upward trend, the divorce rate showed some signs of decrease. It was 1.26 in 1988; then, once again, it began to increase in 1991. There were 179,198 divorces in 1992, the highest

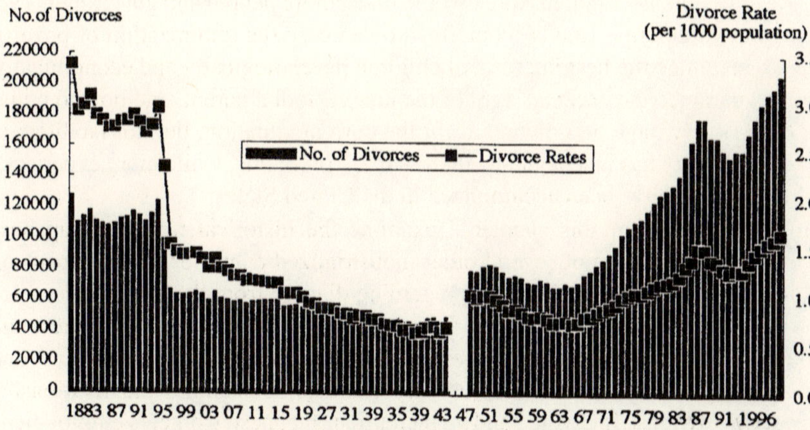

Source: 1883–1943, *Nihon Tokei Nenkan and Nihon Teikoku Jinko Dotai Tokei,* Bureau of Statistics, Cabinet; 1947–1996, *Vital Statistics,* Ministry of Health and Welfare.

Figure 1. Number of Divorces and Divorce Rates in Japan, 1883–1996

number to that point in history. Since then, both the number of divorces and the crude rate have increased every year. This rise is referred to as "the second divorce boom." In 1996, the number of divorces was 206,966 and the crude rate was 1.66.

The crude divorce rate is affected by the proportion of married people to the total population, and by the constitution of age groups. Thus, the current increase in divorce is partly a result of demographics; specifically, the presence of baby boomers who were born between 1947 to 1949, and their children who have started to marry in recent years. The number of divorces increased most in two categories of marriage duration: divorce within five years of marriage, and after more than 20 years of marriage. The probability of first marriages ending in divorce was one in every ten couples in 1965, but it had increased to one in every five couples by 1990 (Ministry of Health and Welfare 1996).

Divorce Rates in Other Industrialized Countries

In the 1960s and 1970s, the divorce rate increased sharply in many other industrialized countries, as well (Figure 2). In the United States, the divorce rate has continued to increase gradually since the middle of the nineteenth century (no records exist before that) except for temporal fluctuations during wars, economic depression, and the 1950s. It was 2.22 in 1962, and reached 5.27 in 1981, when one out of two couples would eventually divorce. The situation is the same in Europe. The gradual upward trend that started before World War II began a steep increase in the 1960s, except for a temporal fluctuation just after the War.

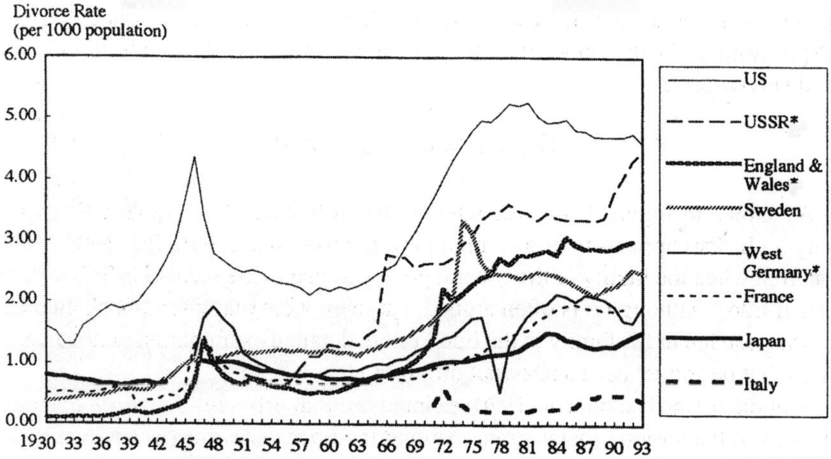

Source: U.N. *Demographic Yearbook.*
* Data from Russia Federation after 1991; data from United Kingdom (including Scotland and Nothern Ireland) after 1971; including data from former West Berlin between 1957 and 1990; data from Germany after 1991.

Figure 2. Divorce Rates in Industrialized Countries, 1930–1993

The long, upward trend of divorce in industrialized countries was brought about by changes in the family system as a result of industrialization. With increasing industrialization, more and more people obtained jobs as employees. People could earn money, not as family units, but as individuals (Goode 1963). Ability, not family line, was valued. During the 1960s and 1970s, more and more women entered the labor force and attitudes toward the family changed. In the United States, the first factor had a more significant effect. The proportion of married women with children under six who participate in the labor force more than doubled from 19 percent to 43 percent (Cherlin 1981/1992). Working outside the home does not necessarily correspond to divorce. However, increased economic opportunity for married women with children raises the possibility of making a living after divorce. In Europe, the effect of changing family norms and values was more influential (Roussel 1988). According to Roussel, "The image of family life is no longer organized around the purpose of founding a stable family; and marriage is not now a primary institution, but it is becoming simply an emotional relationship" (p. 51).

In order to deal with the rapid increase and changes in content of divorce, many countries reformed divorce laws in the late 1960s to 1980s. Before then, a spouse who was responsible for breaking up the marriage could not ask for a divorce, even if the marriage had completely collapsed. In addition, grounds for divorce were limited to infidelity, abuse, abandonment, and so on. Without these grounds the couple could not get divorced, even if both sides agreed to do so.

The courts always had to decide guilt. To reform this inelastic procedure, many countries adopted no-fault divorce laws, which grant divorces without placing responsibility. At the same time, they tended to promote the rights of children of divorced parents.

Divorce in the Edo Period

As shown in Figure 1, divorce rates in the early Meiji Era (1868–1911) were very high. This appears to be a continuous trend from the Edo Period (1603–1867, a period when the family of the general governed in a castle located in Tokyo, then called Edo). Although it is often argued that there were many send-back divorces by the husband or his family in the Edo Period, detailed examination of documents from that period reveal a somewhat different view.

Tsubouchi and Tsubouchi (1970) pointed out that different dynamics operated in each of the various social strata. Among the upper *samurai* (warrior) stratum, who regarded *ie* (the traditional Japanese family) as very important, divorce was infrequent. If a divorce was sought, the husband's family and the wife's family discussed the matter with each other, and with relatives of both sides. The head of each family then had to submit an application to his respective master. In the application, it was noted that after thorough discussion they agreed to divorce because there was no destined relation between them. However, it was difficult to send a bride back to her family as there was a tendency toward hypogamy in *samurai* marriages. One got a bride either from the same level of the family, or a slightly higher level.

Among the lower *samurai,* the political and social meaning of marriage was not so highly valued. Moreover, the tendency toward hypogamy among the *samurai* class created difficulties for daughters of lower *samurai*. They had less chance of finding a spouse; and, even after marriage, they might be easily replaced by other candidates if the husband's family did not like them. One of the teachings of Confucianism was that in general, a woman should follow the ones before her. This was used as a reason to send a wife back to her family. Therefore, send-back divorces were relatively common among the lower *samurai* stratum.

Divorces were further different for common people who comprised most of the population. In their case, divorce was concluded when the husband delivered a letter of divorce to his wife. This *Mikudarihan* was a letter of three-and-a-half lines, named after a common form of the divorce letter that was said to be copied from a divorce letter that appeared in old Chinese stories. This type of divorce is often considered an autocratic divorce decided only by the husband's side because only he could write the letter, and he often wrote that he would divorce her as he pleased. However, based on an analysis of 550 letters of divorce and related documents, Takagi (1987) suggested that the letter was used mainly as a proof of the end of the relationship and as a certificate for the wife's remarriage. The divorce was also concluded after a long discussion with go-betweens, relatives, neighbors, and village headmen who tried to coordinate the results for both sides. It was considered a husband's good sense and consideration for his former wife not to write reasons for divorce in detail, regardless of faults on any side.

Among the common people, there were a significant number of runaway wife divorces. Especially in farm families, the labor of young women was valuable. Therefore, a runaway wife was accepted by her family of origin without much struggle and she had a chance of remarriage. It is true that in some regions, such as villages in the northeastern districts where winter is severe, many send-back divorces occurred. However, even in these cases, the image of divorce or divorced women and the value of virginity were very different from those held by the *samurai* class. In sum, a variety of divorces occurred during the Edo Period, and it was not an unusual event in the society.

Divorce in the Transition from the Meiji Era to the Present Day

Early in the Meiji Era, it became a family head's duty to register all family members at a public office. However, in reality old customs persisted. As shown in Figure 1, the divorce rate decreased sharply after the enforcement of the civil law in 1898. Under the law, one had to marry by law and divorce by law. The system of divorce by agreement was introduced in which a couple who sought a divorce had to submit a formal paper with their signatures and seals. A long-term downward trend in the divorce rate ensued. When the rate began to increase in 1963, changes had occurred in the context of divorce.

As Western thought and morals flowed into the nation, the Meiji government tried to spread the moral sense of the *samurai* class to the whole population. The morals of the *samurai* class were facilitated through school education. In the area of family law, the *ie* system of the *samurai* class was regarded as a model. The morality of *samurai* marriage, in which a virtuous woman would never remarry, and the morality of a modern Western marriage, which emphasized a more equal lifelong bond between a couple, both helped to deter divorce, although the weight of each differed depending on the region and time period.

With increasing industrialization, there were more opportunities to choose one's occupation and to move from one's residence. Many of the second or third sons of farm families began to live independently and move into the cities. The proportion of nuclear families in cities increased, and a sense of citizenship grew stronger. The divorce rate was higher in rural areas than in cities during the 1920s and 1930s (Kawashima and Steiner 1960). Even among extended families, whose proportion was higher in rural areas, the number of send-back divorces decreased in response to the changes in values and social circumstances. The average age at first marriage, for both husband and wife, had increased gradually up until 1942. As women entered into marriages with more maturity, the occurrence of send-back and runaway divorces decreased somewhat.

While the number of divorces had decreased, a new type of divorce appeared in urban areas, especially among families of employees—that of a couple making a personal decision to divorce because they were uncongenial or incompatible with each other. This was similar to the type of divorce that had been increasing since the 1930s in Western societies. Following the radical changes in values that occurred after World War II, the ideology of the conjugal family and individualism gradually penetrated the nation. When industrialization and urbanization accelerated and job opportunities for women expanded in the 1960s, this type of divorce increased.

The Divorce Process

Under the Meiji civil law, there were only two ways to get a divorce: a divorce by agreement or a divorce judgment by a court. After the reformation of the law in 1947, family courts were founded and two other methods of divorce were added: a divorce through mediation and a divorce judgment by a family court. As the proportion of divorces through mediation increased, the proportion of divorces by agreement decreased from almost 100 percent to 90 percent. This occurred because wives who were almost forced to agree on unfair divorce conditions by the husband's side avoided divorce by agreement and asked for mediation in a family court. Currently, more than 70 percent of applicants for mediation are wives. Older wives tend to use mediation more than younger ones. In nearly half the cases of mediation, the couples agree to divorce. The process tends to take more time when the couple has more property and more children, although it also depends on their personalities (Yuzawa 1995). The proportion of divorce through mediation was 8.6 percent of the total in 1994.

In cases where both sides agree to divorce, but differ in conditions, family courts sometimes judge the case. However, if either partner objects to the judgment within two weeks, it loses effect. The number of divorces by family court judgment is very low; there were only 72 cases in 1994.

One can also ask for a divorce from a district court, if it cannot be obtained through mediation or by family court judgment. The district court grants a divorce if the grounds for the divorce fit one of the following five criteria, which are coded in the civil law: spouse committed an infidelity; intentional abandonment by spouse; spouse has been missing for more than three years; spouse has a severe mental illness with no chance of recovery; and other important reasons for being unable to continue the marriage. The proportion of this category of divorce was only 0.9 percent of the total divorces in 1994.

Changes in Reasons for Divorce

Not only the procedures, but also reasons for divorce have changed. According to a survey of women who had divorced by agreement (Bureau of Women and

Adolescents 1961; conducted in 1960), major reasons for choosing divorce were "economic problems" (38%; multiple choice), "incompatibility" (35%), and "husband's having a lover" (29%). There were also reasons relating to a conflict with the husband's relatives: "on bad terms with his parents or siblings" (13%) and "out of harmony with his family custom" (9%). In 12 percent of the cases, "abuse by husband" was given as a reason.

Another survey on divorce by agreement which was conducted by the Ministry of Health and Welfare in 1968 (Statistics and Information Department 1970) showed a similar tendency, but with some differences. In this survey both ex-wives and ex-husbands were surveyed. According to wives' answers, the most frequent reason for divorce was "economic problems" (27%), followed by "becoming disgusted with husband's personality" (24%). "Infidelity of husband or of herself" was the major reason in 20 percent of cases. The proportion of "on bad terms with his parents and other relatives" (9%) was close to the proportion of "uncongeniality in sex life" (7%). Wives gave similar answers in a survey conducted ten years later in 1978 (Statistics and Information Department 1980).

On the other hand, husbands gave reasons of "becoming disgusted with wife's personality" (24%) most frequently, followed by "infidelity of himself or his wife" (22%), and "economic problems" (20%) in 1968. The proportion of "personality" and that of "infidelity" increased in 1978 (31% and 28%, respectively).

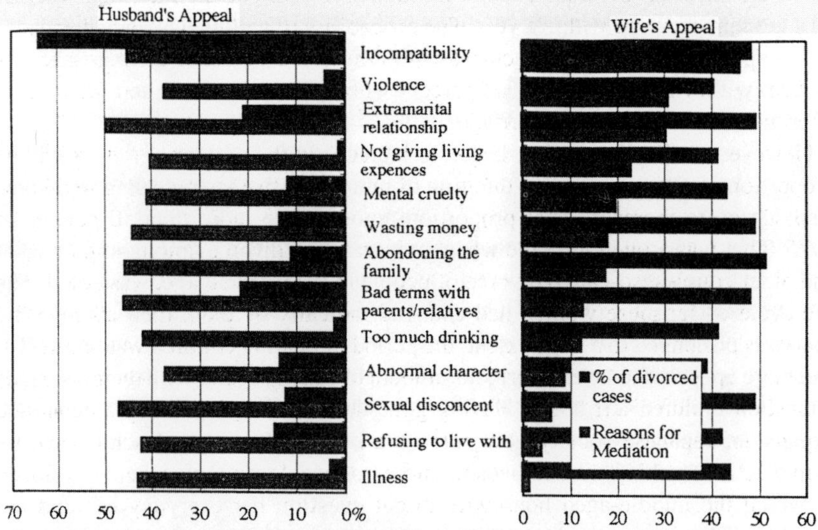

Source: *Annual Report on Judicial Statistics 1995*, Supreme Court.

Figure 3. Reasons for Divorce Mediation and Percentage of Cases Concluded to Divorce, 1995

Women were more likely to be the first to speak of getting a divorce. According to husbands' answers, that percentage rose from 51 percent to 55 percent in the period from 1968 to 1978. The percentage of cases in which husbands spoke first decreased slightly from 36 percent to 35 percent. All the other categories also decreased slightly: husband's parents (2% to 1%); wife's parents (6% to 5%); other (5% to 3%). An interesting discrepancy is that women gave a higher figure of wives' speaking first: 65 percent in 1978.

Among couples who ask for divorce mediation, the most frequent reason is "incompatibility" (Figure 3). That proportion is increasing every year. A noticeable reason for divorce mediation is "husband's violence," which appears in one-third of the cases. Comparing reasons given by wives and those by husbands, one notes several differences. "Being violent," "not giving living expenses," and "drinking too much" appear much more often in wives' appeals. "On bad terms with parents and relatives" and "refusing to live with" appear more often in husbands' appeals.

It tends to be easier to get a divorce in cases where a couple no longer share feelings, as in "being incompatible," "having extramarital affairs," or "abandoning the family," than in cases where a spouse is violent or drinking too much.

The Duration of Marriage

More drastic change is observed in the timing of divorce (Figure 4). In 1947, 61.2 percent of all divorces occurred within five years of the marriage whereas 14.1 percent occurred within a year. The probability of getting a divorce decreased as time passed: 23.4 percent occurred within five to ten years of the marriage, 8.5 percent within 10 to 15 years, 3.8 percent within 15 to 20 years, and 3.1 percent after more than 20 years of marriage.

However, as the divorce rate began to increase in the middle of the 1960s, the proportion of marriages with a duration of more than five years ending in divorce also started to increase. That proportion amounted to more than 50 percent in 1975. That was around the time when an increase in divorce among middle-aged and older couples, so-called divorce of people in the matured age, was noted. The life cycle of Japanese women had changed radically by then: their average life span was prolonged to a large extent, the period of bearing children was shortened, and there appeared to be an enormous amount of time to spend with their husbands after their children left home. Having gone through decades of rapid economic expansion, Japanese society now entered a period of stable economic growth. While active working women became more noticeable, an atmosphere developed in which the middle-aged housewife might question her everyday life and her marriage. This is commonly referred to as "wives in their age of melancholy."

On the other hand, the proportion of divorce among young couples has decreased. Although the expression, "Narita divorce," (divorcing at the Narita International Airport on the way back from a honeymoon overseas) is very

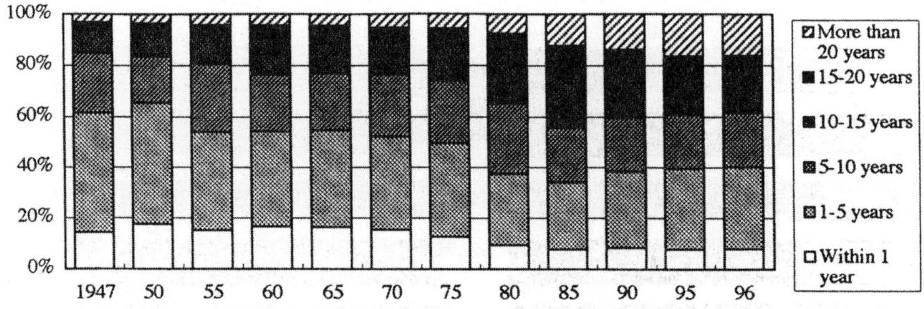

Source: *Vital Statistics*, Ministry of Health and Welfare.

Figure 4. Divorces by the Duration of Marriage, 1947–1996

appealing, the proportion of honeymoon divorces has decreased. In 1996, the proportion of couples who divorced within five years of the marriage was 40.1 percent (7.7% of couples divorced within one year), as compared to 21.3 percent within five to ten years, 12.9 percent within 10 to 15 years, 9.4 percent within 15 to 20 years, and 16.3 percent after 20 years of marriage (8.8% within 20 to 25 years, 4.6% within 25 to 30 years, 1.9% within 30 to 35 years, and 1.0% after 35 years of marriage). Although the principle of divorce decreasing with the duration of marriage—the longer they live together, the less the probability they will divorce—has been maintained, the distribution has changed drastically.

Changing Social Norms on Divorce

As both the number of divorces and the rate of divorce increase annually, people become more tolerant about divorce (Figure 5). The increase in divorce and the change in social norms affect one another. An increase in divorce leads to more tolerance, and this change in social norms leads to an increased divorce rate. In the current period of a rapidly rising divorce rate, a gradual attitudinal change has been observed. In 1979, one out of four women either agreed with or understood the idea that one had better seek a divorce if unsatisfied with one's spouse. In 1984, one out of three women agreed with this idea. The attitudinal change of women was greater than that of men during the same period. Although the nuance of questions and categories of answers differ in surveys taken after 1984, so that it is not possible to make a strict comparison of figures, there has been a sharp increase in the proportion of both men and women who either agree or somewhat agree with this idea.

Teenagers are more sensitive than their parents in their attitudes toward divorce. In a survey of third year junior high school students and their parents in Hyogo Prefecture (Hyogo Institute for Study on Family Issues 1990), 82.2 percent of mothers, most of whom were in their early 40s, answered that they agreed to the

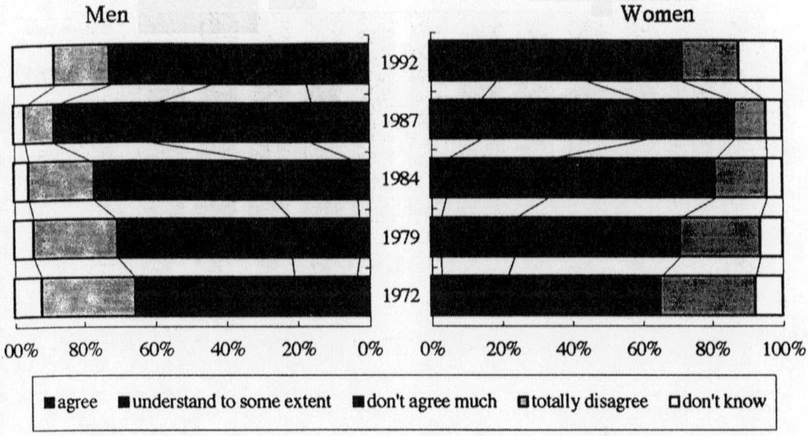

Note: In 1992, categories of choice were "agree," "somewhat agree," "somewhat disagree," "disagree," and "don't know."

Source: *Jyosei ni kansuru Seron Chosa* and *Danjyo Byodo ni kansuru Seron Chosa*, Prime Minister's Secretariat.

Figure 5. Changes in Attitudes on Divorce, 1972–1992

idea of getting a divorce if the couple do not love each other any more. On the other hand, their children—who were much more open-minded about one's lineage or cohabitation than their parents—were somewhat careful about divorce; the proportion of those who agreed with the idea was 63.3 percent for female students and 50.1 percent for male students. At their age, the idea of marriage may still be viewed with admiration. In addition, many of the students might have answered the question from the perspective of children who were involved in the divorces of their parents.

Divorce and Women's Employment

In the period of increasing divorce, more and more Japanese women were entering into the workforce as employees, rather than as helpers in the family business. The advancement of married woman was particularly noticeable. In 1960, only one out of ten (9.2%) married women were employed; by 1995, more than one-third of married women (36.2%) were working as employees (Bureau of Women 1990, 1996). If one plots the proportion of women in the labor force for each age, an M curve appears. However, in the 1970s and 1980s, the bottom of this curve gradually rose. In 1995, 21.2 percent of married women with children under three, and 35.7 percent of those with children aged four to six were employed.

Table 1. Divorce Rates by Husband's and Wife's Occupation (per 1,000 Married Population)

	Husband's Divorce Rate			*Wife's Divorce Rate*		
Occupation	*1980*	*1985*	*1990*	*1980*	*1985*	*1990*
Total	4.91	6.20	5.24	4.90	6.20	5.24
Working total	4.98	6.06	5.58	4.37	5.65	5.41
Professional/technical	4.69	4.70	5.17	5.31	6.04	6.12
Managerial	1.01	2.70	2.71	2.03	5.13	5.71
Clerical	5.30	5.46	6.12	5.45	5.55	6.06
Sales	6.85	7.71	5.98	2.03	8.48	7.74
Service	13.32	16.65	17.32	10.12	12.52	11.68
Protective service	2.74	3.40	3.38	9.11	18.46	17.96
Agricultural/forestry/fisheries	1.74	2.75	1.68	0.75	1.05	0.61
Transport/communications	8.14	9.66	10.20	5.60	7.30	11.82
Craftsmen/production process/labours	4.61	5.21	4.52	2.27	2.81	2.37
Mining	6.95	19.17	—	—	74.22	—
Not classifiable	—	172.86	51.14	—	83.15	45.14
Non-working total	4.51	7.63	3.21	5.40	6.75	5.09

Note: Figures for 1980 are from Kanai, H. Divorce by Occupation. *Kosei no Shihyo*, (15, 1984, p. 16). Figures for 1985 and 1990 are calculated based on data from *Special Report of Vital Statistics* and from census in 1985 and 1990.

The question in this case is whether or not the increase in women's employment caused the increase in divorce. Table 1 shows divorce rates for different occupational categories of wives and husbands in 1980, 1985, and 1990. Divorced wives and husbands were categorized according to the occupations in which they had been engaged before they stopped living with their spouses. A comparison of the divorce rates of working wives and non-working wives in 1980 shows that the divorce rate of working wives was much lower than that of non-working wives (4.37 versus 5.40). This difference was also observed in 1985 (5.65 versus 6.75) when the peak of the first divorce boom, which began in 1964, had just passed. Moreover, divorce rates were relatively low for the occupations that a majority of married women choose. Close to 30 percent of married working women were engaged in factory or labor work, and more than 20 percent in clerical work. The divorce rate for factory or labor workers was 2.81, and that of clerical workers was 5.55 in 1985. Therefore, wives' employment in itself does not necessarily relate to divorce.

However, it is also necessary to focus on the relatively high divorce rates of service workers (12.52) and sales workers (8.48). These occupations have contributed to the expansion of job opportunities for married women. In essence, a feeling of "since I am working I will be able to survive after divorce" might be raising the rate of divorce. At the same time, it should be remembered that only these occupations are widely available for wives who seek jobs in middle age. According to a survey on divorce by agreement in 1978, about one out of six women who started to work after divorce entered into either sales or service occupations. It is reported that wives who consider divorce tend to shift from a status of housewife to full-time worker, or from part-time worker to full-time worker in preparation for life after divorce (Ogawa and Ermisch 1990).

The latest available data, which were collected in 1990, a year before the beginning of the second divorce boom, showed a somewhat different picture. While divorce rates for both working and non-working wives decreased from 1985, the divorce rate was higher for working wives than for non-working wives (5.41 versus 5.09). In addition, while divorce rates for wives who were engaged in sales, service, or factory or labor work decreased, the rate for wives who were engaged in professional or technical, managerial, clerical, or transport or communications work had continued to increase since 1980. It remains to be seen whether these recent trends appear again in the latest census (1995), which was conducted during the current phase of an increasing divorce rate.

AFTER DIVORCE—DEPARTURE FROM DIVORCE

Divorce is basically the decision of a couple. However, their decision influences not only their own lives, but inevitably those of their children as well as the children's attitudes about life and relationships. In this section, I look first at the impact of parental divorce on children, and then examine single-mother and single-father households. After examining recent trends in remarriage, I return to the question I presented at the beginning of this chapter: does the current increase in divorce reflect a breakdown of the family system in Japan? At the end of this section, I take up issues involving a revision of the family law, currently under investigation by the Japanese government.

Divorce and Children

As the number of divorces has increased, the number of minors involved in the divorce of their parents has also increased (Figure 6). In the 1950s and 1960s, the majority of divorced couples had either no children or only one child. Many wives must have foregone divorce, as they could not manage single motherhood with more than two children. However, the prospect for life after divorce became gradually brighter as job opportunities for women, and day-care centers for children, increased, and a child support allowance to divorced mothers was provided. Consequently, there was an increase in divorces among couples with more than two children. According to a public opinion survey in 1986 (Public Information Section 1987a), 54 percent of both men and women agreed that one cannot get a

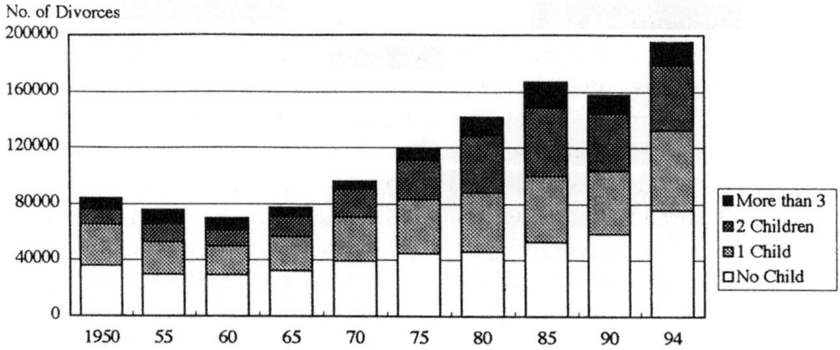

Note: Children under 20.

Source: *Vital Statistics*, Ministry of Health and Welfare.

Figure 6. Divorces by the Number of Siblings, 1950–1994

divorce if there are children. In a more recent opinion survey (Institute of Population Problems 1991; conducted in 1990), 45 percent of men and 50 percent of women agreed that maintaining a loveless marriage will do more harm than good to children.

Under Meiji civil law, minor children basically belonged to the father, both within marriage and after divorce. When the law was reformed in 1947, custody was left up to each couple. Since that time, the number of cases in which mothers received custody of their children has increased gradually (Figure 7). In the middle of the 1960s, when the economic circumstances of divorced mothers had improved, more mothers than fathers received custody of their children. Recently, in three out of four cases, children of divorced parents were living with their mothers. However, there is a marked increase in the number of cases in mediation, where couples dispute furiously over the custody of their children.

Psychological Effects on Children

The psychological effects of divorce on children have not yet been thoroughly investigated in Japan. The accumulated research in the United States has been very helpful on this point. One study shows that children reared in single-parent families develop better than those reared in unhappy families (Hetherington, Cox, and Cox 1982). Although almost all children experience an initial period of great emotional upset following a parental separation, most return to a normal developmental course within one or two years following the separation. However a small number of children experience some long-term psychological problems as a result of the breakup, which may persist into adulthood (Wallerstein and Kelly 1980).

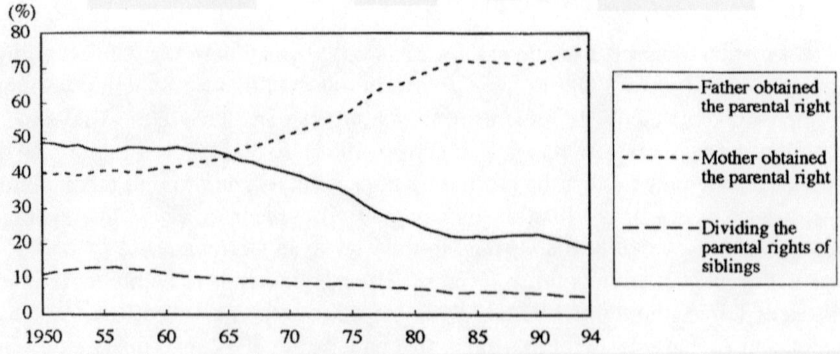

Source: *Vital Statistics,* Ministry of Health and Welfare.

Figure 7. Percentage of Divorces by the Parental Right of Children, 1950–1994

What would help children's adjustment to parental divorce? Children do better when the custodial parent, usually the mother, can reestablish an orderly and supportive household routine. Children also do better when there is less conflict between their parents. However, it is less clear whether children do better, on average, when they have a continuing relationship with both parents after the separation (Cherlin 1981/1992).

Visitation Rights and Joint Custody

In the United States, visitation rights are usually approved as the right of a non-custodial parent and the children. Moreover, after the Joint Custody Law was enacted in the 1980s, in most states, mothers and fathers share parental rights and responsibilities as a general rule, if both sides dispute custody.

Japanese family courts have tried to promote visitation rights of the non-custodial parent during mediation since the 1960s. However, it has never become popular. According to a survey on the relationship between children and their non-custodial parents, the proportion of single-mother families where children kept in contact with their fathers was 29 percent, and that of single-father families where children kept contact in with their mothers was 15 percent (Madoka 1985). It should be noted that most respondents of that survey were members of an active group of divorced women. Therefore, these figures seem to be lower among divorced families in general.

A female lawyer who had been working with a number of divorce cases noticed the following tendency. When asked with whom they wanted to live, a considerable number of children in recent cases answered, "I want to live with either of my parents who will stay at this house."

Divorce Agreements and Financial Provisions

"If I were to divorce, I would ask for 32,000,000 yen [about U.S.$280,000] for consolation." This was the average figure in answers that 580 single, working women and female college students gave in a survey conducted in the Tokyo metropolitan area (*Asahi* newspaper, January 30, 1989). In reality, even in the case of divorce through mediation, the proportion of cases that had any contract on dividing marital assets or consolation money was 56 percent in 1995. The average amount of money that husbands paid to their wives in these cases was 4,052,000 yen, far below the aforementioned figure. The amount tends to be higher in marriages of longer duration: 1,914,000 yen for marriages that lasted less than five years, and 7,276,000 yen for marriages that ended after 20 years. This latter figure is still far below women's expectations. In the case of divorce by agreement, women obtain much less. According to a survey in 1978, 50.7 percent of husbands gave some of their assets to their wives. This figure is higher than that for 1968. However, in the majority of cases, the monetary value of the assets was less than 1,000,000 yen.

Living with Mother after Divorce

According to the annual Basic Survey on the National Life, the number of mother-headed households was about 335,000 in 1965 and about 482,000 in 1995. Because the number of all types of households has increased, the proportion of mother-headed households has remained at about 1.3 percent. Another type of single-mother household is living with grandparent(s). Including this type, the number of single-mother households increased about 789,900 in 1993. Causes of single-mother households have changed. In the 1950s, eight out of ten women lost their husbands as a result of illness, war, or traffic accidents. The proportion of mothers who separated from their living husbands has gradually increased, and in 1993 accounted for 73.2 percent (divorced = 64.3%, never married = 4.7%, other = 4.2%).

The average age of single mothers was 40.2 years in 1993 (Bureau of Children and Families 1995). Of these, 56.8 percent had more than two minor children. The economic situation of divorced, mother-headed families is better today than in the past; however, it is still not easy. Although 89.1 percent of divorced mothers have a job, and 55.3 percent of them are full-time employees, their average annual income, including tax, in 1992, was 2,020,000 yen. That was much less than that of widowed mother families (2,540,000 yen). In fact, that figure is less than one-third (31.2%) of the average income for all types of households (6,480,000 yen).

Despite the parental obligation to support minor children even after divorce, in three out of four cases fathers have never paid child support. Only 14.9 percent of single-mother families receive it now, and 16.4 percent received it in the past.

According to a survey on divorce and single-mother families in 1986 (Japan Research Institute on Child Welfare 1987), the amount of child support that was actually paid to those families is less than 20,000 yen a month in about half of the cases, and 20,000 to 40,000 yen in one-third. More than 80 percent of those families receive child support allowances from the state.

In the case of divorce through mediation, a majority of couples have developed a contract regarding child support payments, and the proportion of such couples has gradually increased, from seven out of ten in 1987 to eight out of ten in 1994. The average amount of the payment for child support has also increased from 37,000 yen a month in 1987, to 44,000 yen a month in 1994. However, this payment still covers only a part of the expenses for children, as single-mother families pay about 33,000 yen (1988) for school expenses alone in a month.

Seventy-one percent of single mothers have some difficulty, and one-third have economic problems (Figure 8a). Sixty-four percent of single mothers are concerned about their children. Although mothers of intact families are also concerned most about education and schooling of their children (Bureau of Children and Families 1992), the burden of school expenses is much heavier for single-mother families (Figure 8b).

According to the survey on divorced and single-mother families, about 70 percent of them made a stable living in such severe circumstances. The survey revealed two driving forces: an independent attitude of divorced mothers, and

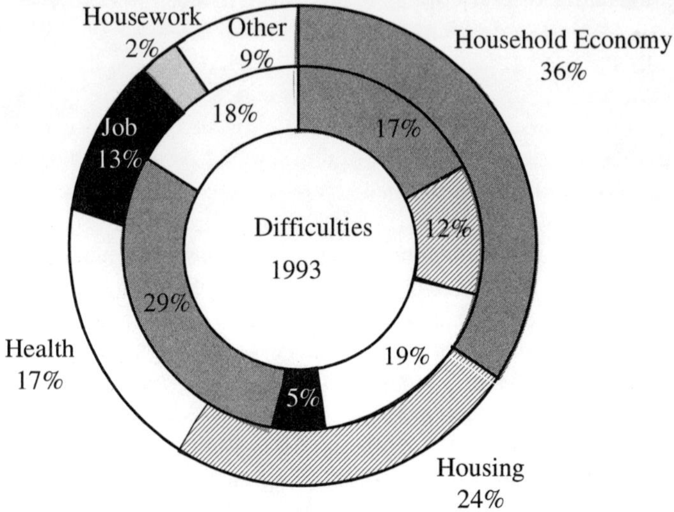

Source: *Zenkoku Boshi Katei nado Chosa 1993*, Ministry of Health and Welfare.

Figure 8a. Difficulties Single Mother (outer circle) and Single Father (inner circle) Faces

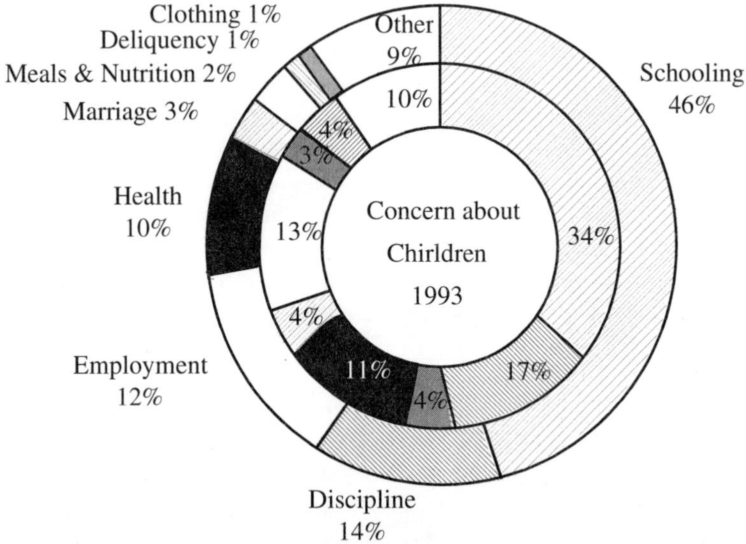

Source: *Zenkoku Boshi Katei nado Chosa 1993*, Ministry of Health and Welfare.

Figure 8b. Concerns About Children: Single Mother (outer circle) and Single Father (inner circle)

support from their relatives. Receiving support from relatives just after divorce was a significant factor in restoring the life of single-mother families.

Living with Father after Divorce

The proportion of children who lived with their fathers after divorce decreased after the 1950s, then stabilized around 22 percent in the 1980s. This percentage has gradually decreased again in recent years (see Figure 7), although the overall number of children living with their fathers has been increasing as the total number of divorces increases. The number of single-father households was about 64,000 in 1975, and about 84,000 in 1995. The proportion of single-father households to the total number of households has been 0.2 to 0.3 percent for the past 20 years. Many single fathers also live in three-generation households. Including this type, the total number of single-father households was 157,300 in 1993.

The first national survey on single-father households was conducted in 1983 (Bureau of Children and Families 1995). Even at that point, there were more divorced, single-father families (60.1%) than widowed, single-father families. In the third national survey conducted in 1993, the proportion increased to 62.6 percent.

The average age of single fathers is 44.2 years old. Some 55.7 percent of them live with more than two minor children. Ninety-three percent of divorced fathers have a job, and 71.7 percent of them are full-time employees.

Their average annual income, including tax, in 1992, was 4,230,000 yen. Although this figure is much higher than that of single-mother families, it is two-thirds (65.3%) of the average income of all kinds of households.

Six out of ten single fathers (62%) have some difficulty, and one-third of them struggle with housework (see Figure 8a). While most single mothers do housework by themselves, single fathers take care of housework on weekdays in less than half (46%) of the cases. In 36 percent of the cases, grandparents take care of it, and in 12 percent of the cases, children take charge of it (among single-mother households, 3 percent of children take care of housework.) This is not only because fathers are not as accustomed to housework as mothers are, but also because fathers have to conform to the Japanese business customs that often ask male workers to work overtime, to work on holidays, or to make business trips. If a man refuses, it often leads to a decrease in his income, or even to the loss of his job. According to a survey by the Association of Single Father's Welfare in the suburban city of Hirakata, three out of ten fathers have changed jobs after becoming a single father, and this proportion is higher for fathers with younger children: 40 percent for fathers with children under three, and a little less than 20 percent for fathers with school-age children (Hirano et al. 1987).

Studies indicate that 59.8 percent of single fathers are concerned about their children. Just like single mothers, they are most concerned about the schooling of their children. However, they are more concerned than single mothers about their children's discipline, meals and nutrition, and clothing and belongings (see Figure 8b). While 71.2 percent of single mothers have someone to consult with their problems, that proportion is only 46.5 percent for single fathers. Both single fathers and mothers often consult with their relatives, friends, and neighbors. However, single fathers rely on public offices more than single mothers.

Remarriage

Some people who had difficulty getting a divorce may decide to remain single. On the other hand, some remarry with the hope of building a successful marriage the next time. The meaning of "success" in marriage varies among people.

Figure 9 shows the proportion of remarriages to the total number of marriages for men and women. In 1952, when the effects of World War II still lingered, the proportion of remarriages for either husband, or wife, or both was 18.8 percent. It gradually decreased to 10.8 percent in 1971, when the first generation of baby boomers created a marriage boom. After the boom, the number of first marriages consistently decreased, while the number of remarriages gradually increased, reaching 18.6 percent by 1989. As children of the baby boomers entered the marriage market, the proportion of remarriages once again decreased after 1990.

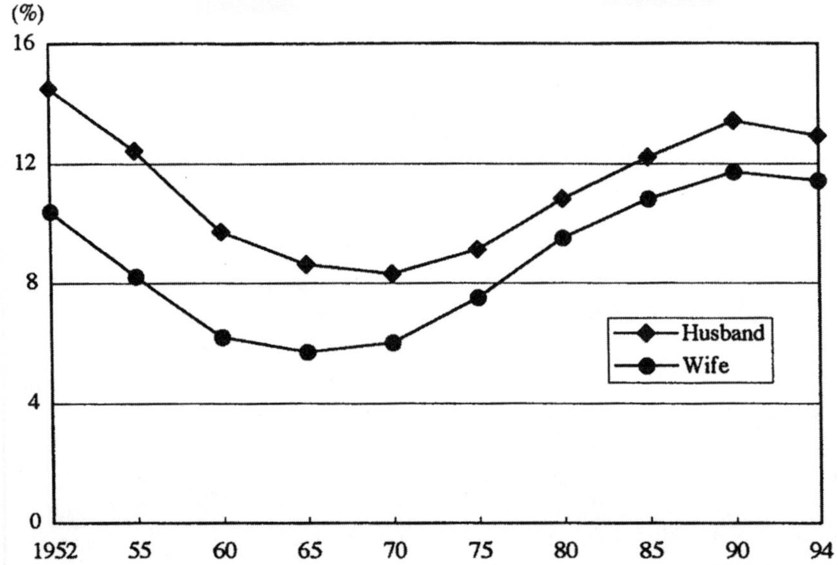

Source: *Vital Statistics*, Ministry of Health and Welfare.

Figure 9. Percentage of Remarriages, 1952–1994

However, the total number of remarriages in 1994 was 139,911 which exceeded the figures for 1952 and 1989.

Furthermore, the increase in women's remarriage rates was greater than that of men's (see Figure 9). Therefore the gap between the proportion of men and women decreased. The probability of remarriage for a divorced man was 88.9 percent in 1965, and decreased to 73.2 percent in 1990; that of a divorced woman was 46.2 percent in 1965, and it increased to 61.6 percent in 1990 (Ministry of Health and Welfare 1996). Two kinds of unions have increased noticeably: first marriage for husband, remarriage for wife, and remarriage for both husband and wife. Proportions of these two types of unions and that of the traditional type of remarriage (remarriage only for husband) were, respectively, 2.8 percent, 3.1 percent, and 4.9 percent in 1971, and 6.5 percent, 5.0 percent, and 6.4 percent in 1994. The negative image of divorce is gradually declining, as are the double standards with regard to sex.

What do divorced people think of remarriage? According to a survey on divorce by agreement, more than half of divorced men had no plans to remarry at that time, but hoped to remarry in the future; on the other hand, more than half of divorced women did not want to remarry. Their intention seemed to be influenced by the presence of children. Compared with divorced men who were not living with children, a larger proportion of single fathers hoped to remarry. However, compared with divorced women who were not living with children, a larger proportion of single mothers did not want to remarry. This difference can be traced to the everyday lives of single mothers and fathers. Despite economic hardships, single mothers enriched their lives with jobs, friends, and children. Single fathers tended to become more perplexed with housework and communicating with their children, both activities which they were unaccustomed to handling.

This finding, however, appears to be changing according to a survey on remarriage and old age (Madoka 1991). While the proportion of divorced women who hope to remarry is still small, the proportion who want love and companionship has increased. Furthermore, there were numerous responses to the survey from single fathers who seemed to enjoy life with their children.

To remarry means to form a new family. Thus, many people have to coordinate new family relationships with old ones. This coordination is especially important in the cases where the husband or wife has children who may be living with or separate from them.

Breakdown of the Family System in Japan

In the preceding sections, I have analyzed divorce in Japan from various aspects: as a historical trend, in comparison with other countries, through institutional changes, through attitudinal changes, in relationship with women's employment, through the effects of divorce on children, and through remarriage. I raised the question of whether the current increase in divorce reflects a breakdown of the family system in Japan. I would answer no. While the probability of first marriage ending in divorce has increased (8.9% for men and 10.1% for women in 1965, versus 19.6% and 18.6% in 1910; Ministry of Health and Welfare 1996) and the probability of ever marrying has decreased (89.9% for men and 92.0% for women in 1965, versus 81.6% and 89.9% in 1985; Takahashi 1989), most Japanese men and women still marry, and 73.2 percent of divorced men and 61.6 percent of divorced women will remarry. In addition, divorced mother families and divorced father families gradually restore stable family life with the support of private or public networks.

Therefore, the current upward trend in divorce in Japan does not mean that the Japanese family system has broken down. It does mean that people's idea of the family has changed. In fact, the Japanese people are becoming more aware of the value of the family. In the 1993 survey on national character, 42 percent of the randomly selected respondents chose the family as the most important thing in their lives. The proportion was 11 percent in 1958, and 33 percent in 1988 (Institute of Statistical Mathematics 1994). In addition, one out of ten people chose their children as most important. On the other hand, only 4 percent of them chose their job as most important, and the proportion of people who chose the state or the society was 1 percent. Moreover, Japanese people are now more concerned with the substance (relationship) of the family than its form. In 1953, 73 percent of the respondents agreed to the idea that if one has no children, one would adopt one to maintain the family line. That proportion decreased to 22 percent in 1993. We have seen this kind of attitudinal change toward marital relationships, too. People have become more concerned about their relationship with their partner, rather than just maintaining the appearance of marriage. In sum, the Japanese family system has not broken down, but people's ideas about family and individuality have been changing, leading to more variations of the family.

Revision of the Civil Law

Under these circumstances, more people came to feel that some provisions in the present civil law are inappropriate. Meiji civil law was revised in 1947, and enforced on January 1, 1948. Most of the framework of the *ie* (traditional family) system was abolished then, except for inheritance of properties related to the care of the family grave. For a certain period (probably up to the mid-1960s), the 1948 version of civil law, the so-called new civil law, had led the attitude of the Japanese people on family issues. Today, however, more than 40 years later, it does not always fit current realities. It has even been argued in the Supreme Court that some provisions are against the Japanese Constitution.

In January 1991, a Civil Law Section of a Legislative Council began to deliberate on a revision of the civil provisions on marriage and divorce. Although a central issue of the revision is whether spouses should be allowed to keep their own family names after marriage, the following six points are also the focus of discussion regarding revision of the divorce law:

1. Specification of conditions under which no-fault divorces are granted by courts.
2. Provision of a way to confirm an agreement on divorce by both sides. At present it is possible for one party to submit a report of divorce to a register of a municipal office without the knowledge of the other party. If the other party can expect the possibility of the submission, there is a way to counteract this by submitting an application for not receiving a report of divorce to the office in advance. The effect of this letter lasts for six months. However, if a report of divorce has already been accepted, this letter has no effect. In such a case, the person has to appeal to the court to cancel it.
3. Clarification of the rule on dividing marital assets at the time of divorce. At present, a wife's contribution, especially in cases in which she has been a housewife, tends to be undervalued.
4. Establishment of an obligation for a non-custodial parent to pay a child support to a custodial parent and establishment of a means to ensure its fulfillment.

5. Promotion of the joint rights of parents, and the joint custody and the visitation rights of a non-custodial parent in cases where such arrangements most meet the interests of their children.
6. Abolishment of the provision that only women are prohibited from remarrying for six months after divorce (currently, men can remarry anytime after divorce).

After discussing these measures for five years, involving repeated hearings from concerned parties such as district courts, family courts, organizations of lawyers, academic organizations, women's organizations, and consumer's organizations; conducting a public opinion survey; and asking citizens to send their opinion through fax, the Civil Law Section of the Legislative Council submitted a proposal of an outline of the revision of the civil law to the minister of Justice in February 1996. Major points of the proposal related to divorce are as follows:

1. Courts should admit a separation of a couple for more than five years as a ground for divorce. However, if either of the couple or their children will be put in a remarkably severe situation, mentally, socially, or economically, this ground might not be admitted. On the other hand, one of the present grounds for divorce—spouse has a severe mental illness and has no chance of recovery—should be eliminated.
2. In order to guarantee a just distribution of marital assets, all the assets considered in the distribution should be listed clearly. If each party cannot clarify its respective contribution for obtaining the assets, they should be divided equally.
3. In order to guarantee the best interests of the children, divorcing parents should make an agreement on necessary matters relating to the custody and expenses for bringing up children.
4. In order to guarantee the liberty of women to remarry, the period during which women cannot remarry should be shortened to 100 days.

The proposal for revision was not submitted to the National Diet in May 1996, however, because of a furious dispute over family names. A significant number of congressmen felt that the Japanese family system would suffer if spouses were allowed to keep their own family names after marriage. Discussion of the revised provisions on marriage and divorce has continued sporadically in Japan since this proposal was tabled.

As was shown by the case of Meiji civil law and the cases of other countries, changes to the law will surely affect the people's notion of the family and cause changes in the trend of divorce. On the other hand, as is clearly shown by the process of the revision of the family law, such changes would never have happened without the emergence of attitudinal changes toward the family in Japan.

REFERENCES

Bureau of Children and Families, Ministry of Health and Welfare. 1992. "Jido Kankyo Chosa Kekka no Gaiyo 1991" ("Outline of Results of a National Survey on Environment Surrounding Children in 1991"). Unpublished manuscript.

______.1995. "Zenkoku Boshi Setai nado Chosa Kekka no Gaiyo 1993" ("Outline of Results of a National Survey on Single-Mother Househoulds and Others in 1993"). Unpublished manuscript.

Bureau of Women and Adolescents, Ministry of Labor. 1961. *Kyogi Rikon no Jittai: Chosa Kekka Houkokusyo* (*The Actual State of Divorce by Agreement*). Tokyo: Bureau of Women and Adolescents, Ministry of Labor.

Bureau of Women, Ministry of Labor. 1990. *Fujin Rodo no Jitsujyo 1990* (*The Actual State of Women in Labor Force in 1990*). Tokyo: Printing Bureau, Ministry of Finance.

______. 1996. *Hataraku Jyosei no Jitsujyo 1996* (*The Actual State of Working Women in 1996*). Tokyo: Japan Institute of Workers' Evolution.

Cherlin, A. J. 1981/1992. *Marriage, Divorce, Remarriage*. Cambridge, MA: Harvard University Press.

Goode, W. J. 1963. *World Revolution and Family Patterns*. New York: Free Press.

Hertherington, M., E.M. Cox, and R. Cox. 1982. "Effects of Divorce on Parents and Children." Pp.233-288 in *Nontraditional Families*, edited by M. Lamb. Hillsdale, NJ: Lawrence Erlbaum Associates.

Hirano, T., T. Oka, H. Machino, and Y. Akasaka. 1987. *Fushi Katei: Kurashi no Jittai to Toujisya Soshiki eno Michi* (*Single-Father Families*). Kyoto: Minerva Syobo.

Hyogo Institute for Study of Family Issues. 1990. *Katei Kyoiku Kadai ni kansuru Chosa Kenkyu Houkokusho* (*A Report of a Survey on Themes on Home Education*). Kobe: Hyogo Prefectural Office.

Institute of Population Problems, Ministry of Health and Welfare. 1991. *Jinko Mondai ni kansuru Ishikichosa Houkoku 1990* (*A Report of an Opinion Survey on Population Problems*). Tokyo: Institute of Population Problems, Ministry of Health and Welfare.

Japan Research Institute on Child Welfare. 1987. *Rikon to Boshi Katei 1986* (*Divorce and Single-Mother Families in 1986*). Tokyo: Japan Research Institute on Child Welfare.

Kanai, H. 1984. "Shokugyo betsu ni mita Rikon" ("Divorces by Occupation"). *Kosei no Shihyo* (*Journal of Health and Welfare Statistics*) 15: 14-20.

Kawashima, T., and K. Steiner. 1960. "Modernization and Divorce Rate Trends in Japan." *Economic Development and Cultural Change* 2: 213-237.

Madoka, Y. 1985. *Rikon no Kodomo Report: Rikon no Oyako Kankei ni miru Gendai Kazoku Mondai* (*A Report on Children of Divorced Parents*). Tokyo: Fujita.

______. 1991. *Saikon Jidai* (*The Age of Remarriages*). Tokyo: Chikuma Syobo.

Ministry of Health and Welfare. 1996. *White Report on Health and Welfare 1996*. Tokyo: Institute on Health and Welfare Affairs.

Ogawa, N., and J. F. Ermisch. 1990. "Women's Career Development and Divorce Risk in Japan." Paper presented at the Nihon University International Symposium on Family and the Contemporary Japanese Culture: An International Perspective, Tokyo.

Public Information Section, Prime Minister's Secretariat. 1987a. *Kazoku/Katei ni kansuru Seron Chosa 1986* (*Public Opinion Survey on the Family in 1986*). Tokyo: Printing Bureau, Ministry of Finance.

______. 1987b. *Jyosei ni kansuru Seron Chosa 1987 (Public Opinion Survey on Women in 1987).* Gekkan Seron Chosa. Tokyo: Public Information Section, Prime Ministers Secretariat.

______. 1993. *Danjyo Byodo ni kansuru Seron Chosa 1992 (Public Opinion Survey on the Equality of Men and Women in 1992).* Gekkan Seron Chosa. Tokyo: Public Information Section, Prime Ministers Secretariat.

Research Committee on the Study of the National Character. 1994. *A Study of the Japanese National Character: The Ninth Nationwide Survey.* Tokyo: Institute of Statistical Mathematics.

Roussel, L. 1988. "Die soziologische Bedeutung der Demographischen Ershüetterung in den Industriеländern der letzten zwanzig Jahre" ("Sociological Meaning of Demographic Quakes in Industrialized Countries in the Latter Half of 20th Century"). Pp.39-54 in *Die "Postmoderne" Familie ("Postmodern" Family),* edited by K. Lüscher, F. Schultheis, and M. Wehrspan. Konstanz: Universitatsverlag.

Statistics and Information Department, Minister's Secretariat, Ministry of Health and Welfare. 1970. *Divorces: Special Research of the Socio-Economic Aspect and Vital Statistics 1968.* Tokyo: Health and Welfare Statistics Association.

______. 1980. *Divorces: Special Research of the Socio-Economic Aspect and Vital Statistics 1978.* Tokyo: Health and Welfare Statistics Association.

Takagi, T. 1987. *Mikudarihan: Edo no Rikon to Jyoseitachi* (*Women and Divorce in the Edo Period*). Tokyo: Heibonsya.

Takahashi, S. 1989. "Effects of the Japanese Mortality Declines on Life Cycle Variables." *Jinko Mondai Kenkyu* (*Journal of Population Problems)* 1: 19-33.

Tsubouchi, Y., and R. Tsubouchi. 1970. *Rikon: Hikaku Shakaigakuteki Kenkyu* (*Divorce: A Comparative Sociological Study*). Tokyo: Sobunsya.

Wallerstein, J. S., and J. B. Kelly. 1980. *Surviving the Breakup: How Children and Parents Cope with Divorce*. New York: Basic Books.

Yuzawa, M. 1995. *Zusetsu Kazoku Mondai no Genzai* (*Illustrations of Family Issues at Present*). Tokyo: Nihon Hoso Syuppan Kyokai.

CHANGES IN GREEK MARRIAGE AND DIVORCE LAW

THE IMPACT OF THE FAMILY LAW REFORM OF 1983

Aspasia Tsaoussis-Hatzis

INTRODUCTION

The relation of law and social change involves an inquiry into the relation between changes in legal rules and the changes that take place in a given social structure. These changes could be economic and technological innovations, social and demographic changes, or ideological and cultural processes, which take place within a particular social system without changing its fundamental structure. At the same time, legal rules often act as accelerators, if not vehicles, of such changes.[1]

Departing from systems theory,[2] we understand changes in marriage and divorce law as dynamic processes of normative "recasting" of relevant ideas, behaviors, and expectations. Our approach emphasizes the pedagogical capacity of norms and the power of ideas in (re)ordering social relations.[3] The Greek legal reform constitutes an illustrative example of how a change in law can effectively change the interaction among laws, ideas, and human conduct.[4]

In this light, family law becomes *constitutive* of new social expectations rather than functioning as a mirror of existing social norms. This is a view of family law as a law that is a powerful creative force of social conditions. Moreover, the power of the norms conveyed by family law lies in their implementation regardless of judicial intervention.[5] Legal rules, as models of behavior that are socially conveyable, do not necessarily depend on the particularization of the *eventus judicii.*[6] Under this perspective, the primary function of family law is not the regulation of intrafamilial relations, but the social orientation of the behavior and the action of social actors (who are all "family members," in one context or another).[7]

A PORTRAIT OF THE GREEK FAMILY

Today, the average Greek family has the basic characteristics of the nuclear family found in all contemporary states of the Western world. This family system emphasizes monogamy, neolocal residence, a modified-extended kinship linkage, bilateral descent and inheritance, egalitarian decision making, and relatively free choice of mate.

However, it departs from the nuclear family model in certain important ways. In Greece, the transition from agrarian to urban to postindustrial society was completed over the course of only three decades.[8] Thus, the Greek family has inherited, and tends to regenerate, traditional social values and mores. It is not only a forum where these values are reproduced, and, in a certain sense, recycled, but also a place where new values appear and coexist with the older ones. This is true of all family forms the world over, but the incredible speed with which the Greek society transgressed family types makes the changes in the Greek family much more dramatic.

At the core of the Greek family's value system there is a nexus of close kinship ties which have been shaped by the turbulent history of the Greek people. The contemporary Greek family is the last stop in a long line of kinship groups that have had to fight for their survival (both literally and metaphorically). Over the centuries, the members of the traditional extended family had found haven, support, protection, and stability in the family unit.[9] Indeed, families developed intricate and highly effective support systems for their members.

For many generations, moving away from the extended family resulted in the social, if not the physical, death of the straying members (Mousourou 1989, pp. 23-24). In these settings, where the group predominated over the individual, marriage was a lifelong union of two families. Therefore, the selection of a mate was a matter for the parents to decide. It was considered disgraceful for a young man or woman to express any preference, and a preferred mate was willingly given up if the parents strongly objected (Kouretas and Papathomopoulos 1968).

The extended family was not the only form of family in the Greek mainland. It was, however, the predominant family structure in most communities, such as those in the region of Mani, the nomadic societies of the Vlachs, and the pastoral communities of Macedonia and Thrace (Nitsiakos 1991). Therefore, the linear schematic representation of the history of the Greek family as "past-extended/present-nuclear" is just an oversimplified way of describing the transition from communal to conjugal forms of family life. It is more accurate to see the Greek extended family as an institution characterized by elasticity (undergoing successive phases of growth and decline) and differentiation (varying by geographical region).

Today, the transition from extended to nuclear family has been completed. The contemporary Greek family operates within a social environment characterized by individualism, which fuels the dominant culture, permeates all social institutions, and defines most socially significant behavior. The modern family, having lost many of its functions, no longer subscribes to the ideals of traditional familialism. But it generates its own ideals, different from those that prevail in other settings.

Compared with other countries, in Greece the average family of procreation is not isolated, but in constant interaction with the parents' families of orientation, and more generally with their extended families. This family may be typified as modified-extended, but it does not fall neatly within this typology.

What sets apart the Greek family is its espousal of an "ethic of care" (Gilligan 1982). The Greek family is unique because it is a *nurturing* family. This means that relatives play a great part in the provision of personal care, child care, and financial, emotional, and moral support.

A great many Greek parents raise their children in a spirit of self-sacrifice, placing emphasis on personal care and education. But what distinguishes them from parents in other industrial societies is the kind of care they offer to *adult* children. Indeed, Greek parents consistently provide support (especially economic support and accommodation) to their adult children, helping them not just in times of crisis (unemployment, divorce, etc.) but regularly and regardless of their children's age or marital status.

This helps explain why single young Greek adults rarely live on their own. It also partly accounts for the increased rates of unemployment among young adults of higher socioeconomic status. This "delayed adulthood,"[10] which is a predominant characteristic of post-1960s generations, spins off into other phenomena, like delayed employment, delayed marriage, and the overall delayed assumption of all adult responsibilities. Over-protective parents raise over-dependent children, and it is this prolonged dependence that often makes it difficult for some children to later become reliable spouses and responsible parents.

Greek parents tend to maintain close ties with their own elderly parents, looking after them by providing them with food, shelter, and care in their old age, and often taking them into their home, especially if they are ill or widowed (Pitsiou 1986). In return, many grandparents give their pension benefits directly to their (adult) children, thus providing significant financial assistance. But more importantly, Greek grandparents eagerly assume many child-rearing duties (such as child-sitting), actively participating in their grandchildren's upbringing.[11]

It is true that in countries where public provision for the elderly is lacking, there is usually an increased number of older people living with relatives (most often their adult children) (e.g., Gaunt 1983, p. 250). But the emergence of this phenomenon in Greece is mainly attributed to closer family ties.

Regardless of its origins, this care towards older people is certainly the most traditionalistic element of Greek family life. Another such element is the gendered nature of the personal care provided within and by the family. In Greece, as in most Western states, women are much more likely to provide care for immediate and more distant family members than are men (Parker 1985; Ungerson 1987). Therefore, it would be more accurate to speak of a "gendered modified extended family," composed of a male breadwinner and his female dependent, who is expected to provide the unpaid labor involved in caring for the children, the sick, and the elderly (e.g., Land 1978).

To sum up, Greek families tend to be child-centered and, in another sense, woman-centered. Indeed, women and children play a major role in family life: women because of their increased responsibilities and children because of their increased rights.[12]

THE REFORM OF GREEK FAMILY LAW

The 1975 Constitution of Greece established the principle of gender equality.[13] As more Greek women entered the labor market and abandoned their tradition-assigned roles, legal theorists and practitioners became increasingly aware of the need to modernize family law. This long-awaited modernization came about in 1983, when the Greek Parliament ratified Law 1329 by a unanimous vote.[14] Law 1329/1983 radically amended many provisions in the Family Law section of the Greek Civil Code (GCC),[15] in the introductory law of the Civil Code, in Commercial Law, and in the Code of Civil Procedure.

The law's basic goal was to provide the legal framework that would permit the full implementation of gender equality in the aforementioned fields of law. Gender equality, as applied to intrafamilial relationships and family-related institutions, brought about a literal transformation of Greek Family Law.[16] But, more importantly, it transfigured the substance, the function, and the very ideological core of the law.

The amended statutes uprooted obsolete social practices as they had been crystallized in legal discourse and set new normative standards for social actors. New social attitudes were shaped and family life was regulated in an equitable fashion. This was accomplished by infusing a sense of fairness into the family and making the equitable relationship the new norm in all intrafamilial relationships. All vestiges of the patriarchal family were removed and replaced with institutions that were congruous with the contemporary notion of the egalitarian family.

Marriage and divorce law were *rationalized* and liberalized, not just in form but in essence. The rights and obligations of spouses under marriage and divorce were grounded on new legal bases. Men and women were exposed to new values and were shown new, democratic ways of regulating their relationships.

The new Family Law was progressive, responsive to the needs of women and children, and characterized by an increased degree of social awareness. The law also aspired to "fulfill and continue Greek tradition."[17] This continuation of Greek tradition should be seen primarily as the (successful) attempt to shake off the last vestiges of Roman law and thus lay down an unadulterated form of Greek law, with provisions that can be traced all the way back to ancient times.

But the new law also broke away from tradition, by leaving out all those "traditional" institutions that had become oppressive and obsolete.[18] It weeded out the "bad" elements from traditional family life and kept the "good" elements, changing them, enhancing them, or redefining them through the use of modern legal language.

And here lies the greatest asset of the new Law: its transformative-pedagogical function. The law envisioned the Greek family as the ideal field in which citizens could learn to apply democratic principles: dialogue, participation, consensual decision-making, formal equality, search for a common interest, and so forth (Rodios 1991). By participating in the democratic function of the family, all its members, and especially children, gain a sense of purpose and increase their social consciousness. Thus, they are socialized and politicized as democratic individuals.

MARRIAGE

In the Greek legal order, marriage is protected distinctly from (and more comprehensively than) the family.[19] In modern Greek law, marriage is founded on monogamy.[20] As a legal institution, marriage is not directly associated with child-bearing, since procreation is treated as just one of the consequences of marriage. By protecting marriage, the legislators of the Greek Constitution of 1975 intended to regulate sexual relations in the context of permanent cohabitation, and not to establish a particular social or legal type of family.

The Constitution protects the nuclear family as a group of individuals who are basically connected by reason of a mere biological fact (human reproduction through the sexual contact of two persons of different gender). In this sense, it is protection of the natural or de facto family; that is, the family that is created regardless of (and outside of) marriage. Of course, the Constitution also protects the so-called legitimate family, as this has been standardized in social life and legal text. But exactly because marriage is no longer a *sine qua non* element of the constitutional safeguarding of the family (because marriage and family are each safeguarded separately), the Constitution covers both the legitimate and the formerly "illegitimate" family.

Therefore, the Constitution does not *impose* the legitimate family as the sole, exclusive or obligatory framework of sexual relationships, of physical reproduction of human beings, and of cohabitation. But it does *propose* marriage—and the legitimate family—as an autonomous legal good and as a model of behavior, enveloping it with privileged treatment.

In Greek family law, marriage is a contract (Deliyannis 1986, pp. 56-61; Karasis 1994, pp. 20-31). Of course, it is a *sui generis* contract, because it departs from all other contracts regulated by Civil Law, in form, content and purpose. In rough terms, the marriage contract does not establish an ordinary contractual relation between the parties, but initiates conjugal living, which is an exercise in sharing and compromising one's individual interests.

Capacity

The age of majority was previously fixed at 21 years. Under the new law, the age of majority has been reduced to 18 years (GCC 127). The age limit for full legal competence has likewise been reduced to 18. Pursuant to these changes, the general provision regulating capacity to marry stipulates that to contract marriage, both parties must have completed their eighteenth year of age (GCC 1350 § 2, 1351). Persons under 18 can enter marriage only by court order and only if marriage is to be expedited on serious grounds.[21] Over the past few years, the average age at which Greek men and women enter into marriage has increased.[22] However, there is a steadily rising number of teenage marriages, as well as a growing number of teenage mothers.[23]

Formation of Marriage

The enactment of Law 1250/1982 brought about a much-needed modernization of Greek marriage law. The law's most significant innovation was the introduction of *the civil marriage* as an option for solemnization. Thus, for the first time in centuries, the civil marriage was recognized as legally equivalent to religious marriage. This put an end to the Church's exclusive jurisdiction over marriage[24] and reestablished it as a private matter.

The spouses have a choice between a civil marriage before a civil official and a religious marriage before a priest of the Greek Orthodox Church, or by an ordained minister of any other creed or faith known (recognized) in Greece (GCC 1367 § 1). The spouses may also choose to have both a civil and a religious wedding ceremony. Regardless of the type of ceremony chosen, certain conditions must be met in order for a marriage to be legal.[25]

Furthermore, the new marriage law has abolished certain impediments previously recognized by law.[26] Most of these impediments gratified prejudicial attitudes and reinforced the tenacity of certain well-established stereotypes about "morally correct" ways of family life. On a practical level, by imposing arbitrary obstacles to marriage, these impediments fostered the growth of "illegitimate" marriages, families, and children.

Under the new law, there are only four impediments to marriage: (1) the legal existence of a prior marriage, which has not yet been dissolved or annulled by irreversible judicial decision (GCC 1354); (2) the spouses-to-be are blood relatives of lineal consanguinity or of colateral consanguinity up to the fourth degree (GCC 1356); (3) the spouses-to-be are relatives of lineal affinity or of colateral affinity up to the third degree (relations by affinity are preserved even after annulment or divorce) (GCC 1357 and 1464 § 2); and (4) adoptive parents or their direct descendants and adopted children may not marry each other, even after the adoption has been annulled (GCC 1360).

If a marriage is celebrated under impediment, then it may be declared null and void by a decree of nullity (GCC 1372-1373). Furthermore, a marriage is voidable if it has been celebrated on grounds of fraud (GCC 1374) or threat (GCC 1375).

Finally, Law 1250/1982 brought about significant changes in procedural marriage law. For example, it simplified court proceedings by abolishing provisions that had generated practical difficulties and it removed bureaucratic hindrances toward the acknowledgment of marriages celebrated by Greeks living abroad.

But the law's greatest modernizing element was the institution of the civil marriage. This change was received with mixed feelings in the Greek legal world and in Greek public opinion (as this was reflected in the press of the time). But the reaction of the social actors at whom this legal change was aimed was most telling. Statistics showed that for six months after the enactment of the new law, civil weddings comprised 13.87 percent of the total number of weddings. In the following year, the percentage took a plunge to 7.66 percent.[27] Since 1990, over 90 percent of all marriages are religious.

Those who choose to get married "in city hall" have various reasons for doing so: ideological, financial,[28] social, even religious, in the sense that they are followers of different faiths or creeds.[29] But most couples who choose civil marriage have personal reasons for avoiding the extravagance of the church wedding: some are middle-aged and are getting married for the first time, still others are divorced and wish to remarry (but have adult children and prefer a more understated wedding to the religious ceremony).

We observe that the secularization of marriage, which was unquestionably a legal breakthrough, was "absorbed" into the mainstream of Greek customary tradition as just another tendency. Most Greeks obviously regard civil marriage as too "impersonal," because in their minds it is likened to an administrative procedure. Thus, the majestic religious ceremony easily becomes the favored choice. Also, in Greek public opinion, the choice of wedding is somehow correlated with age: the young and never-married choose religious marriage, the not-so-young and divorced choose civil marriage.

The Legal Consequences of Marriage

Personal Relationship of the Spouses

Greek marriage law is grounded on the fundamental principle of the equality of the spouses before the law. The ramifications of this principle in the realm of the spousal relationship include the principle of companionship and reciprocity, the principle of free distribution of spousal roles, and the principle of free will (in matters that are irrelevant to marital life).

The primary mutual obligation of the spouses arising from marriage is cohabitation (GCC 1386). This obligation includes a material element and a moral element. The material element is the obligation of the spouses to be physically living together under the same roof. The moral element is actually an array of reciprocal moral-social obligations that are born from marriage. These are, indicatively, the obligation of love, fidelity,[30] mutual trust and respect, copulation, and reproduction.

Secondly, the spouses are required to jointly make decisions on all matters pertaining to conjugal life (GCC 1387 § 1), under two important restrictions: that, in ordering their life in common, the spouses do not impede each other's professional or other activities and do not violate each other's "spheres of personality."

Before the reform, the mutual obligations of the spouses were not so mutual. First and above all, the husband was proclaimed the head of the household. He made every decision that was directly or indirectly related to marital life, as long as the decision did not constitute an abuse of right. The wife managed the household, but subject to the provision declaring the husband's supremacy (that is, in the narrow margin left by the husband's decision making).[31]

It is obvious that pre-reform marriage law imposed a system of hierarchy on spousal relationships. The family was to be run in a manner little different from the one used in the army. The wife was relegated to the private sphere and was symbolically given a "managerial" status, which corresponded to no real power over anything or anybody.[32] The law established that initiative and leadership were for men, and passivity and inferiority for women.

It has been argued that such regimes have served to actually protect women, because they are effective in shielding them from an outside world that is generally hostile toward them, or at best gender-specific. But this kind of paternalism is harmful for women in the long run because it helps to embed negative stereotypes about women in the social fabric.

Furthermore, the law should not only be about what is effective, but also about what is fair and just. Positive stereotypes of women need to be "constructed" and integrated in legal statutes so that they can be transmitted and "taught" to the social actors through the implementation of rules. This is the only way a legally fabricated "ideotype" of human behavior can ultimately become social knowledge.

The spouses have a mutual obligation to support and provide for one another and their children. Each of them is required to contribute, to the extent of his or her powers, to meeting the needs of the family. This contribution could be effected by means of the spouses' personal work, income, and property.[33]

The surname used by each of the spouses in marriage is a formal, yet essential, part of their personal relationship. In the regime that preceded the reform, the wife was obligated to assume her husband's last name.[34] Under the new law, the wife keeps her own last name, but may use her husband's last name or add it to her own "in her social relations" (GCC 1388).

The children born to a legitimate marriage may assume the surname of either spouse or a combination of the two surnames. However, the parents-to-be are required to choose the surname of their children before getting married. Curiously, it is not the surname, but the first name of the offspring that creates problems for many Greek couples.[35]

Financial Relationship of the Spouses

Greek marriage law establishes a separate property regime by a default rule,[36] as the spouses may choose a community property system, either before or during their marriage, if they so desire (GCC 1403 ff). The separate property regime has been an integral part of Greek legal tradition since the times of the Roman and Byzantine eras, having survived because it was to a great extent based on dowry-giving, which was abrogated by Law 1329.

For centuries, the dowry had been a fundamental social and legal institution of Greek society. As in most agricultural societies of Europe, it was the characteristic way of devolving property *inter vivos* to women (Goody 1973; Robertson 1991). Although its functional purpose was to ensure a woman's financial independence, in practice it had become a source of inequity and discrimination (Teperoglou 1968; Symeonidou-Alatopoulou 1980).[37]

The Parent-Child Relationship

Before 1983, the parent-child relation was established on a legal basis that was discriminatory against women and ineffective in the protection of children and their rights. The law imposed a rigid hierarchy, in which the father ruled over everyone in the family.

The new provisions have abolished all traces of this *patria potestas*.[38] Parents no longer have authority over, but rather have custody of, their children (ENVE 1989). The rights of parents over their children are functional rights, in the sense that they are exercised in view of serving not parental individual rights, but children's moral and material interests. All aspects of the parent-child relationship are structured around the principle of the child's best interests (GCC 1511) (Tombrou 1983).

Under the new law, parents and children "owe each other mutual help, affection and respect" (GCC 1507). The mutual obligation for help entails all kinds of material and psychological support. The law also establishes a specific obligation of children toward their parents. Children, to the extent that they are members of the parental household and are raised or supported by their parents, are required to offer their parents services toward the household's management or the exercise of their profession (GCC 1508).[39]

The primary responsibility of parents is "caring" for their children.[40] In alignment with the predominant principle of gender equality, any discrimination as to gender in the upbringing and education of children is prohibited. The law also directs parents as to the disciplinary measures they take against their children. Corrective measures are permissible only when they are pedagogically necessary and do not infringe upon the child's dignity (GCC 1518 § 2).

DIVORCE: THE CONSEQUENCES OF MARITAL DISRUPTION

Greece has one of the lowest divorce rates worldwide and the second lowest divorce rate among European Union (EU) member states. In 1978, the divorce-marriage ratio was 1:20, that is, one divorce corresponded to 20 marriages. Today, the ratio is about 1:11.[41]

In Greek society, moral sentiment, religious sentiment, and tradition still play an important role. Thus, the decision to marry, and likewise, the decision to divorce, is arguably *less rational* than in other societies.[42] In regard to personal relations, some social systems (such as the law) generate individualistic values, whereas others (such as the Church) reproduce an ethos of collectivism. Consequently, although there is diminishing social constraint in matters of marriage and divorce, there is increased social emphasis on the family as an institution that should be preserved, cherished, and protected.

This becomes more obvious in rural parts of Greece. In such settings, individuals generally have less propensity to divorce, primarily because rural norms are less supportive of divorce (Kataki 1984, pp. 57-62). Husbands and wives living in urban settings find it more acceptable to consider separation or divorce than do spouses living in rural areas. In the cities, individual morality takes precedent over social morality.

The legal and social treatment of marital infidelity is also characteristic.[43] Today, marital infidelity is certainly one of the primary *causes* of divorce, but no longer constitutes legal grounds for initiating divorce proceedings unless it has presumably led to the "irretrievable breakdown" of the marriage. Moreover, social norms have expanded to include infidelity as socially "excusable" behavior under certain circumstances.

The low frequency of divorce in Greece reinforces the argument that countries with strong religious underpinnings tend to preserve the traditional family unit, because they display greater marital stability. Countries such as Poland and Italy where Catholicism continues to exert a powerful influence on social mores display remarkably low divorce rates (12% and 8%, respectively). Ireland and Malta also present good examples.

Over 98 percent of all Greeks are Christian Orthodox. Religious homogeneity (and the ensuing marital homogamy) has been observed to contribute to marital stability and marital quality. Furthermore, religion effectively counteracts the prevalent individualist spirit, by uniting people and imparting a strong sense of shared religious identity.[44]

Interestingly, the Greek Orthodox Church does not condemn divorce (Christophilopoulos 1946). On the contrary, it has consistently espoused the view that it is preferable to allow the dissolution of marriage when it is incurably impaired than to force the spouses to maintain a relationship that has reached a point where no good can be derived from it. Canon law has generally conformed to this principle, and has established no rules that hinder divorce.[45] However, the Holy Synod of the Church of Greece has in past years denounced no-fault divorce.[46]

Throughout the greatest part of Greek legal history,[47] divorce had been a private affair (Ginis 1960; Deliyannis 1968). Marital dissolution came about automatically, either by mutual consent or unilaterally, without the intervention of any state agent. Starting from the eleventh century AD, ecclesiastical courts were also given the jurisdiction of granting divorces. They ruled in a liberal spirit (for example by allowing divorce by mutual consent) and maintained their jurisdiction until the enactment of the first Greek Code of Civil Procedure, when civil courts gained exclusive jurisdiction over divorce disputes (Deliyannis 1987, pp. 234-238).

The New Divorce Law

Before the reform, divorce law was essentially fault-based.[48] In this system, divorce acted as a sanction of marriage-related offenses. Law 1329/1983 revolutionized divorce law,[49] primarily by introducing the notion of "objective breakdown" and by instituting divorce by mutual consent. Under this law, there are three kinds of divorce: unilateral contested divorce, unilateral contested no-fault divorce, and divorce by mutual consent (Georgiadis and Stathopoulos 1991, pp. 379-456).

Unilateral Divorce

Under the new divorce law, each of the two spouses may initiate divorce proceedings when marital relations have irretrievably broken down, "on grounds which are imputable to the defendant or to both spouses, so that the continuation of the conjugal relationship has become justifiably intolerable for the plaintiff" (GCC 1439 § 1). If the spouse suing for divorce has not produced evidence to the contrary, "irretrievable breakdown" is presumed in four different cases: bigamy or adultery on the part of the defendant, desertion of the plaintiff, or an attempt on his or her life by the defendant (GCC 1439 § 2).

The presence of these four cases of "marital fault" in this new provision has generally been seen by progressive legal theorists (Manesis 1983; Deliyannis 1987) as an adulteration of the "pure" principle of objective breakdown that reformers initially wanted to introduce. The initial reformative goal was to render divorce a mere confirmation of a failed marriage. But this ambition met with strong reaction by conservatives (in political parties, the Church, and so on). The newly elected socialist government had to consider political cost, and so a compromise was seen as the best solution.[50]

The new rules on divorce acknowledge that marital disruption may be attributed to one of the two spouses. But the "ought" conveyed by the statutes is that fault *should not* matter. This was the new ethic inserted and "instructed" by divorce law. This ethic did not replace preexisting values, but rather a chaotic system lacking ethical order and structuring.

Indeed, unilateral divorce was first introduced to the Greek legal order by virtue of Law 868/1979 (Spyridakis 1979), which was structured on the crude assumption that when a marriage fails, one of the two spouses is at fault. The spouse who sued for divorce, had to prove that the defendant (the "guilty party") had actually committed a matrimonial offense. Catching the defendant in the act was the best evidence one could produce in court. But because this was rarely achieved in practice, spouses had to resort to other means in order to gather evidence: private eyes were hired, phones were tapped, and so forth. If all else failed, perjurers were recruited who testified to witnessing a purported matrimonial offense. When both spouses wanted to divorce (usually when neither one or both of them had committed an offense), mutual friends and relatives willingly appeared before the court and perjured themselves in order to help the couple get the divorce.

Of course, judges were well aware of this situation and could do nothing but tolerate these so-called "disguised consent" divorces. As a result, the statute regulating fault-based divorce remained *socially inactive,* because divorces granted pursuant to the statute were actually based on mutual consent.[51] This is a good example of a rule's failure to "tune into" social practice.

Another provision of Law 868/1979 that created many practical problems was its "cruelty provision,"[52] which gave artificial life to many "dead" marriages, in which the spouses lived in separation or had even gone on to create "illegitimate" second families.

The reform brought order to this failing adversarial system,[53] and went even further, acting as a cleansing agent of the diffused hypocrisy conveyed by legal rules.[54] It dissolved the opaqueness of divorce law and introduced the ethics of *individualism* to divorce proceedings, by enacting provisions that valued and prioritized the free will of the spouses as individuals.

No-Fault Divorce and Its Unintended Consequences

When the spouses have been separated for a continuous period of at least four years, irretrievable breakdown is *irrebuttably* presumed, even if the grounds for the breakdown "concern" (are attributable to) the spouse who is suing for divorce (GCC 1439 § 3). This provision introduces no-fault divorce *as a type of unilateral divorce after a period of de facto separation.* According to the letter of the law, "either spouse" may sue for divorce, but in practice it is the "spouse-at-fault" who initiates divorce proceedings.

As expected, this particular type of no-fault divorce was accompanied by much polemic and incited harsh criticism on the part of several legal theorists and practitioners. The irrebuttable presumption lifts all procedural hindrances for the plaintiff, leaving the defendant with virtually no defense.

If the ideological and pedagogical functions of the reform called for such no-fault divorce proceedings (which would provide each spouse with "an easy way out," both substantively and procedurally), then the relevant statutes were effective.

However, they generated unfavorable, unintended consequences, especially in the first years of their implementation.[55] Research[56] has shown that some women, who participated in no-fault divorce proceedings as defendants, consistently suffered great losses from the disruption of their marriage. Preliminary findings revealed that these women could be grouped together into a distinct category, on the basis of certain characteristics that they all shared.[57]

This "class" of disadvantaged, economically powerless, and socially helpless women had somehow escaped the protective shield of Law 1329/1983. These victims of "homemaker marriages" (cf. Stathopoulos 1983; Matthias 1989) suffered the negative unforeseen results of legislation intended to protect all women.

When a traditional marriage ends by divorce, there are losses that are not evenly or at least equitably distributed to the spouses. Research has shown that some women (and their children) bear a disproportionate share of the costs of dissolution. Indeed, if marriage is a contract, then the costs of breakup should be accepted as part of that contract. But fairness suggests that spouses should share the losses that divorce has caused.

Therefore, any future reformative effort should concentrate on the economic aspects of divorce for wives in traditional marriages. Policymakers should consider enacting a new provision that will remedy the financial difficulties of these women. This provision would regulate the specifics of the postdivorce economic well-being of these wives and would be incorporated in the alimony provisions of the Code.[58]

Abolishing no-fault divorce altogether is not a solution.[59] Scholars who employ the tools of economic analysis to the study of marriage and divorce have argued that in regimes where no-fault is not an option, women are presumably better off because they profit substantially from bargaining with their husbands in order to agree to the divorce.[60] Other legal theorists have asserted that no-fault divorce grounds have generally failed to serve the purposes for which they were enacted.

However, in the Greek legal order, it is not the no-fault provision per se that has produced negative consequences. Problems began when the no-fault provision was followed by a harsh ruling on alimony. It is the alimony provisions that give "no-fault divorce" an almost punitive character for divorcing women.

Divorce by Mutual Consent

Law 1329/1983 introduced consensual divorce to the Greek legal order for the first time. The new provisions establish a procedure that is swift and effective. If the spouses have been married for at least one year, they request the divorce by filing a joint petition.[61] When there are minor children from the marriage, the spouses are required to produce a written agreement with which they regulate matters of custody and visitation.

Primarily because of its relatively short waiting period, divorce by mutual consent has become increasingly popular since its legitimization in 1983. Over the past few years, roughly 75 percent of all divorces issued yearly by Greek courts are by mutual consent.[62]

Consensual divorce is kinder to litigants, in many respects, and not solely in terms of the divorce proceedings themselves. More importantly, divorce by mutual consent is, at least theoretically, a boon to former spouses when alimony is awarded. Indeed, as we will see next, when the marriage has ended by mutual consent, the judge cannot consider a spouse's reasons for requesting a divorce as a factor likely to reduce the amount of maintenance.

Alimony

Permanent and Rehabilitative Alimony

Under Greek divorce law, postdivorce spousal support is granted in the forms of permanent and rehabilitative alimony, when either one of the former spouses cannot ensure his or her maintenance from his or her earnings or from other assets.

Permanent alimony is "alimony with no specific time limit," in the sense that it may be granted for as long as the conditions stipulated by law apply, therefore possibly for life. It is granted in three different cases: (1) if at the time that the divorce is granted, the spouse requesting alimony is at an age or in a condition of health that do not allow him or her to begin or to continue the practice of suitable employment; (2) if the said spouse has custody of a minor and is for this reason impeded from finding suitable employment; and (3) in any other case, when the award of alimony is called for on grounds of equity (GCC 1442).

There has been a demonstrated judicial preference for a restrictive interpretation of the relevance of factors such as age and bad health, which has generally tended to curtail the actual amounts awarded as alimony.[63] The courts are even more cautious when awarding lifelong (that is, permanent in the strict sense) alimony. Their restrictive interpretation of the relevant provision is based on the assumption that alimony is granted to the needier of the two spouses as a means of support, and therefore is not intended to turn the needy spouse into a parasite.

Along similar lines, legal theorists (e.g., Deliyannis 1987, p. 318; Kounougeri-Manoledaki 1993, pp. 154-155) agree that the nature of postdivorce spousal support is not punitive or compensatory, but moral. They have argued (Nikolakopoulou-Stefanou 1986; Matthias 1989; Vathrakokoilis 1990) that permanent alimony can turn the obligor-spouse into a quasi-private insurance fund, especially if one takes into consideration that alimony amounts are readjusted periodically. Furthermore, the obligation for permanent alimony may act as a disincentive for remarriage and the creation of a new (second) family.

In theory, neither permanent nor rehabilitative alimony were intended to protect one sex over the other. Postdivorce maintenance is a right afforded to former spouses regardless of sex. However, especially in the first years after the reform, men were overwhelmingly the initiators of divorce suits, so women were disproportionately rendered claimants to alimony.

In practice, permanent alimony has become the protective shield of divorced homemakers. The losses that these women suffer from the dissolution of their marriages have to be somehow offset by the alimony awards they receive. Judicial practice has demonstrated that when judges have the required social knowledge and the degree of awareness needed to understand this trade-off, they do grant alimony awards that "compensate" similarly situated women for their losses.

Rehabilitative alimony is granted when the conditions for permanent alimony do not apply and the claimant cannot find permanent and appropriate employment or needs professional training (GCC 1442 § 3). As opposed to permanent alimony, rehabilitative alimony has served to facilitate employed former wives' achievement of full financial autonomy.

However, it has become evident that the three-year time limit established for rehabilitative alimony is insufficient for (formerly or currently) employed former wives to find adequate means of self-support and child support.[64] Household labor and child-rearing responsibilities leave little time for any "suitable employment." The job market is not gender neutral, part-time employment is not fully established (but see Law 1892/1990), employers in the private sector choose not to hire women with children, and the Greek state is lacking in welfare services. At best, these "rehabilitated" former wives are relegated to minimum-wage jobs,[65] which leave them with very limited choices regarding schooling, child care, and so on. Hence, a below-average standard of living is the best most of them can hope for.

Restriction or Exclusion of Alimony and the Relevance of Fault

The new law posed further limitations to the fundamental right to alimony, and reinserted fault as a determining factor in postdivorce proceedings. Maintenance may be excluded[66] or restricted "on serious grounds, particularly if the marriage lasted for a short period of time or if the beneficiary is at fault for the divorce or has deliberately brought about his or her own state of neediness" (GCC 1444 § 1). These are all cases which other civil-law legal systems describe as comprising "gross unfairness." The new Greek divorce law has stopped at these limitations, while other continental divorce laws have not.

The provision further determines that the right to alimony ceases, "if the beneficiary remarries, or is permanently living with another person in a free union" (GCC 1444 § 2). This is an interesting rule. It is important, because it makes the de facto union a legal equivalent to marriage. But it does so in the context of restricting alimony. In this sense, one could call it "biased," because the equivalency is a legal fiction.[67] It is beyond any doubt that couples living in a free union do not enjoy the legal rights that married couples do,[68] nor do they have similar social standing.

Division of Property and Distribution of Assets

Upon divorce, each spouse is entitled to a share of the marital assets.[69] The size of the share is to be determined by the courts. In Greek legal theory, the establishment of this claim on marital assets is considered a divergence from the separate property regime. This divergence was called for on two grounds: first, because an implementation of the regime in its rigid and absolute form would lead to injustices (Spyridakis 1983, pp. 109 ff; Deliyannis 1987, p. 105; Daskarolis 1992, p. 272), and second, because the goal of all these new provisions was to reinforce the principles that would help a marriage function properly.

The ideal infused by the new law is the financial independence of the spouses (combined with their emotional, psychological, and personal interdependence). But this ideal is "diluted" when the marital relationship is dissolved, because other ideals (like equity) which were less manifest in the duration of marriage, now come into play. In constructing the "model marriage," the legislators made sure that they were protecting marriage itself, as an ideal relationship. This was accomplished by drafting rules that encouraged companionship in husband-wife relations and by removing those elements that could reduce marriage to a mere business transaction.

The claim to a share of marital assets offers (at least in theory) substantial protection to all financially disadvantaged former wives, but especially to divorced homemakers, because it recognizes their unpaid contribution to the growth of marital assets. In other words, it assigns monetary, legal, and social value to household labor. Married women (especially custodial mothers) are further protected by the provision that regulates division of the family home after divorce (GCC 1393).

Custody of Children

The statutes regulating custody of children are grounded on the egalitarian model of the family, in which spouses are considered equal and children are shared responsibilities. From an ideological-pedagogical perspective, the statutes promote "the best interests of the child" as the highest ideal in child custody law.[70]

Greek courts may award postdivorce custody: (1) to either one of the parents,[71] (2) to both parents indivisibly (joint custody),[72] (3) to both parents divisibly (shared custody),[73] or (4) to a third party.

CONCLUSION

Tradition, religion, and morality are still powerful forms of social control in Greece. Over the past few decades, Greeks have displayed an unfaltering allegiance to the ideals of family life. A strong indication of this observation has been provided by the impact of the family law reform of 1983, as despite the instituted changes, there were no sharp increases in the divorce rate (and no sudden decreases in the marriage rate). Put simply, the liberalization of divorce law did not bring about divorce liberality.[74]

However, the impact of a legal reform is not gauged by statistics alone. Rather, legal change entails a broad spectrum of not easily quantifiable effects on the social and economic level. Legal permission to divorce may bring about changes in social norms and values, alterations in attitudes and ideologies, and shifts in distribution of income. These are all changes in the social environment of the legal system that researchers measure in reference to fluctuations of the divorce rate. Undoubtedly, the divorce rate tells us a great deal about this environment, the other social systems in it, and the changes in their functions. But we could draw equally valuable information about this system by assessing the import of legislative change on values, ideals, and beliefs.

In this understanding, we should be interested not just in the number of divorces granted per year, but also in what kind of divorce proceedings were chosen and what kind of marriages were dissolved by divorce. This is a query into whether or not the existing legal rules that regulate marriage and divorce enhance the *quality* of family life.

In our case, it was shown that the choice of most divorcing Greek couples consistently is divorce by mutual consent. We can draw a number of inferences from this statistical evidence: On a procedural level, couples choose consent over dispute. On a psychological level, couples reveal a preference for fast and painless divorce. On another level, couples choose the most efficient solution, because their combined gains from divorce by mutual consent greatly exceed their combined losses.[75] Therefore, on a social level, a new ideal has been established: the ideal of *egalitarian divorce.*

This ideal was literally born out of a (new) legal rule. Originating from the legal subsystem (from the level of statutes), it has infiltrated other subsystems (at the level of social relations). Again, because social actors have adjusted their behavior according to the new rule, and this new behavior has come to be expected (relied upon by others), it has acquired normative significance.[76] Thus, the ideal is embedded in social practice, continuously being reproduced by the legal subsystem as well as by its environment.

It is not important to trace the origins of this ideal. For the sake of our argument, it does not matter why and when it first appeared in social life. What matters is that this ideal was *validated and promoted by divorce law.* As marriage law inserted the value of the egalitarian partnership, divorce law instituted the structures that allowed these partnerships to end in egalitarian fashion.

In the past two decades there undoubtedly have been major role shifts within the Greek nuclear family: the husband is no longer the exclusive provider and the wife is not necessarily the homemaker/housekeeper and primary caretaker of children. At the same time, new forms of family life (such as single-parent families) have appeared. These newly emergent family forms have modified lifestyles and values, as each of them constitutes a new model for sharing intrahousehold work and allocating family responsibilities.

But the fundamental family unit is still the legitimate nuclear family: the family that has grown out of the marriage of the spouses. In a sense, marriage has become more important than family. The marital relationship itself is not exalted or considered a morally superior way for two adults to share a life. So it is not the marital relationship per se that is exalted in the collective (un)conscious, but the married couple's potential to have "within wedlock" offspring.

Most Greeks feel that this is the lifestyle that has the greatest probability of succeeding in accomplishing a family's agreed-upon goals. This also helps explain why Greeks have exhibited a remarkable resistance to "progress" beyond the nuclear family, specifically, to new family lifestyles that abound in other parts of the Western world. We believe that Greeks have enveloped the nuclear family with the protective cloak of traditionalism. In other words, they are trying to perpetuate the "husband-wife-children" scheme, having labeled this as a "new tradition" worth preserving.

This tendency for tradition is what renders the Greek nuclear family different from its counterparts in the European or the American continent. For modern Greeks, living a typical family life does not mean living within an extended family.[77] But many of the *values* inherent in the extended family have been retained.

Today marriage has become less important to Greek men and women than it was two generations ago. Young people no longer consider marriage to be their life's destination. Parents no longer push their adult children to get married, and they certainly do not select their children's mates. They increasingly emphasize the importance of getting a good education or finding a good job.

Men and women generally look for companions. When they do get married, they do not necessarily believe it is for life. When they decide to have children, they do so because they really want to have children. So the predominant *raison d'etre* of the contemporary Greek family has changed: it is not the financial, but the psychological support it is able to offer its members.

Greek Family Law has laid down the institutions that help the modern Greek family achieve its goals. But it has gone further than just establishing the rules of the game and making sure the players play a fair game. The new Family Law has not acted solely as arbitrator or mediator to family-related disputes. It has operated as "pedagogue" to all the participants in family (and family-like) processes. This democratization of family relationships improved the overall efficiency of Greek Family Law.

ACKNOWLEDGMENT

I am grateful to Aristides Hatzis for his insightful comments on an earlier draft of this work.

NOTES

1. An inquiry into this transformative function of the law is an inquiry into the microdynamics of law; whereas the macrodynamics of law are concerned with the role of law in changing the social structure itself. See Papachristou (1984, pp. 137-141).

2. Under general systems theory (Parsons 1951), law as a social system is a subsystem of society. More particularly, the legal system is conceived as a differentiated functional system *within* society (e.g., Luhmann 1988/1989, p. 138). A systems framework helps us to refine our conception of the social character of the law and to formulate a better understanding of the process of legal reform.

3. In discussing the role of ideas in law and society, McIntyre (1994, p. 107) observes that "even in the absence of law, the structure of marital relations reflects ideas about the 'real nature' of men and women."

4. Glendon (1989, p. 10) points out that scholars are usually more absorbed by the coercive power of law, thus minimizing "its persuasive and constitutive effects."

5. The legal system can be conceived as the totality of relations between three subsystems: the subsystem of rules, the subsytem of behaviors, and the subsystem of judicial intervention. For an elaboration of this model (which applies systems analysis to law), see the work of leading legal sociologist Nikolaos Intzessiloglou (1984, 1986, 1987, and 1990).

6. For Carbonnier (1972, pp. 135-136), law is whatever is *"justiciable,"* which is susceptible to an *eventus judicii* (from Latin: the eventuality of a judicial judgment), what can be questioned (*"mise en question"*) by a judge.

7. From the aspect of the law's legitimating function, law is *"legitimans";* that is, a means of orientation of human action toward the accomplishment of a goal which (usually) is the consensus on certain human actions and activities (Ferrari 1989).

8. The adjustment mechanisms of the family had to be working full force in order to make the transition a smooth one. Under a systems theory (as this is applied to family law), the subsystem of the

family is constantly absorbing the "noise" coming from the other subsystems in order to maintain its stability and balance ("homeostasis").

9. Becker (1991, p. 343) explains that the kinship group is important in traditional societies "...because it protects members against uncertainty," and is also generally effective as a sort of "insurance company," enabling its members to monitor each other.

10. The term we coined for this largely uninvestigated phenomenon is characteristic. Some Greek social psychologists have discussed the phenomenon, but their analysis is descriptive and based on sparse empirical evidence, taken primarily from the field of psychiatry and psychoanalysis (Kataki 1984).

11. In fact, Greece is the European Union country with the lowest percentage of senior citizens living in old-age homes. According to recent figures, only about 25 percent of senior citizens live alone and are self-sufficient (Teperoglou 1979; Amira et al. 1986). It is characteristic that 40 years ago, the United States had the same percentage of persons over age 65 living alone (Becker 1991, p. 369).

12. Kataki (1984) aptly describes the Greek nuclear family as "the child's dominion" and the average wife as "the queen of her prison."

13. Paragraphs 1 and 2 of article 4 of the Constitution of 1975 stipulate that (1) all all Greek men and women are equal before the law; and (2) all Greek men and women have equal rights and obligations.

14. The law was carefully drafted and was worded in gender-neutral language. The Special Legislative Committee that drafted the Bill was composed of 15 members; seven men (all of them professors of law) and eight women (seven of whom were lawyers and one a professor of law). The unanimous vote that the proposals of the Committee received was a very rare occasion in parliamentary practice. The president of the Committee, Professor Aristovoulos Manesis, considered the unanimous vote as the best indication that the legislative reform truly responded to the needs and the beliefs of contemporary Greek society (Manesis 1983, p. 7).

15. Greece, like most European countries, uses civil law, which is statute-based and extensively codified.

16. The establishment of gender equality has had a profound influence on the regulation of family-related relationships in other civil-law European countries such as France and Germany (Glendon 1989, pp. 97-102).

17. The preamble to Law 1329/1983 cited four goals that it aspires to achieve: the eradication of all anachronistic and unfair institutions; social liberation; the fulfillment and continuation of Greek tradition (as one of the principal determinants of the Greek social conscience); and the establishment of new models of intrafamily behavior. See Preamble ot the Bill of Law, "Implementation of the Constitutional Principle of Equality . . ." of the Legislative Committee to the Minister of Justice (Nov. 9, 1982), pp. 273-274.

18. Greek legal theorists generally agree that although traditional norms of family morals should be a determining factor of legislative policy, legislators should always be aware of any new forms of social relations and take them into consideration. Otherwise, some have argued, "the pedagogic potential of family law would be utterly negated and this would mean the adoption of a kind of legislative fatalism, since the insistence on the prevalent traditional family mores would condemn family law to stagnation" (Papachristou 1983, p. 119).

19. Under article 21, paragraph 1 of the Greek Constitution, "the family, as a foundation for the preservation and the progress of the nation, as well as marriage, motherhood and childhood are under the protection of the State." Children and adolescents are protected more emphatically: ". . . the State . . . takes special measures for the protection of youth . . ." (Idem, para. 3).

20. Monogamy is a fundamental principle of Greek marriage law. Consequently, polygamous unions of Greek citizens whose religion allows polygamy are not protected by the Constitution. However, the Constitution does protect the polygamous legitimate unions of foreign citizens whose countries have legislation that allows polygamy (Androulidaki-Dimitriadi 1989, p. 126).

21. Under established precedent, it has come to be accepted that such serious grounds are, for example, the minor's pregnancy, or the fact that the couple who wishes to marry has had a longstanding love affair (which can no longer be tolerated in the minor's social surroundings), or the potential endangerment of the minor's health from deferring the wedding date. In reflection of the state's special concern for teenage marriages, Greek civil law contains provisions that facilitate transactions for married teenagers, empowering them to perform many different kinds of legal acts (GCC 137).

22. In 1980, the average age for men was 27.9 years, while in 1992 it was 29.3 years. For women, the respective ages were 23 (1980) and 25 (1992).

23. According to data published in *UNICEF Report: The Progress of Nations* (1996), in the period spaning the years 1990 to 1995, 22 out of 1,000 live births were to mothers in the 15- to 19-year-old age group (the European Union mean being 15).

24. The Church also intervened in matters of divorce, sharing some legal power with the state until 1982, as it was the only authority that could declare the spiritual dissolution of a marriage and thus permit the remarriage of either former spouse. Of course, since the turn of the century, its intervention had been purely bureaucratic and not substantive (with a few exceptions in conservative rural areas). This power extended only to the Greek Orthodox population (98%) and not to the other creeds and churches that had jurisdiction over their followers in divorce matters. Today, The Orthodox Church still dispenses these "licenses to divorce," which have limited applicability in instances when the couple about to remarry wishes to have a religious ceremony.

25. First, prior to the ceremony, a marriage license must be issued by the competent civic authorities. Second, both parties must be present and clearly consent to the marriage. Third, the parties should announce their upcoming wedding by means of public announcement that is to be posted in the city hall (in rural areas) or run in a daily newspaper (in cities). These conditions (GCC 1368), together with the form (civil or religious or both) of the ceremony, are the elements that conclude the civil contract of marriage.

26. These included impediments from (1) different religion, (2) the preexistence of three marriages, (3) relation by affinity, (4) affinity to an "out-of-wedlock" child, (5) spiritual affinity because of baptism, (6) guardianship, (7) prior conviction of adultery, and so on (article 3 of Law 1250/1982).

27. These two percentages were respectively the highest and lowest of all the years following the implementation of Law 1250/1982. In other words, civil weddings had their zenith and their nadir, all in the course of a year and a half. After this initial jolt, the rate stabilized, fluctuating between 8 and 11 percent.

28. Some have argued that church weddings are generally more costly, but this is so only for couples who choose to marry in large city churches and invite a lot of guests. If for example, a couple marries in a small church or a monastery, the cost is minimal (because they have to pay only for the relevant fees, which they would also pay for a civil wedding). Indeed, many couples prefer a mystic rather than a glamorous atmosphere.

29. However, Law 1250/1982 affords couples who are not followers of the same religion the possibility of performing two religious ceremonies (one for each religion).

30. As we will see later, the violation of this obligation is an indication of "irretrievable breakdown" of the marriage. Before 1982, adultery was also a misdemeanor (under article 357 of the Penal Code). This provision was abolished in 1982.

31. In practice, the wife got to decide about what the family's meals would be or what brand of detergent worked best for the family's soiled clothes.

32. The social message the law conveyed to women was "you might not have any rights, but don't complain: you have no responsibilities either."

33. In other words, the spouses have a general obligation to contribute to managing the household. The extent of this obligation is determined "according to the conditions of family life" and it is performed "in the manner required by conjugal life" (GCC 1389).

34. Calliroe Parrain (1888), the first woman journalist in Greece and perhaps the most significant among the first Greek feminists, wrote: "Nowhere in the Bible, nor in the ancient writers, can one find

the inception of this habit, nor is the first woman mentioned with her husband's name, as Eve Adam, or Sarah Abraham. . . . It seems that all this started in the Middle Ages. . . . It is fair to have the spouses use a joint name, like they have a joint heart, a life, and a future."

35. Modern Greek men (unconsciously paying homage to a formerly dominant familialism) are very protective of their family name, as well as their given name. For example, it is customary for the first son born to a family to assume not only the last name, but also the first name of the paternal grandfather. Thus, the family name is preserved and its intergenerational value is certified and ensured. If the first child is a girl, she usually receives the first name of the paternal grandmother. A few decades ago, and especially in rural areas, it was standard practice for the second child to also take its name from the father's side of the family (from a paternal grandparent, a paternal uncle or aunt, and so forth). When a growing number of families began stopping at two children, and equality between the sexes started infusing social institutions, the custom gradually changed. Today, many couples give their first child the first name of either one of the paternal grandparents and their second child the name of either one of the maternal grandparents. Of course, a growing number of couples are disowning this tradition.

36. The "autonomy of assets" of each spouse is guaranteed by the provision that stipulates that marriage does not entail any changes in the financial independence of the spouses in regard to their assets (GCC 1397).

37. Thousands of Greek women were never married, because their families had no real estate property to give to eligible grooms. And since poor men also demanded a dowry (as a matter of fact, it was especially poor men who needed a dowry to make something of themselves or to just lead a comfortable life), these women were never eligible brides, and never married. Under Greek customary tradition, which was prevalent in rural areas of Greece until the early 1970s, brothers could only marry after their sisters had gotten married. So many men either never married, or married when they were already well into their middle age.

38. The new provisions are congruous with international law on the parent-child relation. Greece is a signatory to numerous international agreements on the rights of children. For example, Greece has signed and recently ratified the U.N. Convention on the Rights of the Child of November 20, 1989.

39. This provision is a notable advancement for Greek family law, because it is an enactment of the fundamental principle that with rights come responsibilities. The law recognizes that children are entitled to special protection, but at the same time it assigns children with roles that help them become responsible adults.

40. This basically includes rearing, supervision, education, and breeding, as well as the determination of the child's place of residence.

41. The relevant ratios for the past few years were: 1989—1:10; 1990—1:10; 1991—1:10; 1992—1:12; 1993—1:12; 1994—1:14 (source: Statistical Service of Greece).

42. Rational here is taken to mean "utility-maximizing," as the term is used by "Chicago School"-rational choice theorists (e.g., Gary Becker, Richard Posner, James Coleman, and so on). Becker (1976, p. 14) has declared that "all human behavior can be viewed as involving participants who maximize their utility from a stable set of preferences and accumulate an optimal amount of information and other inputs in a variety of markets."

43. As mentioned, adultery was decriminalized in 1982. It should be noted that very few cases of adultery were actually brought before the courts. However, the threat of filing charges was widely used by divorcing spouses as a bargaining weapon for achieving favorable divisions of property, sole custody of children, and so forth.

44. The collectivist sentiment incited by religion was a significant force uniting the family in ancient Greece (and Rome). Fustel de Coulanges (1956[1874], p. 42) had remarked that "The members of the ancient family were united by something more powerful than birth, affection, or physical strength; this was the religion of the sacred fire and of dead ancestors."

45. Accordingly, ecclesiastical courts have implemented the grounds for divorce laid down by Roman and Byzantine emperors, and have gone so far as to recognize the dissolution of marriage "on grounds of lack of love and affection between the spouses."

46. The official position of the Church was that "automatic" (no-fault) divorce was foreign to Greek Christian tradition, and that it constituted a degeneration of the sacrament of marriage.

47. This was the case in the secular ancient and hellenistic Greece, but also in religion-dominated Byzantine and post-Byzantine times.

48. The fault-based grounds were adultery, malevolent desertion, attempt on the other spouse's life, and irretrievable breakdown of the marital relationship. There were also four "faultless" grounds, which were rarely invoked in practice: insanity, leprosy, presumption of death, and impotence.

49. To gain a perspective of how progressive this new divorce law really was at the time of its enactment, we should point out that in the early 1980s, when most European countries had restructured their family legal systems toward liberalization, only Sweden had instituted a total no-fault system of divorce (whereby either spouse can get a divorce, and neither spouse can prevent the other from doing so). As late as 1987, Finland introduced a similar system. Southern European countries such as Spain and Italy require a judicial separation before a legal divorce (Goode 1993).

50. The resurfacing of fault is the result of this compromise. The law is very crafty in avoiding the word "fault" (and instead uses the vague term "grounds which are attributable to a spouse"), but in practice these "reasons" usually constitute one of the matrimonial offenses recognized as grounds for divorce in the law that preceded the reform.

51. In fact, research revealed that 77 to 80 percent of divorces grounded on GCC 1442 were really consensual (Skorini-Paparrigopoulou 1971, pp. 183 ff.)

52. Under this provision, "there is no right to divorce, if the defendant objects to the dissolution of the marriage because of exceptional circumstances under which the divorce would constitute, either for the defendant or for the children, such extreme cruelty, that in consideration of the living and other conditions of the plaintiff, the preservation of the marriage is required" (article 2b of Law 868/1979). This provision was then introduced to Greek law for the first time, and was patterned after article 240 of the Frnch Civil Code or article 1568 § 1 of the German Civil Code.

53. Perjury, falsification of evidence, and undue animosity between partners are the anticipated by-products of the adversarial divorce system (Rheinstein 1972).

54. Legal theorists who had earlier advocated the introduction of no-fault divorce (e.g., Karakantas 1968) had argued that it would purify social mores, and upgrade marital and family life.

55. Research findings have revealed that there is an identifiable pattern of unintended discriminatory adjudication (Tsaoussis 1993).

56. Unfortunately, very little research on divorce in a sociological and sociopolitical frame of reference has been carried out in Greece. By contrast, legal theorists have published widely on divorce as a legal phenomenon. Most studies deal with the strictly technical-legal and procedural aspects of divorce.

57. They were all past their middle age (and were all socialized within a traditionally male-dominated environment, having been instilled with the belief that "marriage was forever"; they were married for over 15 years on average; they all suffered from various health problems, which rendered them unfil to work or find employment, hence their insufficient earnings. In sharp contrast, their husbands enjoyed financial well-being and chose to share the acquired marital assets with younger women. In many cases these men had accumulated considerable wealth, mainly because of the contribution of their wives either indirectly, through their assumption of full responsibility for the duties of housework and spending time to raise the children, or directly, by way of a sizeable dowry or the full-time employment of the women in their husband's businesses, and so on.

58. In 1987, this type of reform became state law in California, in response to Lenore Weitzman's research on the impact of no-fault divorce (adopted in the state in 1970) on women and children. The Senate Task Force on Family Equity recommended a presumption of permanent support after a marriage of long duration that would meet the economic needs of displaced homemakers and would be viewed as a form of legal insurance, protection, or pension (Weitzman 1985, 1992).

59. All the more since in countries such as the United States, where relevant research has flourished, the scholarly debate over no-fault reforms and their causal relation to the increasing pov-

erty among divorced women is ongoing and as yet unresolved (Weitzman 1985, 1992; Jacob 1988; Fineman 1991).

60. Becker (1991, pp. 334-335) notes that "if men have been more willing than women to divorce partly because they are not given custody of children and partly because they have many opportunities to meet other women while still married, no-fault divorce reduces their incentive to obtain their wives' consent with generous settlements."

61. See GCC 1441 §§ 1-2. Their decision to divorce must be declared before the court, either in person or by special power of attorney, in two hearings which are separated by a period of at least six months. If two years have elapsed since the first hearing, the declaration of the agreement to divorce ceases to be valid. This "grace period" acts as a safety valve which ensures that the couples filing for divorce have made a definite and final decision to end their marriage. In practice, a significant number of couples never appear in the second hearing.

62. In 1991, divorces by mutual consent represented 72 percent of all divorces granted. In 1992, the relevant figure was 77 percent, and in 1993, it was a whopping 81 percent.

63. Under the prevalent tendency shaped by judicial precedent, the factors listed in the relevant provision must constitute a situation of "unsurpassed social neediness" which does not justify the expectation that the financially weakened former spouse will be able to find work.

64. Employed divorced women make up 60.5 percent of the female workforce, while married women with children under the age of 12 represent 49.1 percent of the female workforce.

65. Again, there is some protective legislation in this direction: by virtue of decision 2396/1985 of the Labor Minister, certain provisions in the regulations of the "Social Security Institution" ("IKA," the country's largest insurance fund) have been amended so that women working as domestic help, babysitters, and so on, can have insurance coverage. Despite the enactment of such laws, the economic problems that afflict women remain: with the increased influx of economic refugees into the country in the past few years (and their state of neediness, which has driven them to take any job offered to them), even the low-paid niches in the job market have been filled up.

66. Greek courts have generally demonstrated wariness in excluding maintenance altogether, making good use of the discretion afforded to them by the "either . . . or" wording of the provision itself.

67. This is the case unless the law uses this equivalency to denote that a marriage and a de facto union *should* be comparable situations because they bear many socially and legally significant similarities.

68. Indeed, this is the only point at which the term "free union" is used in the context of Law 1329/1983.

69. When the marriage has been dissolved or annulled or when the spouses have been separated for over three years, and the assets of one spouse have increased since the time of the marriage (whatever they may have acquired by way of gift, inheritance, or bequest or through the disposal of the assets derived from any such cause is not included), the other spouse, to the extent that he or she has in any way whatsoever contributed to this increase, is entitled to demand the restitution of that part of the increase that results from his or her own contribution. It is presumed that this contribution amounts to one-third of the increase, unless a greater or lesser contribution, or no contribution at all, can be established (GCC 1400 § 1). In addition to this claim, spousal compensation may also be sought on the claim of "unjust enrichment" (GCC 908).

70. The court is required to focus on the interests of the child when deciding on the award of custody or about the manner in which custodial rights are to be exercised (Koukouli-Spoliotopoulou 1983, p. 226). Also, depending on the degree of the child's maturity, his or her opinion should be sought and taken into consideration before any decision concerning parental custody is made, insofar as such decision affects the child's interests (GCC 1511).

71. In practice, sole custody is the judges' preferred choice. Although the Greek courts have decidedly moved away from "the tender years doctrine" in awarding custody, the custody of minor children is most frequently given to mothers. However, in recent years there has been a fortification of

male parenting in Greece. Greek fathers, who are generally effective in their traditional "breadwinner" roles, are on average good and responsible parents and in latter years increasingly active in litigation concerning custody of their children (Tsaoussis-Hatzis, "The Best Interests of the Child in Greek Child Custody Law," forthcoming unpublished manuscript).

72. Joint custody was introduced by Law 1329/1983. Greek judges infrequently award joint custody, which appears to be the most reasonable arrangement for allowing children to have access to both parents.

73. Shared custody is infrequent in practice, although it is better suited for serving the ideals of Greek child custody law (e.g., in the case of couples divorcing by mutual consent who were able to agree on a reasonable scheme for allocating duties and responsibilities regarding custody of their children).

74. Therefore, the example of Greece corroborates findings of sociologists in other countries (e.g., Wright and Stetson 1978), who have demonstrated that easier divorce laws have had a minimal effect on the frequency of divorce.

75. This would be the solution that minimizes cost and maximized gains, using a model of divorce based on social exchanges theory. Under this theory, each spouse makes rational evaluations of costs incurred and rewards received in marriage and divorce (Nye 1979).

76. This "dialectic of social norms" is described by McIntyre (1994, pp. 120-131), who notes (p. 102) that relative to law, norms often demand a higher standard of behavior.

77. An indication is that kinship today is considerably less significant than friendship or professional relations.

REFERENCES

Amira, A., E. Georgiadi and A. Teperoglou. 1986. *O Thesmos tis Anoiktis Prostasias ton Ilikiomenon stin Ellada [The Institution of the Open Care for the Elderly in Greece]*. Athens, Greece: National Center for Social Research.

Androulidaki-Dimitriadi, I. 1989. *Kateuthintiries Arhes tou Syghronou Gamou [Guiding Principles of Contemporary Marriage]*. Athens, Greece: Sakkoula Bros.

Becker, G. S. 1976. *The Economic Approach to Human Behavior.* Chicago: University of Chicago Press.

______. 1991. *A Treatise on the Family.* 2nd ed. Cambridge, MA: Harvard University Press.

Carbonnier, J. 1972. *Sociologie Juridique*. Paris, France: Colin.

Christophilopoulos, A. 1946. "I Elliniki Orthodoxos Ekklesia kai to Diazygion" ["Greek Orthodox Church and Divorce"]. *Arheion Idiotikou Dikaiou [Private Law Annals]* 214.

Daskarolis, G. 1992. *Paradoseis Oikogeneiakou Dikaiou, Tomos I [Lectures on Family Law, Volume I]*. Athens, Greece: Sakkoulas.

Deliyannis, I. G. 1968. "To Zitima tis Metarrythmiseous tou Dikaiou tou Diazygiou Ypo to Phos tis Istorikis kai Sygkritikis Episkopiseos tou Thesmou" ["The Issue of the Reform of Divorce Law Under the Light of the Historical and Comparative Review of the Institution"]. Pp. 87-125 in *Epistimoniki Epetiris Sholis Nomikon kai Oikonomikon Epistimon* [*Tribute to Haralambos N. Fragistas*]. Thessaloniki, Greece: Aristotle University of Thessaloniki Press.

______. 1986. *Oikogeneiako Dikaio I: Eisagogi-Systasi tou Gamou [Family Law I: Introduction-Formation of Marriage]*. Thessaloniki, Greece: Sakkoulas Publications.

______. 1987. *Oikogeneiako Dikaio, Tomos II: Syzygikes Sheseis-Diazygio [Family Law Volume II: Spousal Relations-Divorce]*. Thessaloniki, Greece: Sakkoulas Publications.

ENVE [Legal Association of Northern Greece]. 1989. *Themata Gonikis Merimnas [Child Custody Issues]*. Thessaloniki, Greece: Sakkoulas.

Ferrari, V. 1989. *Funzioni del Diritto [Functions of Law]*. 2nd ed. Roma-Bari, Italy: Gius Laterza and Figli.

Fineman, M. A. 1991. *The Illusion of Equality: The Rhetoric and Reality of Divorce Reform*. Chicago: University of Chicago Press.

Fustel de Coulanges, N. D. 1956[1874]. *The Ancient City: A Study on the Religion, Laws and Institutions of Greece and Rome.* Garden City, NY: Doubleday.

Gaunt, D. 1983. "The Property and Kin Relationships of Retired Farmers in Northern and Central Europe." Pp. 249-279 in *Family Forms in Historic Europe*, edited by R. Wall, J. Robin and P. Laslett. Cambridge, UK: Cambridge University Press.

Georgiadis, A. and M. Stathopoulos, eds. 1991. *Astikos Kodix: Kat'Arthro Ermineia [The Civil Code: An Article-by-Article Interpretation]*. Athens, Greece: Sakkoula Bros.

Gilligan, C. 1982. *In a Different Voice: Psychological Theory and Women's Development*. Cambridge, MA: Harvard University Press.

Ginis, D. S. 1960. "Oi Logoi Diazygiou epi Tourkokratias" ["Grounds for Divorce Under Turkish Rule"]. Pp. 239-284 in *Epistimoniki Epetiris Sholis Nomikon kai Oikonomikon Epistimon [In Memoriam of Periklis Vizoukidis]*. Thessaloniki, Greece: Aristotle University of Thessaloniki Press.

Glendon, M. A. 1989. *The Transformation of Family Law: State, Law and Family in the United States and Western Europe*. Chicago: University of Chicago Press.

Goode, W. J. 1993. *World Changes in Divorce Patterns*. New Haven, CT: Yale University Press.

Goody, J. 1973. *The Character of Kinship*. Cambridge, UK: Cambridge University Press.

Intzessiloglou, N. 1984. *L'Apport de l'Analyse Systemique Dans Le Domaine Juridique*. 2nd ed. Lille, France: ANRT.

______. 1986. "Essai d'Identification de la Totalité Sociale du Phénomène Juridique en tant que Système." *Rechtstheorie* 10: 271-279.

______. 1987. "Essai d'Élaboration d'un Programme de Recherche Scientifique Interdisciplinaire pour l'Étude du Phénomène Juridique." *Revue Interdisciplinaire d'Études Juridiques* 19: 43-82.

______. 1990. *Koinoniologia tou Dikaiou, Tomos 1os: "Epistemologike Problematike: To Antikeimeno tes Koinoniologias tou Dikaiou kai e These tes ste Nomike Episteme" [Sociology of Law, Volume I: "Epistemological Problems: The Object of Sociology of Law and its Standing in Legal Science"]*. Thessaloniki, Greece: Sakkoulas.

Jacob, H. 1988. *Silent Revolution: The Transformation of Divorce Law in the United States*. Chicago: University of Chicago Press.

Karakantas, S. 1968. "To Koinonikon Diazygion Os Problima tou Syghronou Dikaiou." ["No-Fault Divorce as a Problem of Contemporary Law"]. Pp. 177-279 in *Epistimoniki Epetiris Sholis Nomikon kai Koinonikon Epistimon [Afieroma eis Charalambon Fragista: Tribute to Prof. Charalambos Fragistas]*. Thessaloniki, Greece: Aristotle University of Thessaloniki Press.

Karasis, M. D. 1994. *Gamos kai Oikogeneia os Dikaiikoi Thesmoi [Marriage and Family As Legal Institutions]*. Thessaloniki, Greece: Sakkoulas Publications.

Kataki, C. D. 1984. *Oi Treis Tautotites tis Ellinikis Oikogeneias: Psychokoinonikes Diergasies [The Three Identities of the Greek Family: Psycho-social Processes]*. 4th ed. Athens, Greece: Kedros.

Koukouli-Spiliotopoulou, S. 1983. "Zitimata Schetika me ti Goniki Merimna kai Synafeis Rythmiseis" ["Child Custody Issues and Relevant Provisions"]. *Dikaio kai Politiki [Law and Politics]* 4: 213-238.

Kounougeri-Manoledaki, E. 1993. *Oikogeneiako Dikaio, Tomos IÃ: Diazygio [Family Law, Volume Ic: Divorce]*. Thessaloniki, Greece: Sakkoulas.

Kouretas, D. T. and E. A. Papathomopoulos. 1968. "Psychopathologia tou dia Synoikesiou Gamou en Elladi" ["Marriage by Arrangement in Greece and Its Psychopathology"]. Pp. 527-537 in *Epistimoniki Epetiris Sholis Nomikon & Oikonomikon Epistimon [Scientific Review of the School of Law & Economics] (Harmosynon Dimitrio I. Karanika [Tribute to Prof. Dimitrios I. Karanikas])*. Thessaloniki, Greece: Aristotle University of Thessaloniki Press.

Land, H. 1978. "Who Cares For the Family?" *Journal of Social Policy* 7: 257-284.

Luhmann, N. 1988-1989. "Law as a Social System." *Northwestern University Law Review* 83: 136-150.

Manesis, A. I. 1983. "Prologos" ["Preface"]. *Dikaio and Politiki [Justice and Politics]* 4: 5-7.

Matthias, S. 1989. "Diatrofikes Axioseis Meta to N. 1329/1983" ["Claims to Alimony after Law 1329/1983"]. Pp. 29-46 in *Pente Hronia Efarmogis tou Neou Oikogeneiakou Dikaiou [Five Years of Implementation of the New Family Law].*

McIntyre, L. J. 1994. *Law in the Sociological Enterprise: A Reconstruction.* Boulder, CO: Westview Press.

Mousourou, L. M. 1989. *Koinoniologia tis Syghronis Oikogeneias [Sociology of the Contemporary Family].* Athens, Greece: Gutenberg.

Nikolakopoulou-Stefanou, H. 1986. *Oi Synepeies tou Diazygiou [The Consequences of Divorce].* Athens, Greece: Sakkoulas.

Nitsiakos, V. G. 1991. *Paradosiakes Koinonikes Domes [Traditional Social Structures].* Athens, Greece: Odysseas.

Nye, I. F. 1979. "Choice, Exchange and the Family." In *Contemporary Theories About the Family, vols. I and II*, edited by W. R. Burr, R. Hill, I. F. Nye and I. L. Reiss. New York: Free Press.

Papachristou, A. 1983. "Isonomia ton Fylon kai 'Paradosi'" ["Equality of the Sexes Before the Law and 'Tradition'"]. *Dikaio kai Politiki [Law and Politics]* 4: 117-125.

______. 1984. *Eisagogi stin Koinoniologia tou Dikaiou: 1. To Theoritiko Plaisio [Introduction to Sociology of Law: 1. The Theoretical Framework].* Athens-Komotini, Greece: Sakkoulas.

Parker, G. 1985. *With Due Care and Attention.* London, U.K.: Family Policy Studies Centre.

Parrain, C. 1888. *Efimeris ton Kyrion* (*Ladies' Journal*) May 30, 1888.

Parsons, T. 1951. *The Social System.* Glencoe, IL: Free Press.

Pitsiou, E. 1986. *Life Styles of Older Athenians.* Vol. 1. Athens, Greece: National Center for Social Research.

Rheinstein, M. 1972. *Marriage Stability, Divorce, and the Law.* Chicago: University of Chicago Press.

Robertson, A. F. 1991. *Beyond the Family: The Social Organization of Human Reproduction.* Berkeley, CA: University of California Press.

Rodios, N. D. 1991. "To Protypo tis Dimokratikis Symbioseos tou Neou Oikogeneiakou Dikaiou" ["The Model of Democratic Cohabitation of the New Family Law"]. Pp. 591-601 in *Nomos: Epistimoniki Epetiris Sholis Nomikon & Oikonomikon Epistimon [Law: Scientific Review of the School of Law & Economics] (Haristiria ston Ioanni Deliyanni [Tribute to Prof. Ioannis Deliyiannis]).* Thessaloniki, Greece: Aristotle University of Thessaloniki Press.

Skorini-Paparrigopoulou, X. 1971. *Oi Logoi Diazygiou kata tin Ellinikin Nomologian [Grounds for Divorce Under Greek Judicial Precedent].* Athens, Greece.

Spyridakis, I. S. 1979. *To Diazygio Kata to Nomo 868/1979 [Divorce Under Law 868/1979].* Athens, Greece: Sakkoulas.

______. 1983. *Oikogeneiako Dikaio [Family Law].* Athens, Greece: Sakkoulas.

Stathopoulos, M. 1983. "Diatrofi kai Anexartisia tis Diazeugmenis Gynaikas" ["Alimony and Independence of the Divorced Woman"]. *Dikaio kai Politiki [Law and Politics]* 4: 102-108.

Symeonidou-Alatopoulou, C. 1980. "I Exelixi tou Thesmou tis Proikas stin Ellada, 1956-1974" ["The Evolution of the Institution of the Dowry in Greece, 1956–1974"]. *Epitheorisi Koinonikon Ereunon [Journal of Social Research]* 36-37: 322-340.

Teperoglou, A. 1968. *I Proika: Ena Koinonikooikonomiko Problima [Dowry: A Socioeconomic Problem].* Athens, Greece: National Center for Social Research.

______. 1979. "Open Care for the Elderly in Greece." *Epitheorisi Koinonikon Ereunon [Journal of Social Research]* 35: 196-207.

Tombrou, A. 1983. *I Telesi tou Gamou kai oi Sheseis Syzygon kai Goneon-Teknon [The Ceremony of Marriage and Spousal and Parent-Child Relations].* Athens, Greece: Tzakas.

Tsaoussis, A. 1993. "Divorce Under Law 1329/1983: A Sociological Approach of Judicial Precedent." Department of History, Philosophy and Sociology of Law, Aristotle University, Thessaloniki, Greece. Unpublished master's dissertation.

Ungerson, C. 1987. *Policy is Personal: Sex, Gender and Informal Care*. London, UK: Tavistock.

UNICEF. 1996. *The Progress of Nations*. New York: UNICEF.

Vathrakokoilis, V. 1990. *To Neo Oikogeneiako Dikaio [The New Family Law]*. Thessaloniki, Greece: Sakkoulas.

Weitzman, L. J. 1985. *The Divorce Revolution: The Unexpected Social and Economic Consequences for Women and Children in America*. New York: Free Press.

______. 1992. *The Economic Consequences of Divorce*. Oxford, UK: Clarendon Press.

Wright Jr., G. C. and D. M. Stetson. 1978. "The Impact of No-Fault Divorce Law Reform on Divorce in American States." *Journal of Marriage and the Family* 40: 575-580.

PART II

INTIMATE RELATIONSHIPS IN MARRIAGE

PART II

INTRODUCTION

Issues of intimacy, sexuality, and love are central to any discussion of marriage. The intimate relationship can nurture the spirit, improve self-esteem, and provide an environment that shields us from the harsh realities of the world. The chapters in this part of the book explore the complexities of intimate relationships in marriage. Barbara Vinick leads off with the chapter "Sexuality Among Older Couples: Perceptions of Spouse and Self," where she documents that sexual expression remains an important element in long-term marriage of older adults. "Mutuality" affects quality of marital relations as do measures of self-satisfaction and perceptions of spouse's satisfaction. Affectionate behavior, as an expression of intimacy, presents as the most critical element in the perception of marital happiness for older couples.

Kate Davidson extends our examination of intimate relationships in "Marital Perceptions in Retrospect: A Study of Older Widows and Widowers" as she describes how older people perceive their marriages after their spouses have died. This group overwhelmingly perceived their marriages to be happy, and for this generation, both widows and widowers perceived high satisfaction due in large part to the patriarchal hierarchy that was maintained between couples. Of particular interest was the fact that wives' economic dependency was fostered while they met the emotional needs of husbands. Conflicts within these marriages were reduced by adherence to "appropriate" gender behaviors, while at the same time, positive perceptions were heightened.

In our next chapter, "Perceptions of Intimacy and Support, Social Patterns in Contrast: European Americans and Chinese Americans," Deborah Woo continues the examination of perception of intimacy within marriages by focusing both on issues of conflict and on factors that lead couples to perceive their marital relations as "special and close." This study finds that while there were differences in perception between Chinese and European Americans, respect for individual autonomy and decision making, self-sacrifice, and providing an outlet for one another's feelings are important factors in building intimacy in relationships.

Sandra Lee Browning and R. Robin Miller study marital intimacy in the black community with "Ain't It a Crime: Black Women's Perceptions of Intimacy with Black Men Involved in Crime." With black males having a one in four chance of being under the supervision of the criminal justice system within their lifetime, black women are faced not only with a limited marriageability pool, but also with an unpredictable future with their black men. A neglected aspect in the social science literature is how the black male's involvement in the criminal justice system affects his marriageability, and the perceptions black women hold about his involvement in crime. This research investigates the perceptions black women have about black male involvement in crime, dating, and marriage.

Rounding out this section is the work, "Familial Values, Gender Politics, and Meanings of Infidelity: A Cross-Cultural Comparison of Extramarital Affairs," by Jui-shan Chang, which directs the examination of intimacy to include meanings of romantic love, infidelity, constructions of manhood and womanhood, and the gender politics involved in marital and extramarital relations. Through depictions of extramarital affairs reported in print, differences between Chinese and Western understandings of intimacy are found.

SEXUALITY AMONG OLDER COUPLES

PERCEPTIONS OF SPOUSE AND SELF

Barbara H. Vinick

SEX AND THE ELDERLY

I don't know if I'm oversexed, but I'm a lover. I like to pet, kiss, hug. I have more fun out of (that) . . . than the ultimate end. . . . (S)ome people want sex and forget the rest of it . . . and I think that's wrong.

—Seventy-three-year old retired salesman

Generally (my husband) gets into bed first and watches television. I come up a little later. We have twin beds, but I get in with him and we snuggle together. It's so nice to be close. It's beautiful.

—Sixty-seven-year old woman remarried three years

These sentiments, expressed during interviews for a study of remarriage in old age (Jacobs and Vinick 1979), illustrate realities of marriage rarely encountered in the research literature on aging, where intimacy between older spouses has been a neglected topic. Researchers, perhaps uncomfortable discussing sexual behavior and feelings with older people, have generally avoided such topics, even as part of written questionnaires. As a result, remarkably little knowledge exists concerning sexuality and intimacy, within and outside of marriage, among people who are no longer young.

Kinsey and associates' pioneering efforts concerning male and female sexuality (1948, 1953) are still cited as authoritative sources on the sexual "outlets" of older men and women, including heterosexual intercourse, self-stimulation, and homosexual behavior. Based on small samples of men and women over age 60 as part of much larger study populations, Kinsey's research first documented the continuing sexual ability and interest of older men and women, less than among younger people, but not necessarily terminated, even at very old ages. The image of old age as devoid of sexual activity and interest was cracked, if not demolished.

Masters and Johnson (1966) were the first to report the "mechanics" of sexuality in later life. Based on small numbers of men and women whose physical sexual responses were measured in a laboratory, findings showed slower sexual arousal, and other aspects of sexual "performance" that changed with age, including the need for greater and more direct sexual stimulation. The research results added evidence for the continuing capacity of elders for sexual behavior and pleasure.

Researchers at Duke University obtained information about sexual interest and activity from a group of 270 men and women who were aged 60 and older when the study began in the 1960s. Repeated interviews with the participants throughout the decade supported the Kinsey group's findings about the decline in frequency of sexual intercourse among men, but one in five men were still sexually active (engaging in intercourse) in their 80s. Sexual activity was less among women—one in five in their late sixties—although the decline with increasing age was not as great as for men (Pfeiffer, Verwoerdt, and Davis 1972). Using the same sample of elders, George and Weiler (1981) had similar findings, although they reported less decline in sexual activity and more stability over the six-year period that they tracked.

Sexual interest, implicitly defined as interest in intercourse, showed much the same pattern, as men professed more interest than women at similar ages. Among men, the healthier were more interested, whereas among women, those who had enjoyed sex in the past were more interested (Pfeiffer et al. 1972).

Other studies further undermined the notion of the sexless older years. The Starr-Weiner report (1981) indicated startling retention of sexual interest and involvement among the 14 percent of senior center participants who agreed to be interviewed, as did the well-publicized Hite report on women's sexuality (1976) based on 3 percent of readers of a popular magazine, and Brecher's (1984) results of a survey among readers of a consumer publication. While it seems likely that those who participated in those studies were more "sexy" than average, as Starr (1985) has contended, their main value has been in portraying "the potential and possibilities" of sex in the later years.

SEXUALITY IN OLDER MARRIAGES

Since the mid-1980s, publications concerning sex in old age have been devoted largely to summarizing the sparse information that exists (Riportella-Muller 1989; Whitbourne 1990; Schlesinger 1996; Weg 1996). The long-awaited national survey of sexual behavior (Laumann et al. 1994) unfortunately ignored people aged 60 and older.

The Duke study found marital status to be a particularly important determinant of sexual activity, as well as sexual interest, among women. Whereas unmarried men were just as active sexually as the married, married women reported greater sexual activity and interest than widows and the never-married (Pfeiffer and Davis 1972). Older men have greater "opportunity" for heterosexual behavior, as they are far outnumbered by older women. Also, societal approval is greater for relationships between older men and younger women than vice versa. Still, the finding may be somewhat outdated as older women, following the lead of younger generations since the 1970s, have become less restrictive about sexuality outside of marriage.

Although the psychological benefits of all types of intimacy, including caressing and self-pleasuring, have been acknowledged (Comfort 1976; Traupman, Eckels, and Hatfield 1982; Weizman and Hart 1987), with a few exceptions, frequency of intercourse has formed the basis for research findings on sexuality in long-term marriages. A study of married elders that utilized a large national sample indicated, for example, that 53 percent of married people aged 60 and older had "had sex" with the spouse in the last month—comprising 65 percent of those 60 to 65, and 24 percent of those over 75 (Marsiglio and Donnelly 1991).

A few research findings have suggested that marital happiness and sexual activity may be related among long-married couples. An early study in Sweden (Persson 1980) found that 70-year-old women who were still active sexually (a third of the sample) more often judged their marriages to be happy, although the same was not found among the 70-year-old men in the sample. Perhaps not surprisingly, another study of couples aged 40 and older found that those who had sexual problems were more likely to report marital problems. However, the association was less strong among couples aged 60 and older than among the younger couples studied (Murphy, Hudson, and Cheung 1980).

A more recent report (Ade-Ridder 1990) indicated more frequent sexual activity (defined as intercourse) among older married couples who rated their marital quality higher, although couples who had given up intercourse altogether did not report lower marital happiness, findings not easily interpreted. Retention of sexual interest among both men and women was related to earlier interest, supporting previous research results (Kinsey et al. 1948; George and Weiler 1981).

THEORETICAL ORIENTATION AND RESEARCH QUESTIONS

A classic description of marriage is that it consists of separate "his" and "her" versions (Bernard 1972). That observation is consistent with the *social interactionist* theoretical perspective. Based on the work of an influential social philosopher (Mead 1934), the interactionist approach emphasizes the importance of individual perceptions and understandings of experiences and events. Interactionists would point out that people who have been exposed to the same set of circumstances or who have shared the same activity may nevertheless have very different interpretations and meanings attached to the same "objective" facts and experiences, even married couples.

From that perspective, it follows that the frequency of intercourse is less interesting than the *perception* of each spouse concerning his or her satisfaction with frequency. As depicted in the Woody Allen film *Annie Hall*, perceptions may be quite different. (Annie to therapist: We do it all the time—three times a week. Woody to therapist: We do it hardly at all—three times a week.) Some studies of younger couples have indicated considerable disparity in husbands' and wives' perceptions of joint family activities, including decision making (Ross and Mirowsky 1984), contraceptive practices (Neal and Groat, 1976), family division of labor (Shaw 1988), and sexual behavior (Julien et al. 1992).

Similarly interesting from an interactionist perspective is the "accuracy" of each spouse concerning the other's views. Fields (1983) found that happier married couples had greater "congruence of perceptions" regarding each other's personal characteristics, such as "argumentative," "humorous," and so on. Another study found greater marital satisfaction among couples whose independent reports of their interactions were more alike (Christensen, Sullaway, and King 1983). Such agreement may be a product of effective communication between husband and wife, a widely recognized ingredient in satisfying marital partnerships (Cutrona 1996).

As we have seen, most of the research on sexuality among older married couples has been limited to sexual intercourse. Research among young couples has indicated, however, that husbands and wives value different aspects of sexual expression in marriage (Denny, Field, and Quadagno 1984; White et al. 1986; Hatfield et al. 1989). Those results suggest that the terms of intimacy need to be broadened in research concerning long-married couples, as well.

Overall, this research is based on a concept of sexuality as an element in a relationship between two people, as more than a physical act that affects only the actor, a conceptualization that has been overlooked in much previous research. Sexual behavior is viewed as a type of interaction in marriage—special, no doubt, but subject to questions that have been asked about other types of behavior that affect both marital partners. Thus, this study addresses the following questions:

- Do husbands and wives desire change in their partners' sexual behavior? In other words, are they *satisfied* with current levels of sexual expression?
- How *accurate* are the partners in knowing whether or not their husbands and wives desire change?
- How do satisfaction and accuracy relate to *marital happiness*?
- *Which forms of sexual expression* are most important to marital happiness in long-term marriages?

SAMPLE

The sample for this report is composed of 217 couples, married an average of 35 years (ranging from 15 to 46 years), who were recruited for a study of retirement attitudes and adaptation. All of the husbands are participants in the Normative Aging Study, a study of the normal aging process among men, founded in 1963 by the Veterans Administration in Boston. The all-male participants in the study had no evidence of disease when they were selected, and have generally remained healthy after more than 30 years. The couples on whom we are reporting here are a subgroup of those men and their wives.

The average age of husbands in the sample for this study is 61.5 (range = 54 to 78). Forty-two percent are retired, evenly divided among present or former white- and blue-collar workers. Over 90 percent white, they represent a variety of ethnic groups, residing in cities, suburbs, and rural areas, generally within 50 miles of Boston. Wives are aged 45 to 77, with a mean age of 59; 42 percent are in the workforce. Because the couples were not recruited specifically for a study of sexuality, they may be more typical of the general population than if they had volunteered for research concerning sexual behavior.

MEASURES

Data on sexual expression in long-term marriages comes from the Areas of Change (AC) instrument (Weiss and Birchler 1975; Margolin, Talovic, and Weinstein 1983) administered to each spouse as part of a questionnaire battery following a separate face-to-face interview. The AC contains 34 items of shared and separate behaviors in marriage, such as "keep the house clean," "get together with our friends," "work at a job or hobby," "show appreciation for things I do well." Each of the items is prefaced by "I want my partner to:" and followed by a scale indicating amount of change desired in the partner's performance (much less, less, somewhat less, no change, somewhat more, more, much more). This chapter reports analysis of three items pertaining to sexual expression. "I want my partner to:"

- have sexual relations with me
- pay attention to my sexual needs
- show me affection

The second half of the AC contains the same 34 items of behavior in marriage, prefaced by "It would please my partner if I:" (in this case) "had sexual relations with him/her", "paid attention to his/her sexual needs," "showed him/her affection." These are followed by the same scale indicating amount of change desired. It was therefore possible to match one spouse's *actual* desire for change in the partner's behavior with the partner's perception of how much change he or she *thinks* the spouse desires in that area.

DESIRES FOR CHANGE COMPARED WITH SPOUSES' PERCEPTIONS

For this section of the analysis, the response categories "much more," "more," and "somewhat more" were collapsed into one "more" grouping, as were the three "less" responses. Those who reported that they desired "no change" are designated "satisfied" in Figure 1; lack of desire for change was equated with satisfaction with the status quo.

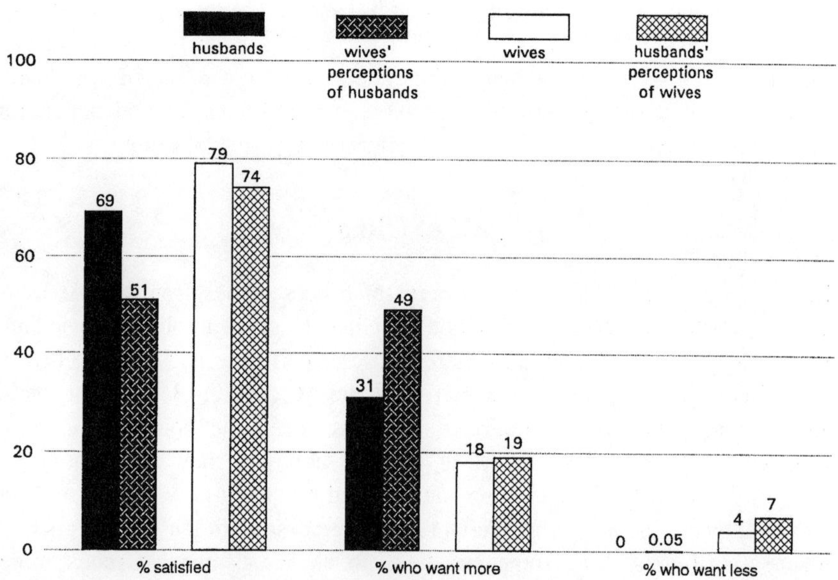

Figure 1. Sexual Relations with Spouse

As Figure 1 indicates, women were slightly more satisfied with the current level of sexual relations than were their husbands—79 percent of women versus 69 percent of men. However, almost twice as many men as women wanted an *increase* in sexual relations—31 percent and 18 percent, respectively. Husbands were quite accurate in their perceptions of wives' desires. Wives, however, tended to *over*estimate husbands' desire for increase in sexual relations; half thought that husbands wanted more, when fewer than a third actually did. Wives also *under*estimated husbands' satisfaction; only half thought that husbands desired no change (were satisfied) when more than two-thirds were actually satisfied with the current level.

The item "pay attention to my sexual needs" showed almost the same pattern as "have sexual relations with me." Wives were more satisfied, while underestimating husbands' satisfaction, and overestimating husbands' desire for change. Correlations between the items about "sexual needs" and "sexual relations" were very high—between .80 and .90—suggesting that respondents were equating one with the other.

The responses to the items "I want my partner to show me affection" and "It would please my partner if I showed him/her affection" displayed a rather different pattern. As seen in Figure 2, men's and women's responses were similar: They were equally satisfied, and both tended to overestimate spouses' desire for more affection and underestimate spouses' satisfaction with the amount of affection shown to them.

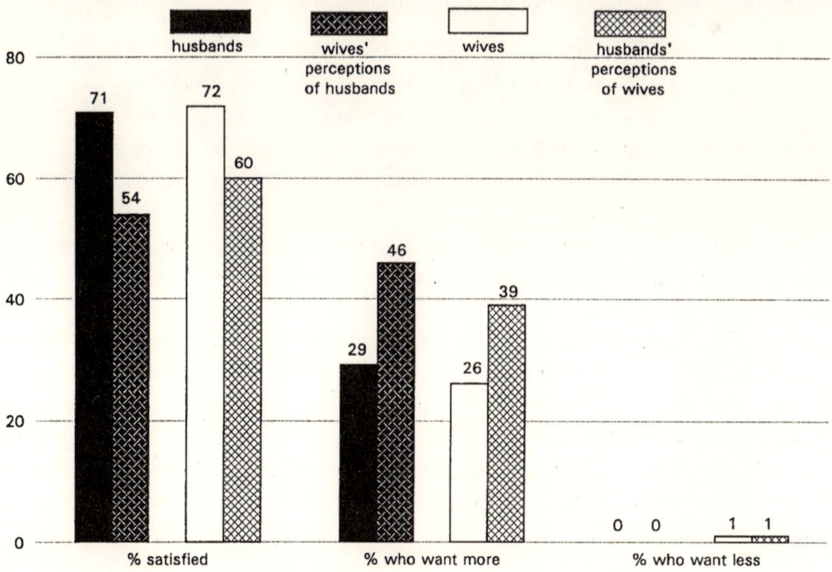

Figure 2. Showing of Affection

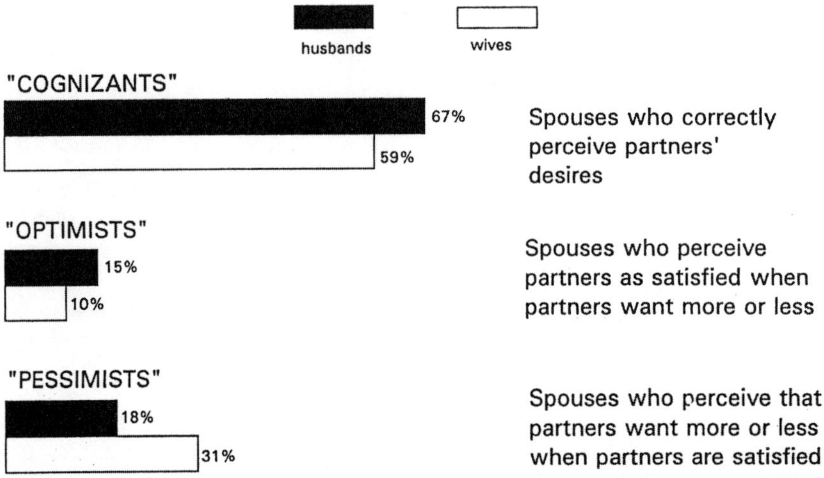

Figure 3. Accuracy of Spouses' Perceptions

In sum, wives were somewhat less desirous of change in husbands' sexual behavior than were husbands in regard to wives' behavior. Wives were more satisfied with the status quo, especially when compared with husbands' wishes for more sexual relations, and more attention paid to their sexual needs.

ACCURACY OF HUSBANDS AND WIVES

How accurate were husbands and wives in perceiving their partners' desires? Figure 3 summarizes the three areas concerning sexual relations, attention to sexual needs, and affection.

When scores were matched, husbands correctly perceived wives' desires for increase, decrease, or no change in the three behavioral areas 67 percent of the time. The corresponding percentage of wives, labeled "cognizants," was 59 percent. The main difference in husbands' and wives' perceptions can be seen in the numbers of "pessimists"—individuals who thought that the spouse wanted change when, indeed, he or she was satisfied (wanted no change)—31 percent of wives versus 18 percent of husbands. At the same time, 15 percent of husbands were "optimists"—those who thought that the spouse was satisfied when, indeed, change was desired—compared with 10 percent of wives who thought likewise.

RESPONSES ON THE AREAS OF CHANGE INSTRUMENT AND MARITAL HAPPINESS

An important aspect of the research concerns how husbands' and wives' responses on the AC questionnaire related to their assessments of marital happiness. In another part of the interview, the respondents were asked to rate their marriages on a seven-point scale from "extremely unhappy" to "perfect" (Fincham and Bradbury 1987). In order to determine if responses on the AC varied along with ratings of marital happiness, responses on the AC were recoded to form a four-point scale reflecting the amount of change desired in the three areas, from most change ("much more" and "much less") to "no change."

Table 1 indicates substantial and significant negative correspondence between desire for change and marital happiness. In other words, husbands and wives who wanted more change in each of the three areas rated their marriages less happy. Similarly, husbands and wives who perceived their *spouses* as wanting more change rated their marriages less happy. It is interesting to note that husbands' marital happiness was most closely related ($r = .48$) to their *own* desire for affection from their partner, and that wives' marital happiness was most closely linked to their perception of the *husband's* desire for affection from them ($r = .39$).

Further inquiry concerns the spouses' accuracy and marital happiness. Are people whose spouses are "cognizant" (accurate) in their perceptions happier than those whose spouses are "incognizant" (either optimists or pessimists, as defined previously)? Comparing the mean marital happiness scores (via paired t-tests) of husbands and wives with cognizant and incognizant spouses, the answer is "yes." For example, Table 2 shows that among husbands whose wives were cognizant about whether they wanted more, less, or no change in sexual relations, the mean marital happiness score was 5.24. Among those whose wives were incognizant, the mean marital happiness score was 4.75, a significant difference. Wives'

Table 1. Correlations of Marital Happiness with Personal-satisfaction Items and Perception-of-partner's-satisfaction Items

	Husbands	*Wives*
I want my partner to:		
have sexual relations with me	$-.38^{***}$	$-.18^{**}$
pay attention to my sexual needs	$-.39^{***}$	$-.30^{***}$
show me affection	$-.48^{***}$	$-.33^{***}$
It would please my partner if I:		
had sexual relations with him/her	$-.21^{**}$	$-.25^{***}$
paid attention to his/her sexual needs	$-.29^{***}$	$-.26^{***}$
showed him/her affection	$-.27^{***}$	$-.39^{***}$

Notes: $^{*}p < .05$, $^{**}p < .01$, $^{***}p < .001$.

Table 2. Marital Happiness when Spouses are Cognizant Versus Incognizant of Partners' Desires for Change

	Husbands' Mean Marital Happiness		
	Wives Are Cognizant	*Wives Are Incognizant*	*t*
Husbands' desires re:			
sexual relations	5.24	4.75	2.44^{*}
attention paid to sexual needs	5.25	4.74	2.50^{*}
affection shown to them	5.21	4.78	2.00^{*}
	Wives' Mean Marital Happiness		
	Husbands Are Cognizant	*Husbands Are Incognizant*	*t*
Wives' desires re:			
sexual relations	4.93	4.62	1.76 (ns)
attention paid to sexual needs	4.96	4.54	2.52^{**}
affection shown to them	4.91	4.61	1.03 (ns)

Notes: $^{*}p < 0.05$, $^{**}p < 0.01$, $^{***}p < 0.001$.

marital happiness scores illustrate the same pattern: Among those whose husbands were cognizant of their desires concerning sexual relations, the mean score was 4.93; among those whose husbands were incognizant, the mean score was 4.62, not a significant difference (as regarding attention to sexual needs), but similar to the arrangement of husbands' scores.

CONTEXT OF SEXUAL EXPRESSION

Finally, we wanted to determine the importance of a person's desires for change in sexual expression, as well as his/her perceptions of the spouse's desires for change, in the light of other potentially important factors in the couple's life. Which factors had the most influence in determining marital happiness? Did other factors overshadow sexual expression in "explaining" marital happiness?

To investigate those questions, we used multiple regression, a statistical tool that indicates the contribution of various factors to the prediction of a certain outcome—in this case, the marital happiness scores of husbands and wives in the study. The factors we included in the "model" were the desire-for-change scores in the three areas of sexual expression (sexual relations, attention paid to sexual needs, and affection), the perception-of-spouse's-desire-for-change scores in the three areas, self-rating of health from another part of the interview (poor, fair, good, excellent), rating of the spouse's health (poor, fair, good, excellent), the score on a measure of life satisfaction (Liang 1984), and the spouse's marital happiness score.

Table 3. Predictors of Marital Happiness: Contextual and Sexual Variables—Dependent Variable: Marital Happiness

Variable	*B*	*Variance Explained (%)*
Husbands		
Wife's marital happiness	.49	25
Satisfaction with affection	.29	7
Perception of wife's satisfaction with affection	.19	4
Self-rating of health	.20	4
Rating of wife's health	.13	2
Adjusted R^2 = 40%		
Wives		
Life satisfaction	.50	25
Husband's marital happiness	.35	10
Perception of husband's satisfaction with affection	.22	4
Satisfaction with affection	.17	2
Adjusted R^2 = 41%		

Table 3 shows the five factors that were significant for predicting (explaining) marital happiness among husbands. The most influential factor was the wife's marital happiness. The next most important was the husband's satisfaction (desire for change) in affection shown to him by his wife. That was followed by his perception of his wife's satisfaction (desire for change) in affection shown to her. Health of husband and wife also contributed to explaining a total of 40 percent of the variation in marital happiness from husband to husband.

The factors that were most important for predicting wives' marital happiness were somewhat different from those that predicted husbands' happiness. Of the four significant ones, the most important factor among women was general life satisfaction, followed by the husband's marital happiness. The perception of the husband's satisfaction with affection was next, and finally the wife's own satisfaction with the affection shown to her by her husband.

DISCUSSION OF FINDINGS

The results of the study indicate that sexual expression remains an important element in long-term marital relationships. As among younger men, long-married men appear to retain greater "body centered" orientation to sex (Hatfield el al. 1988), equating sexual behavior with physical gratification to a greater degree than do women (Hendrick et al. 1985). Twice as many men as women desired an increase in sexual relations.

However, even though approximately a third of husbands wanted an increase in "sexual relations" and attention paid to "sexual needs," sexual expression that takes the form of affectionate behavior appears more crucial to marital happiness than sexual intercourse for both older husbands and wives. As indicated by the results shown in Table 3, none of the "sexual relations" or "attention to sexual needs" items proved to be significant predictors of marital happiness among the couples in the study.

Results point strongly to mutuality as a key to marital happiness, which was clearly related not only to self-satisfaction, but also to perceptions of the spouse's satisfaction. As shown in Table 1, both husbands' and wives' happiness was closely linked to their perceptions that their partners were satisfied. The correlations concerning affection among wives (closer correspondence between marital happiness and perceptions of the spouse's satisfaction than their own satisfaction) suggest that older wives are more "spouse-centered" sexually than their husbands, an observation congruent with the literature on gender that emphasizes women's affiliative orientation (Gilligan 1982).

Correlation results found expression in the multiple regressions of Table 3: Among husbands, their own satisfaction with affection was predominant; among wives, husbands' satisfaction took precedence over their own desires for affection. Perhaps women's greater ability to express affection in relationships with friends and other family members accounts for men's greater dependence on affection from their wives; men tend to put all their eggs in one "emotional" basket (Rossi and Rossi 1990).

As summarized in Figure 3, women were somewhat less aware of husbands' desires for change than were men concerning their wives' wishes. That was especially true in regard to sexual relations (see Figure 2), where women were markedly pessimistic (unaware that husbands were satisfied). The explanation may hinge on women's tendency to communicate feelings more easily than men (Cutrona 1996), so that men may have had more information from their partners. Wives' overestimation of husbands' desires for sexual relations may be the result of ignorance of husbands' diminished sexual interest as compared to earlier years of marriage—wives' perceptions may not have kept pace with current realities.

Likewise, husbands' notable underestimation of wives' satisfaction with affection may be an indication of "egocentric bias" in reports of marital interaction. One study found "a regular tendency to overreport one's own and/or underreport partners' contributions," even on negative items, among both husbands and wives (Christensen et al. 1983). Husbands in this study may have accepted too much responsibility for affectionate interaction with their wives, just as women may have accepted too much responsibility for engaging in sexual relations, perceiving that they were behaving inadequately to satisfy the spouse, when, in truth, the partner desired no change in their behavior.

Notwithstanding those explanations, the substantial proportion of individuals who were unaware of their partner's desires for change in sexual behavior suggest that, although desirable, discussions or other communication concerning sexual matters may not be easy for married couples, even those averaging 35 years of marriage. Given the link between marital happiness and satisfaction with sexual expression, interventions to help couples to improve communication in the sexual realm are no doubt worthwhile in long-term marriages, just as they are among the newly married.

THE FUTURE

This report, based on six forced-choice questionnaire items, obviously has only scratched the surface of a multifaceted topic. The perspective on sexuality as marital interaction deserves more attention, as many questions remain unanswered.

A limitation of the present study is the absence of reports by respondents of specific baselines regarding desires for change: Change from what present level of behavior? In that regard, a particularly intriguing set of inquiries involves couples who are at the extremes of the spectrum—those who never engage in intercourse, and those who are unusually active. For example, it has been reported that men usually institute the cessation of intercourse in marriage (George and Weiler 1981). How does that finding coincide with men's greater interest in physical gratification, compared to women of similar ages, even if, on average, men are a few years older than wives?

Information about satisfied couples who are incapable of or uninterested in sexual intercourse would provide a valuable balance to the emphasis on the continuance of sexual behavior into old age. Just as the myth of the asexual later years has contributed to inappropriate advice given to elders, and to their own discomfort as "dirty old" men and women who should give it up (Schlesinger 1996), some couples may benefit from reassurance about declining activity or interest in intercourse, especially if other forms of sexual expression may substitute effectively, as these findings suggest.

But what other forms of sexual expression are satisfying to men and women? Another limitation of the present study is lack of information about the substance and meaning of "affection," which may have different connotations and content for men and women. Presumably, the behavioral manifestations of affection would include endearments, kissing, and hugging, but the content probably differs from couple to couple, from barely perceptible gestures to extravagant ones, depending on personality, disposition, and marital history.

Another issue concerns the health of marital partners. Some medical procedures, medications, and diseases make sexual intercourse difficult, uncomfortable, or impossible (Weg 1996). Although most of the respondents in this study were healthy overall, some must have had disabling conditions. How did that affect reports of satisfaction with levels of sexual expression among husbands and wives?

The preceding questions, as well as many others that could be asked, illustrate the gaps in knowledge that exist concerning sexuality in later life generally, and among long-married couples more specifically. The heyday of research was in the early 1980s, and surprisingly little information has been added to the literature since then. A careful study based on a national sample would be valuable to replicate findings of the past and to explore new issues and avenues. As precursors to that, smaller qualitative studies that allow elders to speak freely on issues of intimacy would give better insights into all the factors involved.

As noted at the beginning of this chapter, a key to better understanding might well be more "sexual interest" and risk-taking on the part of researchers. If posed sensitively in an appropriate setting, queries involving sexuality are accepted by elders and need not be avoided. As potential contributions to pleasure and contentment in the lives of older men and women, various forms of sexual expression should not be overlooked as significant and consequential areas for investigation. Moreover, as the subject of many myths and misconceptions disseminated person-to-person and in the media, sexual expression among older people should be addressed using carefully designed studies that will produce valid and helpful knowledge for future generations of elders.

ACKNOWLEDGMENTS

I am grateful to David Ekerdt for conceptualization and leadership of the Retirement and Marital Quality project, supported in part by the Medical Research Service of the Veterans Administration and a grant from the National Institute on Aging (AG05661), from which data for this paper was obtained. My appreciation also extends to Hadassa Kubat, who provided excellent programming and analytical support. Some material was presented previously at the Gerontological Society of America Annual Meeting, 1994, and at the Midcontinent Region Annual Conference, the Society for the Scientific Study of Sex, 1994.

REFERENCES

Ade-Ridder, L. 1990. "Sexuality and Marital Quality among Older Married Couples." Pp. 48-67 in *Family Relations in Later Life,* 2nd ed., edited by T. Brubaker. Newbury Park, CA: Sage.

Bernard, J. 1972. *The Future of Marriage.* New York: World Publishing.

Brecher, E. 1984. *Love, Sex, and Aging.* Boston: Little, Brown.

Christensen, A, M. Sullaway, and C. E. King. 1983. "Systematic Error in Behavioral Reports of Dyadic Interaction: Egocentric Bias and Content Effects." *Behavioral Assessment* 5: 129-140.

Comfort, A. 1976. *A Good Age.* New York: Crown.

Cutrona, C. E. 1996. *Social Support in Couples.* Thousand Oaks, CA: Sage.

Denny, N. W., J. K. Field, and D. Quadagno. 1984. "Sex Differences in Sexual Needs and Desires." *Archives of Sexual Behavior* 13: 233-245.

Fields, N. S. 1983. "Satisfaction in Long-Term Marriages." *Social Work* 28: 37-41.

Fincham, F. D., and T. N. Bradbury. 1987. "The Assessment of Marital Quality: A Reevaluation." *Journal of Marriage and the Family* 49: 797-809.

George, L., and S. J. Weiler. 1981. "Sexuality in Middle and Later Life." *Archives of General Psychiatry* 38: 919-923.

Gilligan, C. 1982. *In a Different Voice.* Cambridge: Harvard University Press.

Hatfield, E., S. Sprecher, J. T. Pillemer, D. Greenberger, and P. Wexler. 1989. "Gender Differences in What Is Desired in the Sexual Relationship." *Journal of Psychology and Human Sexuality* 1: 39-52.

Hendrick, S, C. Hendrick, M. Slapion-Foote, and F. Foote. 1985. "Gender Differences in Sexual Attitudes." *Journal of Personality and Social Psychology* 48: 1630-1642.

Hite, S. 1976. *The Hite Report.* New York: Dell.

Jacobs, R. H., and B. H. Vinick. 1979. *Re-engagement in Later Life.* Stamford, CT: Greylock Publishers.

Julien, D., C. Bouchard, M. Gagnon, and A. Pomerleau. 1992. "Insiders' Views of Marital Sex: A Dyadic Analysis." *Journal of Sex Research* 29: 343-650.

Kinsey, A. C., W. B. Pomeroy, and C. E. Martin. 1948. *Sexual Behavior in the Human Male.* Philadelphia: W. B. Saunders.

Kinsey, A. C., W. B. Pomeroy, C. E. Martin, and P. Gebhard. 1953. *Sexual Behavior in the Human Female.* Philadelphia: W. B. Saunders.

Laumann, E., J. Gagnon, R. Michael, and S. Michaels. 1996. *The Social Organization of Sexuality.* Chicago: University of Chicago Press.

Liang, J. 1984. "Dimensions of the Life Satisfaction Index A: A Structural Formulation." *Journal of Gerontology* 39: 613-622.

Margolin, G., S. Talovic, and C. D. Weinstein. 1983. "Areas of Change Questionnaire: A Practical Approach to Marital Assessment." *Journal of Consulting and Clinical Psychology* 6: 920-931.

Marsiglio, W., and D. Donnelly. 1991. "Sexual Relations in Later Life: A National Study of Married Persons." *Journal of Gerontology: Social Sciences* 46: S338-S344.

Masters, W. H., and V. E. Johnson. 1966. *Human Sexual Response.* Boston: Little, Brown.

Mead, G. H. 1934. *Mind, Self, and Society.* Chicago: University of Chicago Press.

Murphy, G., W. Hudson, and P. Cheung. 1980. "Marital and Sexual Discord among Older Couples." *Social Work Research and Abstracts* 16: 11-16.

Neal, A. G., and H. T. Groat. 1976. "Consensus in the Marital Dyad: Couples' Perceptions of Contraception, Communication, and Family Life." *Sociological Focus* 9: 317-329.

Persson, G. 1980. "Sexuality in a 70-Year-Old Urban Population." *Journal of Psychosomatic Research* 24: 335-342.

Pfeiffer, E., and G. C. Davis. 1972. "Determinants of Sexual Behavior in Middle and Old Age." *Journal of the American Geriatrics Society* 20: 151-158.

Pfeiffer, E., A. Verwoerdt, and G. C. Davis. 1972. "Sexual Behavior in Middle Life." *American Journal of Psychiatry* 128: 1262-1267.

Riportella-Muller, R. 1989. "Sexuality in the Elderly: A Review." Pp. 210-236 in *Human Sexuality: The Societal and Interpersonal Context,* edited by K. McKinney and S. Sprecher. Norwood, NJ: Ablex Publishing.

Ross, C. E. and J. Mirowsky. 1984. "The Social Construction of Reality in Marriage: An Empirical Investigation." *Sociological Perspectives* 27: 281-300.

Rossi, A., and P. Rossi. 1990. *Of Human Bonding: Parent-Child Relationships Across the Life Course.* New York: Aldine de Gruyter.

Schlesinger, B. 1996. "The Sexless Years or Sex Rediscovered." *Journal of Gerontological Social Work* 26: 117-131.

Shaw, S. 1988. "Gender Differences in the Definition and Perception of Household Labor." *Family Relations* 37: 333-337.

Starr, B. D. 1985. "Sexuality and Aging." *Annual Review of Gerontology and Geriatrics* 5: 97-126.

Starr, B. D., and M. B. Weiner. 1981. *The Starr-Weiner Report on Sex and Sexuality in the Mature Years.* New York: McGraw-Hill.

Traupmen, J., E. Eckels, and E. Hatfield. 1982. "Intimacy in Older Women's Lives." *The Gerontologist* 22: 493-498.

Weg, R. B. 1996. "Sexuality, Sensuality, and Intimacy." Pp. 479-488 in *Encyclopedia of Gerontology, Volume 2*, edited by J. E. Birren. San Diego, CA: Academic Press.

Weiss, R. L., and G. R. Birchler. 1975. "Areas of Change." Unpublished manuscript, Department of Psychology, University of Oregon, Eugene, OR.

Weizman, R., and J. Hart. 1987. "Sexual Behavior in Healthy Married Elderly Men." *Archives of Sexual Behavior* 16: 39-44.

Whitbourne, S. K. 1990. "Sexuality in the Aging Male." *Generations* 15: 28-30.

White, K. M., J. C. Speisman, D. Jackson, S. Bartis, and D. Costos. 1986. "Intimacy Maturity and Its Correlates in Young Married Couples." *Journal of Personality and Social Psychology* 50: 152-162.

MARITAL PERCEPTIONS IN RETROSPECT

A STUDY OF OLDER WIDOWS AND WIDOWERS

Kate Davidson

INTRODUCTION

This chapter focuses on a study of older widows and widowers who had experienced long-term relationships in a post-World War II United Kingdom. Particular attention is paid to the notion of the "companionate marriage," a relational concept that is especially identified with this cohort. The character of the relationship between couples formed during the 1930s and 1940s needs to be viewed in the historical context of the meaning of marriage to preceding and succeeding generations. Most respondents reported having had a happy marriage and all were proud that they had kept their "marriage vows."

The chapter explores what constitutes a happy marriage and argues that for this generation, the patriarchy inherent in these marriages whereby wives subsumed their wishes to the demands of their husbands, is a factor in the maintenance of a long-term marriage. A hierarchical relationship, where boundaries are drawn and observed, fosters reduced conflict and an increased likelihood of conjugal longevity. For the overwhelming majority of these older people, the dissolution of marriage was brought about by the death of their only spouse and so for them, it was indeed a case of "Till Death Us Do Part." The chapter concludes with a brief discussion on the initial bereavement period following a long-term partnership.

DESCRIPTION

People born during the second and third decades of the twentieth century and married in the 1930s and 1940s have experienced, on average, longer lasting marriages than any preceding or succeeding generations. They have also witnessed the most dramatic technological, societal, and cultural changes: from horse and cart to space travel; from total disgrace of illegitimacy to a third of children born to unmarried parents; from a massive colonial empire to legislation on racial equality. In the present-day United Kingdom, data reveal similar proportions of lone parents and remarriage rates as were experienced 100 years ago (CSO 1996). The fundamental difference lies in the cause of the dissolution of marriage: in the nineteenth century, premature death of a spouse was the reason for the break-up and reconstitution of a family, whereas toward the end of the twentieth century divorce is the overriding factor (Burgoyne, Ormrod, and Richards 1987).

This chapter is based on a U.K. study of men and women, born before 1930 (at least 65 years old at time of interview) and widowed for a minimum of two years.[1] Twenty-five widows and twenty-six widowers were interviewed and encouraged to talk about their histories, marriages, present lives, and futures. The sample was not representative in terms of gender distribution. Men were deliberately overrepresented since the purpose of the study was to make a direct comparison between the sexes.

The age range for both widows and widowers in the sample was 65 to 92, but the average age of the widows was 75 years while that of the widowers was 78 years. This is reflected in the average length of their "only or last" marriage as 36 years for the widows and 38 years for the widowers. The men had had the longest as well as the shortest marriages: more men than women had celebrated their golden (fiftieth) wedding anniversary (seven as compared to three), but men also had the shortest "last" marriages because more had remarried after a first bereavement or following divorce. Two of the 25 women had been married twice, one was widowed very shortly after marriage during World War II and one divorced, as compared to nine of the 26 men, two of whom had remarried after divorce and seven after widowhood. One widower has subsequently married for the third time, having been widowed twice.

In the United Kingdom, the visionary Beveridge Report of 1942 and the subsequent postwar health and welfare legislation did much to alleviate the health and financial difficulties of early motherhood and widowhood, but it also served to confirm the place of women as principal caregivers and consolidated the notion of economic dependency on a male earner (Williams 1991; Baldwin and Falkingham 1994). Beveridge, a rather ascetic academic, was far from being anti-female. He entertained the greatest respect for women's intellects and consulted them in the construction of his report. However, he did reflect the nationalist and familist conceptualisation of a (still) colonialist United Kingdom, and his principal concerns were with the maintenance of full, male employment and the production of a healthy new generation. Beveridge made the assumption that "During marriage most women will not be gainfully occupied" (Beveridge 1942 para. 111). Rather, she would have duties that he described as "vital, unpaid service as housewives" (para. 309).

This perception bears a close resemblance to Talcott Parsons's functionalist ideas of marriage (Parsons 1956). There was the notion of men and women being equally valued in a relationship that required them to carry out different tasks or functions for which they held different expertise.

The sample was collected from a variety of sources and was widely heterogeneous in terms of income, health status, housing conditions, and social networks. However, there was one focal point to which all respondents frequently referred as having the most profound influence on their lives, and that was the Second World War, 1939 to 1945. For many women, it legitimized their entry into the labor force, and gave them the freedom that attends financial independence. Some entered the military forces, others worked for government, in factories, or on the land, which took them away from home and direct parental influence. For the men, particularly those who fought, what they witnessed and experienced affected them deeply. Some told of a loss of faith in God, government, and the status quo of a divided and unequal society. The return to peace and civilian life promised a different world and they were determined to be a part of it.

Although women recognized the horror and devastation wrought during the hostilities, many of them reported having a very "good time" during the War and enjoying a freedom hitherto not allowed young women and certainly rarely experienced by their mothers' generation. Through the medium of the cinema and through personal acquaintance with U.S. military personnel, the U.K. population was increasingly influenced by the North American culture of the pursuit of individual satisfaction and fulfilment. For these women, expectations of greater personal happiness characterized their relationships, especially with men of their own generation.

> Mrs. T: The war opened up a whole new world for many of us, for me, and we weren't prepared to be the same as our mothers. (Mrs. CT, aged 75, m 28 years, w 12 years)

Both widows and widowers reported that after the War they knew that "nothing was ever going to be the same again." Nevertheless, there was a concerted political and social movement intent on reconstituting idealized, nuclear family life with the husband and father as breadwinner, wife and mother as homemaker. There was also an overwhelming feeling that the men should be rewarded for the sacrifices made during the War.

> Mrs. B: Men had to be men and women were women and looked after the house and the children. I look back and suppose I spoiled him. He had such a terrible War and I spoilt him. (Mrs. JB, aged 75, m 50 years, w 3 years)

However, postwar wives, buoyed by their wartime experiences of greater equality in the workplace and ability to cope "while the men were away," expected that their marriages were going to be more symmetrical, companionate relationships based on genuine partnership.

For this generation of women, the proliferation of electrical appliances brought about by increased prosperity permitted more leisure time than had been enjoyed previously. This meant that they did not have to contend with as much domestic drudgery as their mothers had. Added to this were major shifts in child-bearing patterns. Since the beginning of the twentieth century the average number of children born in families has been decreasing, and so, consequently, has the period of time between the birth of the first and last child. At the beginning of the century in England and Wales, the mean family size for all women under 45 at time of marriage was 3.53. By the time most of the respondents were married, in 1941, this number had decreased to 1.98 (Allan 1985). With most of the perils of child-bearing removed, a new technology that permitted greater domestic freedom, and a perceived greater sexual equality, a change in marriage dynamics meant that in this new union, neither side should dominate the other, each instead achieving his or her goals with the other's cooperation. The reality experienced by the widows in this sample was somewhat different, as will be discussed.

COMPANIONATE MARRIAGE

In the United Kingdom, the postwar, social reconstruction witnessed a key shift from the idea of marriage as an institution to marriage as a relationship, and from this emerged the concept of companionate marriage (Finch and Summerfield 1991). Although the phrase, "companionate marriage" had been in use since the 1920s, its practice was not widely embraced until after World War II. There was a strong desire to foster the notion of teamwork in a relationship—a notion that had its roots in the shared risk and joys of the community at large during the War. After the disruption and separations, what was perceived as important was a period of stability, and the focus for the provision of this was the family. The new partnership ideology was based on equality, sharing, and the pursuit of personal happiness. Indeed, in these postwar unions, there was a manifestly, less overt, dominant/subservient, power structure between husbands and wives than had been the norm in Victorian marriages (Harris 1983).

The character of the relationship between the widows and widowers in this study needs to be seen in the historical context of the meaning of marriage to preceding and succeeding generations. For their parents, there was less emphasis on communication and shared activities. For their children's (and more so their grandchildren's) generation, there is not only a decreased emphasis on marriage as the normative lifestyle for women, but a huge increase in the prevalence of divorce when expectations have not been fulfilled (Mares 1995).

> Mrs. B: Dad ruled the roost like, and my mum was the sort that men were the be all and end all. You had to do everything for the man. They had to be given their dinner first and their cup of tea first and that. She was a wonderful mother mind, don't get me wrong. It was all the man. The men had to be waited on. I suppose some of that rubbed off on me. I had to have the dinner ready when he come home from work and that. . . . My daughters used to say "Oh, we're not going to put up with that mum, when we're married." And they don't. But they've both got good marriages. (Mrs. MB, aged 70, m 47 years, w 7 years)

Mrs. B learned from her childhood that it was usual for women to look after men quite comprehensively, and that this was the basis of being a good wife. She was pregnant at 15 (at the beginning of the War) and married the baby's father soon after their daughter was born. By the age of 18, she had another child. She was determined to make the marriage "work" and to do so she felt, certainly in the early years, that she needed to emulate her mother. It is interesting to note that although her daughters do not behave in this way, she says "but they've both got good marriages." This would imply that they have good marriages despite their refusal to acquiesce to the overriding wants of their husbands.

The late 1960s and early 1970s, when the respondents' children were forming partnerships, gave rise to the second wave of feminism in the United Kingdom. Although eschewing many of the manifestations of the movement, older women started to question overtly their positions as sole performers of domestic work. They saw their daughters and daughters-in-law negotiating domestic and parenting responsibilities with their partners and compared this with their mother's complete dominance by their fathers.

The widows, in particular, placed themselves in the transitional ground between the marriage experience of their parents and that of their children, in terms of egalitarian decision making and shared labor, both outside and within the home environment. However, the roles within the postwar companionate marriage continued to be traditionally gendered (Morgan 1991), although they were more likely to be perceived as complementary than divided.

> Mrs. M: I mean, my husband never changed a nappy, never bathed a child, never had to cook a meal. And I think it's the same on the other side. The men looked after the money and that sort of thing. Looked after buying the cars and all that. I don't think many women did that sort of thing. (Mrs. SM, aged 65, m 32 years, w 8 years)

This whole domestic ideology was reinforced in postwar Britain: the state closed down most of the nurseries set up to enable women to work for the war effort while the return of the men from the forces resulted in a rapid rise in the birth rate. Furthermore, women themselves who had grown up socialized into believing that they belong in the home, returned together, at least for the short term. Riley (1983) has pointed out that the motives and reasons for women surrendering status as full-time workers were complex and varied. The explanation that they were forced back into the home was, she has argued, as unsatisfactory as the idea that women chose to return to their rightful place as mothers and homemakers. However, after the austere days of clothes rationing, women felt able to reclaim their femininity with the "New Look" fashion of full, long skirts and narrow waists. Many saw domesticity as liberation from factory work and hard, land labor (Roberts 1984).

The respondents in this study, both male and female, were socialized into viewing marriage and having children, in that order, as a life goal. For the vast majority of women, this was viewed as the ultimate goal. All the widows who had children felt that their "proper" role was to be at home while their husbands provided for the family. Although the gendered division of housework persisted through most of the married years, all but two of the widows had worked at some stage of the marriage and had continued to do the vast majority of the domestic labor. The widows strongly voiced the idea that the husband's need to be looked after while he was working was perceived to be the most important factor.

K: Did he do any housework or anything like that?

Mrs. H: Oh no! Oh no! He wouldn't dream of doing anything and I wouldn't dream of letting him.

K: Why not?

Mrs. H: Well, you just didn't dear. You weren't a proper wife if you let your husband do housework and things like that! (Mrs. VH, aged 87, m 33 years, w 32 years)

Those women who had no children, or whose children had not survived infancy, still viewed their role as homemakers, even if they were themselves working.

Widowers did work in the home, but this tended to be house maintenance rather than housework. Men and women were perceived as "experts" in their own fields and as such saw themselves as the better equipped to perform the job, according to the task.

Mrs. H: B used to do everything like that. He wouldn't let me help decorate or anything.

K: Why wouldn't he let you help?

Mrs. H: He thought I would be useless! (Mrs. VH, aged 65, m 31 years, w 13 years)

Mrs. L: He did all that. Electricity, woodwork, painting, decorating.

K: Did you do things like that?

Mrs. L: No. He was so good at it, I thought, let him get on with it. I helped him, I always used to paste the paper for him, ready to hang up and all that sort of thing. But, no, I never did it because he was good at it. No, I liked sewing. I loved sewing. (Mrs. BL, aged 83, m 50 years, w 10 years)

The division of labor and respect for each other's abilities was considered an important aspect of a successful marriage.

Mrs. D: But he respected me because I done all the house and kept the home and I went to work, I had a job because I had only the one girl and she was going to school. Well, he respected that and I respected him. In a marriage, you've to have respect for one another. (Mrs. SD, aged 70, m 28 years, w 22 years)

Mrs. D attributes her happy marriage to how she interpreted appropriate behavior of men and women within the marriage relationship. Socialised by Victorian parents, for this cohort the mutual expectations as to what constitutes "wifely" or "husbandly" conduct is defined by the way "femininity" and "masculinity" are conceptualized.

Mrs. D: I had a very happy marriage. And I had a good husband. He was 11 years older than me. And he was like a father, a friend, a husband, he was everything in one—because I was that much younger than him, and he treated me that way. But he always treated me like a woman. (Mrs. SD, aged 70, m 28 years, w 22 years)

A good husband was a provider and a protector; a good wife cared for him. These were the accepted gender roles of this generation.

HAPPILY EVER AFTER

One of the most common reports about marital satisfaction over time is that there is a U-shaped trend, with satisfaction dipping in the middle years of the relationship and increasing again in later years (Mares 1995). However, as Valliant and Valliant (1993) point out, the studies that have presented these findings have all been cross-sectional, which means that different cohorts were compared, and the sample of older people did not include those who had been divorced. Their longitudinal research indicates that for many couples in long-term marriages, the levels of satisfaction with the relationship stay consistently high. They argue that unhappy couples are more likely to divorce than to stay in the relationship. This research applies to couples in the United States, where historically, the trend of increased divorce rates predates those in the United Kingdom. Divorce for this generation of U.K. widows and widowers was still comparatively rare, both for financial considerations as it could not be afforded, and for social reasons as it was still considered unacceptable (Burgoyne et al. 1987). Some of the respondents did say that there were turbulent years, usually near the beginning of the relationship quite commonly as a result of being separated during the War. Some described difficult years after children were born, mostly because the men felt excluded and sexual relations decreased. However, all of the widowers and two-thirds of the widows who reported very difficult years said that the relationship had improved toward the end of the marriage.

K: How would you describe your marriage?

Mr. B: Well, it was, of course, it was disrupted by the War. Troubles we had were because, you know, you were away three years. But it would have been better if I'd never gone away like that, but there you are. . . . I mean for the last 20 years or so, we were good friends. . . . We were very good friends, we got on very well. (Mr. JB, aged 81, m 53 years, w 4 years)

Most of the people in the sample described their marriages as happy, with ups and downs "like normal people." This description could be termed a "wave effect" rather than a "dip" in marital satisfaction. It is possible that if they did experience a dip, it has subsequently been levelled out in retrospect. Respondents, male and female, interpreted their married lives as "normal" even though they were not perfect unions. This pragmatic approach is the result of coming to terms with the "lived experience" which differed from expectations of a marriage "made in heaven." The men and women are able to locate themselves within society knowing not only how they experienced marriage, but also how other people are likely to experience it. They are then able to adjudge that how they lived was typical.

Mr. S: It was quite happy. There was no problems, like. Obviously there were arguments once in a while—no fisticuffs of anything like that. Just normal, average, like I suppose everybody else's was. (Mr. JS, aged 70, m 34 years, w 6 years)

Mrs. H: We were very happy. We had our ups and downs like everybody. (Mrs. VH, aged 65, m 31 years, w 13 years)

Mrs. H: Happy. Quarrelled, just like other people, only natural. We were very much in love, very fond of one another. (Mrs. VH, aged 87, m 33 years, w 32 years)

Normal married people, according to these respondents, are men and women who do not always live in harmony, but persevere during difficult times and stay together, thus keeping their marriage vows. A "normal" marriage, therefore, is one in which there is not perfect and perpetual harmony. However, the widowers were much more likely than the widows to say they had "very good" or "perfect" marriages.

When I asked widowers to describe their marriages, they tended to say how much they appreciated what their wives had done for them.

> Mr. B: One of the best in the world as far as I'm concerned. A very happy marriage. She always helped me, like. (Mr. CB, aged 85, m 53 years, w 6 years)

> Mr. A: If you love your wife and she do everything, you miss her. I not sorry I marry. Because she was really companion. Yes. And she was cooperative. (Mr. MA, aged 84 years, m 37 years, w 2 years)

> Mr. J: My wife was a Yorkshire and I think they make very, very good wives. I mean they really look after their men. . . . She was a very good wife to me. Excellent wife to me. She used to polish my shoes every morning before I went to work. She did it out of love. You wouldn't get girls these days doing things like that for their husbands. She liked to do things like that. (Mr. EJ, aged 74, m 38 years, w 13 years)

> Mr. B: Very easy to get on with. Yes. Fortunately, very easy. Never told me off or anything like that. Very good. Not like a lot of women. She was very good, very good indeed. (Mr. BB, aged 92, m 49 years, w 9 years)

Lee (1978) examined factors that predicted the morale of older couples and found that for men morale was simply related to the existence of the wife rather than to the quality of the marriage: ". . . it seems to be the case, in terms of morale and related variables, men are more responsive to the simple presence of a spouse while women are more sensitive to variation in the quality of the marital relationship" (Lee 1978, p. 137).

While the men's primary satisfaction was from the temporal presence of their wives, it seems that a successful marriage for the women was centered around what they did with and for their husbands to make them happy, and that the men's primary satisfaction was from the temporal presence of their wives.

Credit for a happy marriage was often given to the dead spouse. Respondents who expressed this were more likely to blame themselves for any turbulence in the relationship. They were able to admit to the "normal" up and down marital relationship and still keep the memory of the spouse as wholesome and blameless.

> Mrs. B: It was nice, it was comfortable. We were friends as well. We were good friends as well. We got on well. We had the odd argument of course, but—in fact, when we were first married, I was a terrible one. If I'd got anything in my hand, I'd throw it! But I got over that as I got older, I mellowed. But

he would laugh it off, you see. He wouldn't get nasty. (Mrs. IB, aged 68, m 42 years, w 4 years)

Mrs. L: Perfect. Well, only yesterday a friend on the phone said to me. I said, "I feel very lonely at times" and he said, "You would do B, 'cos your marriage was made in heaven." I was so happy with my husband. My husband was a real lovely man so—kind and gentle. Never heard him raise his voice, you know. In fact, he should have done, with a wife like me at times. Yes, he should of done. (Mrs. BL, aged 83, m 50 years, w 10 years)

Lopata (1973) has identified this process as "sanctification" of the spouse, where the memory of the late husband or wife has been idealized. Of those who said they had an "exceptional" or "perfect" spouse, the widows were more likely to divulge later in the interview that this was not necessarily so. For example, one widow, whose much younger husband was killed in a car accident, revealed that he was "quite a flirt" and she supposed she is grateful, at least, that she lost him to an accident rather than to a younger woman. Another said she often felt she came third in her husband's affection, after the house and the dog.

Zietlow and VanLear (1991) found that compared with younger couples, long-term couples (married at least 40 years) showed more deference in terms of willingness to tolerate some behaviors, but also to relinquish some behaviors. Long-term couples, they argued, who practiced the most deference were the most satisfied. Interestingly, Zietlow and VanLear note that although couples married during the 1930s and 1940s are usually viewed as operating under a hierarchical, husband-dominant power structure, there were no significant sex differences found between long-term husbands' and wives' use of behavior control. Analysis from this research disputes this finding. Women's acceptance of male domination means that as the couples grow old together, there is less likelihood of conflict in the interest of amity.

Mr. L: I was, had a very, very good relationship with B. She had a few faults, a very few faults. I have quite a lot of faults. And she had far more to put up with, to suffer with me than the other way around. She couldn't stand fools, regularly, but she stood me for all those years. She was the ideal wife for me and we were very happy. Very happy. (Mr. KL, aged 70, m 43 years, w 3 years)

Mares (1995) comments that elderly adults report relatively high, stable levels of marital satisfaction and that their communication styles are more likely to be low-risk and low-disclosure orientated. Common to both genders was the notion that there was a good friendship between them, which strongly relates to the ideology of the companionate marriage.

TOGETHERNESS

For widows, the most common reason given for why they had had a happy marriage, was: "we did everything together"; "we worked well together." When asked what they and their husbands did together, it was revealed that within the domestic sphere they did very little and what they did do together largely depended upon what their husbands liked doing. The togetherness was being in close proximity rather than doing the same thing, but it was recognized as a very important factor for their happiness.

> Mrs. L: Oh yes, we did things together.
>
> K: What sort of things?
>
> Mrs. L: I mean, not shopping, he couldn't stand shopping. But we used to go out for the day for walks and you see, he built a boat, a lovely boat. We used to cruise the Thames, up and down. Every weekend. And he loved fishing, so if he was happy sitting on the bank fishing, I would be happy reading, having a read. (Mrs. BL, aged 83, m 50 years, w 10 years)
>
> Mrs. M: For instance like the [local Arts and Music] Festival, if I were married, I wouldn't be able to spend the amount of time on that, that I do. You see, when the Festival is on, it's every weekend from mid-April right to the end of May. Well, I would never be able to do that. Unless my husband were interested as well. He wasn't. You see, we did everything together. (Mrs. SM, aged 65, m 32 years, w 8 years)

Mrs. M indicates that they would only do things together if her husband was interested.

> Mrs. D: You see, with men about, you have to consider what he wants—if he wants to go out, if he wants to go somewhere. I always went with him. Many a time I didn't want to get up 5 o'clock in the morning to go fishing, but he wanted to 'cos I always say in marriage, if you can't beat 'em, join 'em! (Mrs. RD, aged 78, m 22 years, w 28 years)
>
> Mrs. H: Not having any family, just the two of us. So we were—everything we did, we did together. . . . And then you see, he thought he'd do some sailing. Well, I wasn't too keen but I did it because I would have been more frightened if I was at home on my own, so I went with him. (Mrs. EH, aged 78, m 47 years, w 7 years)

By the wives' deference to the interests of their husband, these quotes demonstrate very clearly the patriarchal relationship within the marriage. Often unspoken at the time, these widows reported that they acquiesced to their husbands' wishes and used phrases in the interview like, "Anything for a quiet life, you know," and "It wasn't worth the hassle," and "I used to just fall in with it" to describe the way they coped with the demands.

> Mrs. B: I let him have all his own way and did everything for him, even though I was working too. (Mrs. JB, aged 75, m 50 years, w 3 years)

The major financial decisions such as relocation, choice of car, and so on, continued to be made by husbands, as had been the norm with the previous generation. Wives were often consulted on what were perceived to be the less important decisions: leisure activities, vacations, furniture, interior decoration.

The women in this sample talked about "his money" and accepted that they had relatively little control over how it was jointly spent.

> Mrs. T: If you say anything, you see, he used to say, "Well, it's my money, it's not your money. It's my life." What do you do then? How do you cope with it? (Mrs. CT, aged 75, m 28 years, w 12 years)

Many had no idea how much money their husbands earned and two had never had a bank account, paid any bills, or knew how to write out a cheque before they were widowed. They did, however, say that they received *their* housekeeping, from *his* money, and they could always ask for more if they needed it. The day-to-day decisions about food purchases and children's clothing were almost entirely made by the women. Edgell's (1980) 1970s research of middle-class couples indicated that husbands usually made the major decisions and primarily consulted their wives for their agreement. One husband said, "By and large I decide nearly everything. Everything that matters, I decide. I do consult my wife and she always agrees" (p. 67).

Most of the widows and widowers reported working together in the garden. Even here, however, there was a division of labor, although it was less obvious initially, because of the spatial proximity of husband and wife.

> Mr. B: She liked to do the flower garden and I did all the heavy digging she wanted, but otherwise, she looked after that top part. I did all the rest, given over to veg and fruit, whatever. (Mr. JB, aged 81, m 53 years, w 4 years)

Many of the couples moved into new houses with gardens, either local authority (public housing) or in the private sector, during the massive rebuilding program of the postwar era. Only one widower lived all his married years in an apartment in central London where he had no access to a garden. Even for those who had moved into established homes, the gardens had been given over to "digging for victory" and had produced mainly vegetables for the previous few years. Now was their chance to establish an area of beauty and leisure, as well as provide an area of play for children.

Mrs. B: Saturday was for the garden. All day when the weather was good. We just worked along side each other really happily. (Mrs. JB, aged 75, m 50 years, w 3 years)

The garden could be viewed as producing the overlap between what went on behind closed doors and what was on view to outsiders, although both remained in the domestic sphere: a focal point where the couple could be *seen* to be working together, making joint decisions about layout and choice of flowers and vegetables.

Mrs. M: When we first moved here, the garden was just like a building site. All stones and rocks and everything. B and I spent weeks clearing it, sieving it, sometimes until well after it was dark. We planned where we would put things. We worked so hard together to get it how we wanted it. (Mrs. JM, aged 67, m 40 years, w 4 years)

The importance of the part played by the garden in the changing emphasis in marital relationships should not be underestimated. Smaller families and increased access to labor-saving devices enabled couples to spend more time together and gardening was socially acceptable as a joint "unpaid" occupation. It is noticeable, however, that the men tended to look after the "useful" part of gardening—the vegetables, and women tended to care for the "pretty" side—the flowers.

Mr. S: We'd be on the garden together on a Sunday or something, you know.

K: You both liked gardening.

Mr. S: Well, she used to look after the flowers, you know. If you couldn't eat it, I didn't grow it, I said. (Mr. JS, aged 70, m 34 years, w 6 years)

Despite the persistent gendered division, the garden is the site where most of the widows in particular, and many of the widowers reported the greatest conjugal contentment.

Mrs. B: So I never learned to change a plug or I've never decorated or any of that kind of thing. In fact, my husband used to do a room each year. Unlike my sons and their wives, they do it together. I wanted to do it together, but he—I mean, I once took a paintbrush and I'd only just started and it was just running down a bit and he said, "Look what you're doing!" And took the paintbrush away from me, and so I never really cooperated. Yes, in the garden, but not in the house. (Mrs. JB, aged 75, m 50 years, w 3 years)

Division of household and domestic responsibilities for older people appears to continue to conform to the traditional gender roles established early in marriage, even after both have retired (Askham 1995).

RETIREMENT AND DOMESTICITY

The socially structured commencement of "old age" for men is generally considered to be 65, when compulsory retirement from the labor force is the norm for most in the United Kingdom. There has been considerable research on men's loss of roles at this time of their lives (Fennell, Phillipson, and Evers 1993). For married women who have worked and retired, the transition is not as dramatic for two main reasons. Firstly, they usually maintain their domestic duties in tandem with paid employment (Ginn and Arber 1991) and secondly, the majority of working wives have occupations in the service and caring industries that are often an extension of the domestic sphere (Beechey 1986). For older women, then, there is a sense of continuity of role after retirement. Most older wives have done, and continue to do, the major share of household duties and caring responsibilities.

> Mrs. B: Saying that, even when he retired, he never cleaned a window, or hoovered and never helped me with the vegetables or anything like that when we were entertaining. He'd clear up. We had a washing up machine then. And he'd do the clearing up but he never—but then again, that was the pattern.
>
> K: There was a distinct division of chores in the house.
>
> Mrs. B: Yes. He thought it was women's work. In the house. Washing, cleaning, and that kind of thing. Even when he was retired, it never occurred to him. (Mrs. JB, aged 75, m 50 years, w 3 years)

Ward (1993) observed that the husband's retirement was associated with an increase rather than a reduction in demands upon his wife's time because he required companionship and entertainment. Research findings on the allocation of household tasks in the context of long-term marriage have proved inconclusive. For example, a retired man may become more involved in activities in the home and any such changes can alter, or reduce the household role of his spouse (Dorfman 1992). Role change may, in turn, affect the self-concepts and psychological adaptations of both spouses (Brubaker 1985). It has been suggested that women view their husbands' increased involvement as intrusive on their traditionally held territory (Keith and Schaffer 1985). There has been some support for the view that older men find the participation in household tasks irksome, perhaps because of its perceived lower status compared to paid employment. On the other hand, there has been some evidence to suggest that some older men enjoy household involvement, although benefits derived varied by the type of task being undertaken (Keith and Wacker 1990). In this sample, the widowers were more likely to say they did their share of housework by washing up and doing some vacuuming, especially after they retired. However, the widows were more likely to say that *all* their husbands did was a bit of washing up and vacuuming, and sometimes very reluctantly.

The widows did say, however, that their husbands did more than their fathers had done in the house, but not as much as their sons (or sons-in-law).

> Mr. M: Oh, we shared a good deal. I used to help with the washing and the washing up and the hoovering. Afterwards, I thought I ought to have done more. (Mr. MA, aged 84 years, m 37 years, w 2 years)

> Mrs. P: And he didn't like housework, but he would help me. He'd say, "Utter waste of time. Why are we doing this?" And I'd say, "Well, because it is nice when it's done." [And he'd say] "Well, I don't think it's necessary." (Mrs. BP, aged 87, m 57 years, w 3 years)

In the majority of cases, of the couples whose husbands had survived into retirement, the women were no longer working. Most of the couples continued to follow patterns established earlier in their relationships—with disparity in time spent on housework and gender-stereotyped division of tasks. This pattern changed only when one of the partners became physically unable to carry out her or his "normal" work.

The "blurring" of roles was primarily one-way: men were more likely to take on the more traditional female household chores, under the tutorship of their wives, than women were to take on the more masculine tasks of house maintenance. For all but a handful of the older widows and widowers, the dissolution of marriage was presaged by illness of the spouse. This illness can be viewed as a bridge, or transition between marriage and widowhood; and for some, this period was more an integral part of their experience of widowhood than of marriage.

DEATH US DO PART

It has long been recognized that widowhood after a long marriage, regardless of its quality, is the single greatest cause of disruption in the life course of an older person (Pollock 1977; Arens 1982-83; Bowling and Cartwright 1982; Stroebe, Stroebe, and Domittner 1988; O'Connor 1994; Stroebe 1994). It is not my intent here to give a comprehensive account of the vast literature that has addressed issues of bereavement; rather, the purpose of this short section is to highlight the similarities between men and women when they first lose the person with whom they have spent the majority of their adult years.

Most of the people interviewed had enormous insight into how and why they were or were not coping with living without a partner. These widows and widowers worked out how and why they felt the way they did. A certain degree of pragmatism, borne of understanding where they fitted in within family, neighborhood, and community, allowed them to realign their lives and act out their roles as "single again" members of society.

Murray Parkes's work on stages of bereavement highlights the feelings of abandonment, anger, and desperation felt by both widows and widowers when first faced with the loss of a spouse (Parkes and Brown 1972). The respondents remember with clarity the devastation and despair they felt. For some, this was tinged with guilt that they "could have done more," or "should have made the doctors do more." One woman had had an angry argument with her husband just before he died, and three years later continued to feel great remorse. Four of the 25 widows confessed to feeling great relief when their husbands died, but still mourned life without the partner they had loved, even if the final years had been difficult, sometimes unbearable for them. They wanted their husbands back, but not as they had been in the final years.

Widows

I didn't think I'd ever live through it.
I was *desperate* when he died.
I lost my whole world.
I almost lost my mind.
When he died I was shattered.
The bottom had fallen out of my world.

Widowers

Why me?
I was floundering.
Loneliness—something I thought I could bear—you can't.
I thought I was going off my nut actually.
I think I was more cross than anything.
I didn't know what to do.
I felt as if I had lost part of myself.

There is much commonality between the genders at this point. This reflects the findings of Hyraks, Kaumonnen, and Paunonen (1997), who concluded that there was no difference between the genders when death of the spouse had occurred less than two years prior to completing their questionnaire. It also concurs with Lund, Caserta, and Dimond's (1993) study, which found that gender is a relatively unimportant predictor of initial grief reactions. For the vast majority, although by no means all, after the disorientation experienced with initial loss when there seems to be no meaning to life, there is a gradual picking up of the threads of "successful living." Successful living, here, is defined by the widowed people themselves, in terms of how they consider they are coping with life. Four of the 26 widowers considered they were not coping at all, neither physically nor emotionally, without their spouses, more than two years after her death. These men made frequent visits to their physicians, were childless, and made constant use of the statutory and voluntary services for older people (mobile meals and day centers, for example). Three of them were over the age of 80 and the fourth had considerable learning difficulties. One of the 25 widows continued to be in deep mourning more than 15 years after the death of her husband who was killed in a automobile accident. She, too, was childless.

Interestingly, nine of the widowers and five of the widows had no surviving children, most having always been childless. The presence of adult children and grandchildren was a much greater factor in self-reported well-being for the men than it was for the women. Those who considered that they were coping well tended to use phrases like "You just have to get on with it," "You take one day at a time," "Life has to go on, regardless," and "You can't just sit there and mope all the rest of your life." Although there was no substitute for a spouse, they reported that friends and family had provided the essential support during those first years of widowhood.

CONCLUSION

To love and be loved is one of the most powerful and fundamental driving forces for sustaining self-esteem and self-identity. For this cohort of older widows and widowers, this two-way process is characterised by a gendered division of mutual dependency: the women's dependency is grounded in their need for the financial security provided by a man; while for the men, the dependency takes the form of practical and emotional support given by a woman. In a patriarchal society, these support systems do not command equal value. The wants and needs of the husband were perceived to be the more important and all but a very few of the widows accepted the status quo until the latter years of their marriage. This questioning coincided with the burgeoning interest in feminism and the refusal of their daughters to tolerate similar domestic divisiveness.

Toward the end of the marriage, there was a much greater domestic input from their husbands, although the perception of the quantity and quality of this involvement differed between the men and the women. Nevertheless, 21 of the 25 widows and 25 of the 26 widowers interviewed, considered they had happy, or very happy marriages. They were proud of having kept their "wedding vows" and there was a general disapproval of the increasing rate of divorce and single parenting of younger generations. When asked what they considered to be the "secret of their successful marriages," all said that they had "worked at it." The women tended to say that they learned how to make their husbands happy, and the men tended to say their wives had been very good to them. For the majority of the couples, the latter years were marked by close friendship despite, and sometimes as a result of, the illness of a spouse. They spoke of a shared sense of humor born of familiarity and perseverance.

Bearing in mind that for this cohort in the United Kingdom, divorce was not an easy option, it could be argued that some of these marriages, given contemporary access to separation and alimony, could have ended in divorce rather than the death of a spouse. It is also possible that the memory of the marriage has been sanitized as a personal coping strategy, or as an account for the research interviewer, or, indeed, for a public account. The women admitted, or were more prepared to admit they had had unhappy periods, but they considered this to be a normal consequence of marriage and they "just got on with it." In much the same way, these women faced life without their spouses with a pragmatism with which they approached and reported their married lives. It would be far too simplistic to say that the longevity of a marriage depends largely upon a woman acquiescing to her husband's desires. Based upon this research there does seem, however, to be a correlation between an adherence to patriarchal values with its standards of appropriate gender behavior, and reported long-term, marital satisfaction.

NOTE

1. In quoting respondents, "m" is used to refer to years married, "w" to years widowed.

REFERENCES

Allan, G. 1985. *Family Life*. Oxford: Blackwell.

Arens, D. A. 1982-83. "Widowhood and Well-Being: An Examination of Sex Differences within a Causal Model." *International Journal of Aging and Human Development* 15: 27-40.

Askham, J. 1995. "The Married Lives of Older People." In *Connecting Gender and Ageing: A Sociological Approach*, edited by S. Arber and J. Ginn. Buckingham: Open University Press.

Baldwin, S., and J. Falkingham. 1994. *Social Security and Social Change: New Challenges to the Beveridge Model*. London: Harvester.

Beechey, V. 1986. "Women and Production: A Critical Analysis of Some Sociological Theories of Women's Work." In *Feminism and Materialism: Women and Modes of Production*, edited by A. Kuhn and A. Wolpe. London: Routledge and Kegan Paul.

Beveridge, W. 1942. "Report of the Committee on Social Insurance and Allied Services." London: HMSO.

Bowling, A., and A. Cartwright. 1982. *Life After a Death*. London: Tavistock Press.

Brubaker, T. 1985. "Responsibilities for Household Tasks: A Look at Golden Anniversary Couples Aged 75 and Older." In *Social Bonds in Later Life: Ageing and Interdependence*, edited by W. A. Peterson and J. Quadagno. Thousand Oaks, CA: Sage.

Burgoyne, J., R. Ormrod, and M. Richards. 1987. *Divorce Matters*. Harmondsworth: Penguin.

CSO. 1996. *Annual Abstract of Statistics*. London: HMSO.

Dorfman, L. T. 1992. "Couples in Retirement: A Division of Household Work." In *Familes and Retirement*, edited by M. Szinovacz, D. J. Ekerdt, and B. H. Vinick. Thousand Oaks, CA: Sage.

Edgell, S. 1980. *Middle-Class Couples*. London: Allen and Unwin.

Fennell, G., C. Phillipson, and H. Evers. 1993. *The Sociology of Old Age*. Milton Keynes: Open University Press.

Finch, J., and P. Summerfield. 1991. "Social Reconstruction and the Emergence of Companionate Marriage, 1945–59." In *Marriage, Domestic Life and Social Change*, edited by D. Clark. London: Routledge.

Ginn, J., and S. Arber. 1991. "Gender, Class and Income Inequalities in Later Life." *British Journal of Sociology* 42: 369-396.

Harris, C. 1983. *The Family and Industrial Society*. London: Allen and Unwin.

Hyrkas, K., M. Kaumonnen, and M. Paunonen. 1997. "Recovering from the Death of a Spouse." *Journal of Advanced Nursing* 25: 775-779.

Keith, M., and R. B. Schaffer. 1985. "Equity, Role Strains and Depression among Middle Aged and Older Men and Women." In *Social Bonds in Later Life: Ageing and Interdependence*, edited by W. A. Peterson and J. Quadagno. Thousand Oaks, CA: Sage.

Keith, P. M., and R. R. Wacker. 1990. "Sex Roles in the Older Family." In *Family Relationships in Later Life*, edited by T. Brubaker. Thousand Oaks, CA: Sage.

Lee, G. R. 1978. "Marriage and Morale in Later Life." *Journal of Marriage and the Family* 40: 131-139.

Lopata, H. Z. 1973. *Widowhood in an American City*. Cambridge, MA: Schenkman.

Lund, D., M. Caserta, and M. Dimmond. 1993. "The Course of Spousal Bereavement in Later Life." In *Handbook of Bereavement. Theory, Research and Intervention*, edited by M. Stroebe, W. Stoebe, and R. Hansson. New York: Cambridge University Press.

Mares, M.-L. 1995. "The Aging Family." In *Explaining Family Interactions*, edited by M. Fitzpatrick and A. Vangelisti. Thousand Oaks, CA: Sage.

Morgan, D. 1991. "Ideologies of Marriage and Family Life." In *Marriage, Domestic Life and Social Change*, edited by D. Clark. London: Routledge.

O'Connor, P. 1994. "Salient Themes in the Life Review of a Sampe of Frail Elderly Respondents in London." *The Gerontologist* 34: 224-230.

Parkes, C. M., and J. M. Brown. 1972. "Health After Bereavement: A Controlled Study of Young Boston Widows and Widowers." *Psychosomatic Medicine* 34: 449-461.

Parsons, T. 1956. *Family, Socialisation and Interaction Process*. London: Routledge and Kegan Paul.

Pollock, G. H. 1977. "The Mourning Process and Creative Organizational Change." *Journal of American Psychoanalytic Association* 25: 3-34.

Riley, D. 1983. *War in the Nursery: Theories of the Child and Mother*. London: Virago.

Roberts, E. 1984. *A Woman's Place*. Oxford: Blackwell.

Stroebe, M. 1994. "The Broken Heart Phenomenon—An Examination of the Mortality of Bereavement." *Journal of Community and Applied Social Psychology* 4: 47-61.

Stroebe, W., M. Stroebe, and G. Domittner. 1988. "Individual and Situational Differences in Recovery from Bereavement: A Risk Group Identified." *Journal of Social Issues* 44: 143-158.

Valliant, C. O., and G. E. Valliant. 1993. "Is the U-Curve of Marital Satisfaction an Illusion? A 40-year Study of Marriage." *Journal of Marriage and the Family* 55: 230-239.

Ward, R. A. 1993. "Marital Happiness and Household Equity in Later Life." *Journal of Marriage and the Family* 55: 427-438.

Williams, F. 1991. *Social Policy: A Critical Introduction*. Cambridge: Polity Press.

Zietlow, P., and C. VanLear. 1991. "Marriage Duration and Relational Control: A Study of Developmental Patterns." *Journal of Marriage and the Family* 53: 773-785.

SOCIAL PATTERNS IN INTIMACY AND SUPPORT

EUROPEAN AMERICANS AND CHINESE AMERICANS

Deborah Woo

> Scientists have long noted an association between social relationships and health. More socially isolated or less socially integrated individuals are less healthy, psychologically and physically, and more likely to die. The first major work of empirical sociology found that less socially integrated people were more likely to commit suicide than the most integrated. In subsequent epidemiologic research age-adjusted mortality rates from all causes of death are consistently higher among the unmarried than the married. Unmarried and more socially isolated people have also manifested higher rates of tuberculosis, accidents, and psychiatric disorders such as schizophrenia. (House, Landis, and Umberson 1994, p. 84)

In the course of everyday living, individuals experience a variety of problems, some easily managed and resolved, others more serious and stress provoking. The ability to cope is mediated by a number of factors, including the nature of the support structures that are available. Indeed, the very health of individuals is powerfully influenced by the quality of these relationships. Thus, in the opening quote, James House and his associates noted the longstanding and cumulative nature of those research findings having to do with the degree of an individual's "social integration." They even go so far as to suggest that, "The evidence on social relationships is probably stronger . . . than the evidence that led to the certification of the Type A behavior pattern as a risk factor for coronary heart disease" (House et al.

1994, p. 89). The value of social support applies not only to familial relationships (nuclear and extended) and marital ties but to friendships and contacts outside the family. Citing a study by Lisa Berkman and Leonard Syme, they report that in a survey of 4,775 adults in Alameda County, California, each of four types of social ties—marriage, contacts with extended family and friends, church membership, and other formal and informal group affiliations—was found to predict mortality, with those persons low on the "social network index" being twice as likely to die as those with stronger networks.

In a separate article highlighting other research findings, David Spiegel (1993) reported that not only is marital status significant but gender is as well, with men being more vulnerable:

> Single, widowed, or divorced men had a death rate twice that of married men; single women had a death rate 50 percent higher than that of married women. The divorced and widowed fared worst of all, especially among men in their 20s and 30s . . . [whose] death rate [was] up to 10 times that of married men their age.

In general, men appear to benefit more than women from social relationships, especially in cross-gender relationships (House et al. 1994, p. 87), and more from marriage than other relationships: ". . . even when single men lived with a family member or friend, their death rate was twice that of married men" (Spiegel 1993, p. 334). Social support also seems to exert a protective effect on those diagnosed with serious illness. Among those with coronary artery disease, "having a spouse or other close confident tripled the chances that a patient would be alive five years later." Cancer patients similarly benefited from social support, with the link "strongest for smoking-related cancers in men and hormone-related cancers, such as breast cancer, in women" (Spiegel 1993, p. 333).

Despite the powerful evidence for social support as a factor in health, less is known about the process by which social support exerts its influence, or under what circumstances the relationship may even be spurious. Thus, House and his associates (1994, p. 84) asked:

> Does a lack of social relationships cause people to become ill or die? Or are unhealthy people less likely to establish and maintain social relationships? Or is there some other factor, such as a misanthropic personality, which predisposes people to have a lower quantity or quality of social relationship and to become ill or die?

Where there is a clear link between positive support and health, what is the process by which support exerts its influence? Is it because such structures make available material and emotional resources that might otherwise not be available, thereby buffering and moderating the negative impact of stress? Or perhaps when people feel cared about, they are also more motivated to take better care of themselves? Whatever the case, the proliferation of support groups in the United States is unquestioned and, according to a landmark study, has become one of the principal ways in which the emotional needs of individuals have been addressed (Wuthnow 1994).

While questions about the nature of the "causal link" have been posed by researchers in the past, the present study began with several assumptions: (1) that social support takes variable forms; (2) that what constitutes meaningful forms of support is dependent upon a background of experiences, including how cultural expectations are situated within a societal context; and (3) that in the American context, social support issues would need to come to terms with an overarching ideology of individualism. As the population that might be seen as most representative of this distinctly "American" brand of individualism, white Americans of European ancestry constituted an important comparison group or backdrop against which to view other subgroups within society, in this case, Chinese Americans. These cultural differences between Chinese American and European American respondents are noted. However, culture is fluid, flexible, and adaptive more than simply "great traditions" handed down intact from one generation to another.

I began the study without a priori assumptions about the nature of intimacy. Rather, those interviewed were asked to identify all those relationships they could describe as subjectively close or emotionally satisfying and then to discuss what meaningful support looked like from "inside" at least one of those significant relationships. The discussion, therefore, did not begin by examining relationships of formal obligation, even though the marital relationship is one such crucial form of support. For one, the initial task of the study had been to examine perceptions of "mental health problems." It soon became apparent, however, that the imputation of a problem hinged on the perceptions of those close to a person. For this reason, the study shifted to perceptions of intimacy and social support as a way of capturing those factors that made them meaningful, valued, or cherished, whether or not they fell within traditional societal and cultural expectations of what these should be. Specifically, people were asked to comment at length on two major things: (1) those aspects that made the relationship "special and close," and (2) situations of typical, recurring, or significant conflict. The focus on conflict or problems reflected the fact that even the most ideal of relationships can turn sour, if not oppressive. Hence an understanding of how problems are met and handled in the "best" of relationships was worth exploring.

A total of 40 interviews (mental health and social support interviews combined) were completed over the course of three years, 13 of these with European Americans and 27 with Chinese Americans, all then living in the Bay Area, although the majority had resettled here from out of state. Subjects were chosen through an informal network, their ages ranging from 24 to 66 years, with the majority in their 20s and 30s. Their occupational background or training was wide-ranging. Thus, for example, European Americans included an optometry student, freelance photographer, sportswriter, naval officer, schoolteacher, and psychologist in training. Chinese Americans were equally diverse, including an X-ray technician, grocery clerk, engineer, stock runner, meter maid, and campus minister. This chapter reports on the major themes that emerged as findings around social support issues.

THE DYNAMIC TENSION IN COMPETING VALUE ORIENTATIONS

A single society and culture can create needs that are complementary or conflicting. There are needs that are inculcated as extensions or manifestations of a dominant cultural orientation. Thus, the American scenario is striking in its professedly wholehearted commitment to the nurturance of individual identity and freedom. At the same time, this very emphasis on individualism is undercut by competing societal pressures, which make individuality less a reality than an ideal. These competing pressures can arise from a variety of sources (e.g., degradation of the individual in the workplace, media representations that homogenize cultural images, or the reduction of individual needs to capitalist priorities). Indeed, it is possible to talk about needs that emerge precisely because of this imbalance. Thus, for example, where individuality as a valued way of being promotes separateness from others, there is an attendant longing for social connectedness, for a sense of "familyness," as one interviewee called it. In other words, the experience of freedom that is relatively unfettered may create in its wake a certain need for structure, guidance, or support, if not some broader political call for unity and cooperation.

In general, an important feature of intimacy among European Americans reflected this tension between individualism and belonging. The same might be said to be true for Chinese Americans, although the dynamics of this stress were in the opposite direction. For the former, there was a natural symbiosis or alliance between "personal" values and those that were socially promulgated. For Chinese Americans, this was not the case and was most apparent where acculturation produced generational conflicts between parents and offspring. In such instances, the younger generation often legitimated their individualistic aspirations by drawing upon competing values derived from the dominant culture.

While generational conflicts are classical occurrences in immigrant families, they are not necessarily direct culture clashes between parents representing "old values" and their children the "new values." For a 28-year-old woman it was an internal conflict between the desire to broaden her existing circle of friends (for the sake of *proving* her own independence) and deep feelings of loyalty for the unconditional support her old set of friends had already given. The value of having new friends, in her words, lay in it being "an indication that I am more independent of someone, somewhat less reliant on my own ethnic group, my own social circle. . . . It's kind of proof to me that I've accomplished something on my own rather than have it [i.e., the friendship circle] develop all around me. . . ." Despite such ambivalence, the Chinese American respondents in general took for granted the fact of fellowship and community and the notion that one's identity was also derived thereof (i.e., identity as realized through a set of reciprocal obligations and commitments). This attitude was found not only among the first generation but, in some cases, even among third-generation Chinese Americans.

EUROPEAN AMERICANS

In general, the qualities that these respondents saw as making for intimacy were those closely tied to personal efforts toward realizing one's individual identity, if not freedom or independence. Individual autonomy, in fact, seemed to be taken for granted. Individual identity, however, was not, and the desire to achieve and establish this identity was not only an underlying need but also a possible source of frustration. Thus, for example, one respondent pinpointed the source of his mother's unhappiness in the fact that her role as a mother did not allow her to realize certain personal goals or ambitions.

> . . . people that meet my mother, nine times out of ten . . . generally see one side of my mother. They see the side of her that is tremendously warm and personable and that selfless side of her that makes a very good mother, which is all very very real. . . . They don't see the side that has a lot of unfulfilled dreams. It's a basic dissatisfaction and unhappiness with her life. They also don't see that side of her that takes out her frustrations on members of the family. . . . You know, in the course of her life she's sacrificed a lot but hasn't done enough to realize her own goals and ambitions.

Chinese Americans, by contrast, often referred to parental self-sacrifice as a sign of love and devotion, with few references to frustrations related to self-realization. Rather, parental fulfillment was bound up less with the idea of individual success than with the success of one's own children.

In a cultural context where the demands of independent living dominate, interpersonal intimacy or attraction is typically inspired by perceptions of the other as (1) a strong and independent person, (2) someone who liked adventure, excitement, or new experiences, and yet (3) one who provided or shared a sense of "familyness." These qualities or attributes were not only the most salient characteristics mentioned in relationships that were valued as close and supportive among white Americans, but they also stood in striking contrast to those mentioned by Chinese Americans.

Implicit Role Models and Mentors

The notion of the individual as a self-governing entity is a taken-for-granted assumption, whether or not it is fully realized. Youthful rebellion is fueled and legitimated by this notion, such that parents often counter by calling, in turn, upon some form of "individual responsibility." This call, however, runs up against a powerful ideology that claims to affirm the person guided by "conscience" and inner dictates rather than by deference to social forms or externally imposed rules. The right to question or challenge authority, in other words, is itself highly valued.

Thus, for example, the intimacy that a 25-year-old man shared with his sister was dominated by his perception of her as "real strong." One of the particular strengths he pointed to was her ability to stand up and say what's on her mind, and specifically, to confront their parents in ways he would never have entertained. In addition to interpersonal "spunk, "she was said to display an "incredible" energy or "drive" when it came to work, another source of his admiration. Asked what he gains from the relationship, he responded:

> She's real strong. She has the ability to say what she feels without having second thoughts about it or thinking about the consequences. I admire that. I really do. She could tell my parents things that I'd never dream of telling them. And she has a drive that's incredible. She'll work and then she'll work some more. . . . She's also a very strong-willed person. I like strong-willed people. . . . She's got opinions, is not afraid to voice them, and I'm not afraid to tell her different. We have a strong relationship because we do differ sometimes. . . . We give and take and "dare" and things like that.

About his former girlfriend, a 27-year-old man recalled fondly how being "stubborn and independent" was one of her many endearing characteristics.

> There were so many things. She's [a] very open, unassuming sort of individual. . . . She was also very stubborn and independent. She finished junior college and then left all her friends and went to Boston because she felt she should have the experience of a large city, even though she hated cities. She went and did it completely on her own, got a job, and took care of herself.

Similarly, the following account captures the interpersonal attraction based on perception of the other as one whose maturity is reflected in self-sufficiency. Thus, a 29-year-old woman expressed her admiration and respect for her friend's "strength," particularly her incomparable capacity to reconcile the peculiar tension or conflict that women experience with what it means to be "feminine" in this society and yet at the same time display a certain degree of independence or self-sufficiency:

> She was more independent than any other woman I had ever gotten to know. I had a lot of respect for her, for that combination of things she was—that she could be so warm and generous, very feminine, and yet at the same time, she was competent in taking care of herself. [What she valued most in the relationship was] . . . her strength. I think she's a strong woman.

Insofar as relationships fulfill complementary needs, the ability to grow or learn from the other is implicit, such that the idea of the other being a strong role model was a common one. Thus, one person spoke to the fact that his status as a "lowly undergraduate" was complemented by his friend's greater experience and training as an older graduate student, someone whom he found himself "modeling after." Alongside the inclination toward independence and individuality, in other words, is a corresponding search for companionship that lends support to individual endeavors.

Based on the interview data, it also seemed that European Americans experience greater problems achieving a satisfactory sense of mutuality in this regard for at least three reasons: (1) different pulls created by different individual goals; (2) fear of dependency; and (3) social insensitivity, disregard, or irresponsibility, the negative underside of individualistic tendencies.

While friendships are often initiated because of some "common" interest, their longevity depended on the ability to sustain this mutuality when individual growth takes each person in different directions. Examples included "philosophical friends" who experienced a "rift" in their relationship when one could not share the other's same intense need to "seek the truth." In another instance, stylistic differences were a source of difficulty, as when one person felt his friend was "kind of conservative" (not in political sense) or "too much of a conformist." Ironically, while a spirit of individualism purportedly espouses tolerance for differences, including individuality in its variety, these examples suggested a low level of tolerance where close friendships are concerned. Crafting, maintaining, and enhancing the bond of intimacy or mutuality seemed to require the patient and careful interweaving of idiosyncratic tendencies. It appeared, in other words, that stability may be more uncertain when cultural pressures individuate and fragment these experiences, interests, views, dispositions, and activities.

Fears surrounding dependency also produced distancing and disengagement. The attraction to "strong" individuals as role models thus seems to have another value—it is itself an assurance against the other's possible dependency and imposition upon one's time and energy. In one woman's opinion, men offered the more positive role models whereas women seemed to relate on a "needy basis."

> With women, I have to respect them. They have to like themselves, feel good about themselves, otherwise they're just another "case." It's like they're using me. . . . It's women who come to me on a needy basis. Men are more interested and independent. I guess because they happen to be people I respect and look up to. It makes me try harder, to be more self-motivated around strong people.

The opinion that women were more insecure was similarly expressed by a 27-year-old man, who found that the support he had to offer a close female companion interfered with his own work responsibilities and "battle with time."

> I'd like the women I see to have more of a sense of security and independence. That's the main

thing. I just have this battle with time. "A" is very insecure, not so much dependent now, but insecure, which means she requires a lot of attention. And that's kind of hard for me to deal with because I have a lot of things that I really want to do that require my time. I end up resenting "A" at times for taking me away. . . .

In general, an individualistic orientation heightens and exaggerates perceptions of differences and imbalance in the relationships. Intimacy can, on the one hand, be stimulated by a contest of viewpoints and opinions, by the fact, as one respondent put it, that "we like to differ." On the other hand, even among relative "equals," there is sensitivity to potential sources of asymmetry. Thus, the mere admission or expression of feelings of closeness can elicit not only discomfort, but the fear, as another person explained, "that he might be depending on me too much." While unreciprocated feelings can evoke discomfort, even in contexts where an individualistic orientation does not prevail, it seems to be the peculiar feature of an individualistic orientation that fears of dependency surface when reciprocity or mutuality in feelings are in question. As the following person pointed out, uncertain commitment and the failure of social support to materialize reinforced her own resolve to be more independent.

I've had such high expectations of so few people . . . and I've been disappointed. . . . One friend in the past disappointed me because she was older, and I expected her to be wiser and more understanding of just the world in general as well as being more self-aware. Another woman I was close to is the sort of person who spreads herself very thin. She was someone I admired, and yet it took me a while to understand that's the sort of person she is, and I should learn to have no expectations. There were a few times when I felt vulnerable and I didn't get the kind of support I would have given her. But then again, I feel that's just part of one's education in life—to be stronger and depend on oneself more.

Commitment and autonomy thus exist in uneasy balance. Disappointment in support from others makes it all the more incumbent "to be stronger and depend on oneself more," a tendency which, by the same token, may also contribute to difficulties "learning to receive help." Asked whom she would call on if an illness forced her to say at home, this same woman responded:

That's a hard one because I *resent* having to call on *anyone*. It embarrasses me to have to call on people to help. I went through this the time I fell and was on crutches and I didn't call "D" [husband], which shocked a lot of people, and I was surprised that they were shocked. I didn't expect him to come rushing home. I *did* go to the people next door since I *couldn't* drive to town and needed to get to work. . . . I was shocked at how many people just volunteered and *wanted* to help. I guess I felt I wanted to reciprocate. If you're on crutches, it's hard to walk. You can't reciprocate. And yet I would feel free to help them if they were in the same situation. It's just a problem I have, learning to receive help.

Finally, relational problems, not surprisingly, stemmed from individualistic tendencies taken to the extreme. Thus, for example, the very respondent who had admired his sister's "strong-willed" nature, her ability to confront their parents and speak her mind, also pointed to behavioral issues that seemed to be an extension of this independent posture, such as her "running away," "being high all the time," and later when married, "dominating" her husband in verbally abusive ways. Another male respondent explained how a former girlfriend almost brought him to a "nervous breakdown" in light of the fact that "she wouldn't take responsibility for her actions." The point here is not that individual irresponsibility is unknown where collective responsibility is emphasized, but that intimacy in the individualistic context includes the expectation that the relationship will *actively* foster individual independence. There was also a lower tolerance for dependency. For this reason, intimate relationships seemed to require greater balance and equality. By contrast, as we shall see, Chinese American expectations surrounding intimacy and autonomy were qualitatively different. For example, a friend who would "listen" and not be controlling was desirable and sufficient in and of itself, without the person also having be a role model for independence.

Attraction to Novel "Experiential" Adventures

A theme articulated by European Americans but generally more peripheral to the expressed concerns of Chinese Americans was the view that one's closest relations offered the promise of adventure, romance, or experiences that reached for the extraordinary, novel, or unconventional in life. Intimacy is expected to strengthen and grow on the basis of opening the self up to new experiences. As one woman explained, her friendship was bonded by such common interests as "Careers, being single. Liking family, liking warmth and closeness and yet having wanderlust, wanting excitement—wanting both those things." In sharp contrast to formal relationships constrained by conventional social roles, intimacy indulged the particularities of the self, inviting exploration into the inner and outer reaches of the total personality. Thus, a male respondent explained how renewing intimacy with his sister would mean rediscovering what she "dreams about, what she considers to be the ultimate in life. . . ." Similarly, another person reflected that an "ideal" relationship would be one in which there would be a really deep, profound, spiritual connection—constantly bringing things to each other."

Despite differences suggesting a range of socio-emotional expectations, intimacy, as ideally conceived, among European Americans, is a confiding, self-disclosing relationship. For one person, this openness would necessarily include a more straightforward expression of negative feelings.

> . . . I guess the thing is more direct expression of our feelings towards each other. We'll talk about . . . how we felt about each other in the past, but the immediate, in vivo, sharing's a little harder. . . . Maybe it would be good if we had a fight. I'd like it if we could express more of our negative feelings towards each other when they're happening.

Similarly, another person noted how greater openness or self-revelation of inner feelings were moments few and far between. His friend's willingness on one occasion to be "vulnerable with me" marked a rare and positive experience, producing a kind of closeness that he wished would occur more often.

> . . . when [he] broke up with [his girlfriend], he called me up on the phone and was crying. It was the first time I ever heard him cry. [He] was being vulnerable with me in a way that he normally isn't. . . . I wish that could happen more. It made me feel closer. . . .

The value of intimacy achieved through self-disclosure may have been experienced or perceived as a very personal or individual matter but the fact that these needs or concerns did not surface in the interviews with Chinese Americans suggests patterned differences that stem from different cultural expectations. Another example of the desire for "expressed" intimacy was indicated by the following person, who not only stated that the full expression is rarely achieved in any single relationship but that this very impossibility leaves one with "a bit of yearning."

> . . . It's not so much that the current relationships aren't satisfactory. If you sort of are not totally ecstatic all the time in your life—and most of us aren't—then the capacity to express various features of your personality are kind of there as a bit of yearning, I suppose, to have something different, or new, or additional.

In sum, while close relationships in general serve a variety of functions, the articulated needs focused on the expressive dimension, specifically emphasizing a need for greater authenticity and self-disclosure. This finding raises questions about the source of such needs. Are they owing to the fact that these emotional needs are somehow inadvertently frustrated, say, by an ethic of rugged individualism? Or, or are these concerns simply more salient because they are a part of a configuration of cherished values? The following section similarly prompts one to ask: does the value placed on family emerge largely because it is a central value? or because it is a shadow of this ideal, in truth a secondary value, less likely to be realized than other cultural aspirations? The answer will imply different solutions.

A Sense of "Familyness"

Despite—or perhaps because of—a powerful pull toward independence and new experiences, the third pattern that emerged in the European American sample was a nostalgia for the warmth, security, and pleasures of family life. Relationships were often colored by yearnings to recapture the same sort of "family things." A female respondent thus recalled the "really nice sense of *familyness*" her friend exuded: "When I think of 'C,' I picture a beautiful, cozy kitchen, with her in it, and being there with her. . . . There's a real warmth there." A male respondent noted that his affection for a particular woman included her family: "The other part is that I fell in love with her family, too." Where expressed, this longing for family seemed to derive from multiple sources—nostalgia for an idyllic past as well as the absence of an ideal childhood, the need to recapture some innocent, spontaneous aspect of the self nurtured in those early years, or else the need for reassuring supervision and guidance that was more available at that stage in the parent-child relationship. Again, while such needs may appear "universal," the degree to which they are generated or satisfied by existing social or cultural arrangements is highly variable.

Twenty years ago, theologian Harvey Cox (1977) suggested that the need for simple human friendship was frustrated by the Western, American way of life. He sought to explain what motivated people to look outside the American scene, specifically toward Eastern or "Oriental" religions, for sources of spirituality. In the end, he concluded that the nature of the attraction stemmed not only from general disillusionment with Western society, but from specific needs and commonalities that drew people to what he called "neo-Oriental movements." According to Cox, motivations clustered around six specific patterns and included the search for (1) simple human friendship or companionship; (2) "real," felt experience, a way to experience God, life, or the world more directly, with an immediacy unmediated by ideas and concepts; (3) an authority that is trustworthy; (4) a more "natural" form of spirituality that is untainted by Western science, technology, and rationalization; (5) an escape from the male domination of Western faiths; and (6) a concern for health, ecology, and conservation of the planet's natural resources. What Cox questioned was the degree to which one could genuinely appreciate these other traditions without the distortion created by psychological individualism and a consumer culture where relationships are premised first and foremost in terms of instrumental exchange and profit. The point that is relevant here is that the emotional, spiritual needs expressed by these individuals may be better understood in terms of larger dynamics that are responsible for a more widespread cultural malaise.

Individualism surfaced as a dominant cultural imperative and as an organizing thread in these interviews on intimacy and support. For this reason, even though 20 years have elapsed since the time Cox offered the above observations and the interviews reported here, the kind of individualism that spurred the search for community during the 1960s is still pervasive. It may take another form of social movement to spur a significant value shift in the dominant culture.

CHINESE AMERICANS

To appreciate the very different responses that questions about intimacy and support evoked among Chinese Americans, one would need to suspend at the outset the idea that social relationships are organized around individualistic needs in the first place. The dominant reality here is that of a social world where individuality has a lesser role, and where the authoritarian structure of family relations constitutes a major backdrop against which intimate relations are evaluated. The individual self or ego is subordinated to the social self, and to the extent that individuality emerges as a separate and distinct entity, it is against a social nexus in which the individual is presumed to be already enmeshed, anchored, or established. Mutuality is often achieved through suppressing individual wants, needs, or desires rather than through each appreciating and bringing out the other's uniqueness. At the same time, social support can be counted upon because individuals have already been socialized to their respective duties and social obligations. The irony or paradox is that one seldom explicitly calls upon these interpersonal resources for fear of abusing what can be taken for granted. Thus, psychologist Benjamin Tong has explained:

> . . . Powerful, untidy feelings are to be managed entirely on one's own, without counsel or solace from other people. Turning to family and friends is a final recourse but here we have a paradox: one can always call on significant others, yet they are in fact to be regarded as the absolute last resort. An alliance or friendship carries with it the right to casually invade each other's privacy but this is not to be abused. A sign of love, therefore, is never having to make demands. One takes care of one's own, beginning with oneself. (Tong, 1978, p. 86)

In short, the result is a form of individual self-reliance that may look as though it were based on "individualistic" assumptions but is actually premised on a different set of assumptions. Individual responsibility, in fact, does not simply include "taking care of one's own, beginning with oneself," but extends to include responsibility for another's welfare. Indeed, as Victor Li notes, in Chinese society, it is one's moral obligation or social duty to monitor or "correct" another's behavior when it deviates.

> If a person truly loves his neighbor or truly is his brother's keeper, then he has a moral and social duty to correct his brother's shortcomings. . . . Similarly, comments and criticisms made about another's life-style or work-style are regarded in the ideal case, as proper exercises of a social duty to one's fellow man, not as invasions of privacy or as meddling in the affairs of another. (Li, 1977, p. 146)

Within such a context, there is the expectation that advice, support, and guidance will be forthcoming, whether or not one's request is explicit. With this in mind, the following were identified as characteristics important to intimacy among Chinese Americans: (1) an ability to simply listen and respect individual autonomy in decision making (i.e., not to abuse one's "right" to intrude upon another's privacy), (2) perspicacity with respect to unstated conflicts or feelings, and (3) extraordinary devotion and self-sacrifice for basic material needs.

Respect for Individual Autonomy and Independent Decision Making

A theme that emerged with some regularity among those interviewed was the view that the "confiding" relationship was one in which a trusted friend not only lends a sympathetic ear but is especially mindful of one's freedom to make one's own decisions. Intimacy that is dear offers measured guidance with respectful reserve. In order to appreciate what this means or why such qualities would be valued, one would have to appreciate the nature of the traditional parent-child relationship. While tension in all families arises over the issue of how much structure or guidance is too much or too little, there are important cultural differences here. In the context of individualism, a European American's individual responsibility as a "free agent" is taken for granted, and anything that would represent constraints upon this freedom (including strict parental discipline) would be seen as inhibiting a child's potential. By contrast, in the Chinese American family, the earliest signs of a child's malleability become the grounds for shaping the young toward certain social ideals. The notion that parents should sensitize themselves to, and cultivate, the innate possibilities within the child's nature is antithetical to this cultural mindset. (While we will shortly see that perspicacity with regard to another's feelings was seen as important to intimacy, parent-child relationships were an important exception in this regard.) If anything, children were expected to be sensitive to *parental* expectations. Thus, one respondent talked about his father as follows:

> It doesn't really matter what my feelings are. It all depends on what his are. It's really an irrelevant question for me. He doesn't have to pick up on how *I* feel. It all depends on how he feels, because when I'm with him he directs my life.

The overarching control which parents try to exert over their children's lives extends far beyond the childhood years. Thus, the following person spoke of his parents' intense supervision as a recurring problem.

> One [problem that keeps coming up] is that my parents still try telling me what to do, even though it's through telephone lines instead of directly. And they still try to run my life indirectly, by saying, "Call this insurance company, call that one," or something like that. That's still a problem.

Many respondents, moreover, pointed out that confiding personal problems to parents merely increased the risk of incurring criticism, blame, or unsympathetic reproach. Thus, the above person continued,

> Plus if I do something wrong, they still berate me. Well, like if I get a roommate, and I complain about my roommate, if I get a roommate who's really messy or something like that—instead of being understanding and saying, "What a shame. It's too bad you're in that situation," they'll say something like, "Why didn't you check this out beforehand?" So you don't receive the sympathy you should be receiving. You receive a lecture you don't need, because you realize what the problem is anyway.

About her mother, one woman similarly explained her own reluctance to confide:

> I don't really like to talk about personal problems with her. She takes them too seriously. Plus in her eyes I tend to be always wrong.

> She had difficulty listening. If no one hears you, that presents a big problem. It was sort of like your word wasn't good enough. What you know isn't good enough.

Another respondent echoed these sentiments when describing his girlfriend's relationship to her parents.

> When her parents would raise an issue for her, she'd just be very quiet. When she'd get in trouble, she knew she could never say anything about it, 'cause her dad was always right. Her dad was always right. So usually she just ate it and kept quiet about it.

Given that the mere expression of a problem was more likely to evoke criticism than empathy, the confiding aspect of intimacy tended to be absent in relations with parents and was thereby the backdrop against which individuals valued the respect for autonomy they did receive in certain relationships. Feelings of being unloved or uncared for are not uncommon. For one person, this feeling stemmed directly from the perception that his father was simply interested in "running our lives."

> It's hard to explain. But I never thought my father cared about us as people 'cause he was always running our lives, telling us to do this, to do that. No one likes being told what to do all the time. . . . I think it's the tradition of Chinese families that the man's the head of the household, and he carries it to an extreme.

It is not surprising, then, that an intimate confiding relationship occurs primarily between peers who share a respect for individual thoughts and feelings. The other's autonomy, moreover, is acknowledged by tempering one's own views or by merely listening. While it may well be argued that listening is a quality "universally" appreciated, the fact the respondents often articulated this in terms of what the other did *not* do is revealing. Describing what made her friendship "special and close," one woman explained:

> It's just the fact that she's there. She'll listen to me and she'll guide me, but she *won't* tell me what to do. She'll let me make up my own mind, which is one of the most important things, I feel. The fact that *I* made my own decisions and she did not really influence me but she helps me open doors to ideas and things like that, and that's really important to me. . . . That's what I like most about her—the fact that she's not pushy. . . .

Similarly, a 30-year-old man explained how his confiding relationship with a friend contained this mutual respect.

> We talk about each other's problems. We respect each other's opinions. I respect it but I don't always take it. I cannot take anybody telling me what to do.

A related factor important to confiding was a self-conscious effort to not "blame" or "judge" the other. For one couple, this involved an explicit, self-conscious effort.

> We try not to blame each other. We try and lay the cards out on the table and say, "This is what I perceive. This is the way it seems to me. I'm not judging you but the way I'm coming about this is through my own point of view. . . . " And you know, [I] *try* not to be judgmental in saying that.

It is noteworthy that respondents who described a more open, confiding relationship with parents were mainly first-generation or foreign-born Chinese. For example, a Taiwanese student enjoyed a relationship of almost limitless rapport with his parents, writing numerous letters to them (as many as 72) over a nine-month period: "My parents know everything. . . . I think I'm a special type. It's not very usual in Chinese students here." Similarly, while Chinese American parents were often described as being narrow and authoritarian, the reports from foreign-born students were significantly different. One described his parents as unusually "broad-minded," adding that "I've always had free choice on the kinds of things I want to do." Career pursuits reflected this freedom, with three first-generation students studying forestry, English, and law. How are such differences to be explained?

One theory is that these more liberal attitudes reflected middle-class mobility. That is, since material security was not a dominant concern, parents were more likely to be supportive of individual career goals, in sharp contrast to those situations where survival issues placed a greater burden on offspring to contribute to the family welfare. Foreign-born students are more likely to come from well-to-do backgrounds in the first place since they would have to have the financial means to study in the United States. Another theory is that these parents were more liberal because they were more educated, modern, urban, or Westernized—and hence, acculturated to individualistic values. A third possibility still is that the liberalness in parental attitudes is consistent with traditional values but is indulgent only of male pursuits.[1]

Perspicacity with Respect to Unstated Conflicts or Feelings

The cultural prescription to "take care of one's own," noted earlier by Tong, may seem to contradict the idea of getting help from a support network. Yet there is a reciprocal side to "taking care of one's own," namely, sensitivity or attentiveness to nuances indicating another might be in need. Given a norm against burdening others with personal problems or demands, much depends on the other's ability and willingness to provide a safe space for venting one's feelings. Perspicacity or sensitivity to emotional nuances was therefore cited as having an important bearing on intimacy. One woman thus explained how her friend gave her the opportunity to express certain feelings, "Just by being perceptive."

> Just by being perceptive. I mean, when we're out together or something and I'm down, she can tell right away that I'm down or something's bothering me. Then she will ask if I want to talk about it. That in itself expresses she does care because she can tell you're in a different state of mind. Your mood's different, and she asks about it. She gives me the option to say something or not. A lot of times when I'm feeling down, a lot of people can't tell. They have to know me really well to tell that I'm down. Even if I try to cover it up, by being jolly or whatever, she can still tell. Some people think I'm always up, but there are a lot of times when I'm down and they don't even realize. But she does. [I do] the same thing [for her]. I can pick up on her moods, just the way she reacts to things, even her speaking voice.

Similarly, a male respondent commented on his girlfriend's ability to "respond to a need before I even express it."

> "D" has a neat way of being sensitive with me. She'll respond to a need before I even express it. . . . My needs [come out] . . . in subtle ways and she picks up on that. But it's trying to alleviate or help a need before I even ask her [that's important]. And usually I won't ask until it gets *desperate*, you know. But she'll usually see and usually ask me to talk about it.

The need for perspicacity is that impassive essential given the exterior that is often assumed as part of "taking care of one's own." The stereotype of "Oriental inscrutability" derives in part from the cultural requirement for emotional self-control.[2] About himself, one respondent said,

> In the beginning, I wasn't so good at being transparent about what was going on. I'd just hold things in and let it brew in there. . . . When I have a problem, I hold it in.

Another respondent similarly described how his parents manage their personal problems by concealing or suppressing their problems.

> . . . they'll keep it in. They'll keep it in. You'll never find out about it. Not usually. My mom will keep something in and it'll just never surface. And she just basically kind of—it just disappeared. She'd never think about it again. I think people like that just become very good at maybe forgetting things. . . .

Given that the suppression of emotions is a form of coping strategy, the ability to detect and elicit discussion of a problem is an important dimension of social support. Thus, one respondent explained how he had to repeatedly prod his parents to open up and talk about their problems.

> I openly tell them, "I can sense there's something going on. You can talk to me. What is it? What, if anything, can I do? I have to know what it is kind of before I can offer anything constructive."

Even for those who have a mutual trust, there is an almost insuperable barrier to just raising issues, especially those that might create direct, interpersonal conflict. About his relationship with his girlfriend, whom he later married, the following Chinese man explained that in the early stages of their relationship, it used to take two hours just to surface the problem.

> It used to take us [so] long to get to the point, maybe two hours before we could talk about it. . . . It took a long time to drag things out of her. I just had to keep saying, "Let's talk about it. Let's talk about it. What good is it not talking about it? . . ." She'd do that with me, too.

One major reason why disclosure is so difficult has to do with anxiety stimulated by the possibility of confrontation or disagreement. While difficulties with disclosure were noted by European American respondents, the barriers did not seem as great, nor did the threat to social harmony seem a major source of anxiety. Expressions of even negative emotions seemed to defuse any accumulating tension. By contrast, emotions among Chinese Americans seemed much more contained and suppressed, such that their release had more explosive, volatile, and potentially irreparable, implications. Despite stereotypes of family harmony among Asian Americans, insiders to the culture frequently pointed to the heated nature of arguments, once these disagreements were allowed to surface.

> . . . there are a lot of arguments, a lot of arguments. It seems like that's the way our family deals with each other—by arguing. It might be a Chinese phenomenon, because I notice that with a lot of Chinese families. Same way. You don't sit down and rationally discuss things. You spout off stuff and you argue.

Whereas intimacy in the European American sample seemed to carry with it injunctions about "truth" and "honesty" that enjoin one to "tell it like it is," intimate relations among Chinese Americans were characterized by cultural barriers to their verbal expression. For one, anxiety over how to gauge the other's reaction could itself make it a delicate issue to talk about certain problems. As the following person explains,

> I don't cover up because I think he will overreact or react in a negative way. But I cover up because I don't know how he *will* react. . . .

Secondly, there is little cultural socialization or preparation for the articulation of inner feelings. As the following woman explained, anger or happiness were more likely to be expressed than sadness, depression, or emotional need. Neither her mother nor her father seemed capable of verbalizing the latter type of feelings. In general, a major obstacle to confiding, apart from fear of possible criticism or blame, is inexperience or lack of knowledge as to how to verbalize certain emotions.

> If you talk to her, she *will* talk to you. She just doesn't give her "inside" feelings. When she's angry or happy, she lets you know it! But when she's hurt or sad, she doesn't express it. Neither did Dad. She can't verbalize certain emotions or inner feelings like, "Gee, I feel so depressed today." Maybe she considers it a weakness. She's *very* sensitive, too. I don't think she ever had the opportunity to express that part of herself even to Dad, since he was never any big talker either.

For those wishing to converse in a more open manner, disclosure would entail a capacity to translate thoughts and feelings into language.

> If I were to start my own family, I'd do things differently. I'd try more openness. When there is a problem that can be verbalized, or even if it's not quite verbalized yet, to discuss it and try to make it more concrete.

In short, while an expressive, confiding relationship might be similarly valued among Chinese Americans, the threshold for this kind of direct communication is relatively higher, especially when there are problems of a direct and interpersonal nature. For example, one woman who said she could "express any kind of feeling" to her friend "and it will be understood—even negative," was hard-pressed to convey a recurring annoyance over the other "being slow" when the two had an appointment to be somewhere. It is her own helplessness in a situation that might be viewed as a relatively small or trivial problem that is striking:

> . . . you can only say so much sometimes. 'Cause even though you can be very close to another person, we don't always agree with everything, and we don't think they're doing the right things all the time. Sometimes they do piss you off, but it's really hard to speak up sometimes . . . and it's very sensitive in the sense of having to speak up. I don't know whether it's just petty things, or I'm just making a mountain out of a molehill type of thing. You might as well just ignore it, but you can't ignore it because it happens. It's a thing that you just don't talk about because—I don't know if I can explain it anymore. It's like everybody has quirks, and even though you know that . . . it really bothers you. . . .
>
> [She bothers me by] being slow. Like you know that you're supposed to be at a certain point in fifteen minutes, but the person just dawdles and dawdles, and you're just sitting there and fifteen minutes have passed by already and you're sitting there waiting. It bothers the *shit* out of me. But what can I do?. . . You can't talk about it, but you feel it though.

Not surprisingly, if trivial problems can present such difficulties for communication, serious problems, which have deeper, emotional dimensions to them, are even more likely to present a challenge for those who are part of this close network of social support. The following two examples are illustrative, respectively, of how matters of individual crises can be overlooked and interpersonal tensions exacerbated when those involved do not perceive a safe or comfortable environment for airing and dealing with these problems. Whereas in the past mediators, go-betweens, or a host of other family members might have played an important role in neutralizing conflict or otherwise aiding in communication, acculturation and and urban American lifestyle have undermined these traditional support systems.[3] In either case, it was commonly acknowledged that direct communication of problems was frequently undermined by fears that their expression was inappropriate or potentially productive of conflict. Thus, one respondent explained that his father's exterior was so impenetrable at times that it would be difficult to "tell if there is a problem." By the same token, he admits not knowing how to help even if he were aware of some "major crisis in his life."

> My father's like Spock in *Star Trek*. It's the same, identical situation. He has the logic of a Vulcan. But deep inside himself, he has the emotions of a human being. And protocol deems that I treat him the same way. . . .
>
> If he needs help physically, such as moving something, there's no question about it. He just says, "Do this, do that." But if it's a problem such as an emotional problem or something like that, he holds it all inside. You can never tell if there is a problem. He never says a thing. I really can't tell. If he loses his account as a broker, and he feels he's going to go bankrupt or something, I couldn't tell the difference. He won't mention it. And even if I know there's a major crisis in his life, I can't approach him. I can't show that I do care. I just let him take care of it in his own way. If he needs any help, he could ask me, but he never does.

The dissolution of close friendships can be attributed to some extent to the inability to open up communication when there are problems. Thus, another respondent sadly recounted how a close friendship of ten years mysteriously deteriorated. The change occurred after he had been away in the service for a few years. All attempts to get his friend to say what was bothering him were of no avail.

> Things were just left unsaid. When we had something to talk about in the past, we would always talk about it. But I guess there was so much distance and so much left unsaid since my return. . . . I think I made honest efforts, attempts to patch up things. I told him many a time that if he ever had anything on his mind, he could feel free to talk to me about it because I felt as though I was really being open to him. But every time I'd ask him, he'd lie to me. He would say nothing was the matter, everything was "cool," and treat me like a brother again. Then two days later, same old shit.
>
> Probably the last time I spoke to him, I told him I always felt like we were brothers, and that I no longer felt we were that close, and I was always hoping we could [find] out what the f---was going on with each other, and see if we could pull it back together. But, you know, after that conversation, I never heard from him again.

From the preceding examples, it is apparent that communication skills take enormous amounts of patience, sensitivity, and, above all, time to develop. Several factors tend to inhibit disclosure, among them, an inability to gauge the other's reaction, uncertainty or fear over possible criticism or blame, unpleasant memories of such direct exchanges in the past, and simply a lack of cultural competence in the delicate expression of feelings that might potentially be construed or interpreted negatively.

Extraordinary Devotion and Self-sacrifice for Basic Material Needs

In general, intimacy in Chinese families tended not to be expressed verbally, even when these emotions were positive. More often than not, direct expressions of caring take the form of behavioral demonstrations related to the provision of material, practical, or basic needs. Confiding, for the most part is related to practical, concrete concerns around which some kind of action or behavioral remedy can be applied. Thus, the woman who complained that her mother had "difficulty listening" noted—somewhat to her own surprise—that they were most comfortable discussing problems in the area of "school" or "work."

> I can share school problems and work problems. She's really sympathetic to things like that. That's funny. Things like that she listens to. In general, I share whatever problems I've had during the day with her. We do talk but to sit down and have a real *heavy* conversation . . . not usually.

Similarly, a Chinese American man noted that the "mechanics of everyday living" provided an area of communication where both he and his parents had some competence.

> The main thing is that I can still go talk to them. I can still ask them for advice on the mechanics of everyday living. I consider them to be very good at that, very adept at that. They ask me for help on financial things because of my work experience and things like that.

Even for the more "psychologically minded," the idea of delving into the interior self contained too much uncertainty. Moreover, according to a 24-year-old man, with experience as a resident advisor of students, there are dangers to delving too deeply.

> I think sometimes psychological help is more trouble than it's worth. They can make things more complicated than they are. For example, they bring up things best left alone, things we weren't previously conscious of or aware of.

In general, the dominant cultural attitude is that most problems are exterior in nature and are best handled that way.

> . . . It's just things that occur, okay, that makes for problems. It's always exterior. It's like money problems, you know, boss problems, work problems. I think exterior pressures that come on that make each of us have these problems. I don't think it's anything in terms of psychological problems or anything like that.

Close personal relations were generally based on a form of constancy and devotion expressed through attention to basic, concrete, or material needs rather than through overt, emotional displays of affection. Respondents who specifically noted the absence of verbal or physical expressions of caring were more likely to be carrying with them expectations drawn from observations of caring expressed in other cultural contexts. Thus, the following person described how he had felt unloved, especially after seeing how the parents of his more "American" friends were more openly demonstrative, especially with respect to verbal or physical displays of affection.

> . . . if my friends' parents were to express love, they would do that by expressing it both verbally and physically, the touching, the caressing, the saying "I love you." My parents would never do that. I want that, because I'm probably sitting more on the "American" side of the fence a lot more in my own identity than I am on the "Chinese" side. I asked my mom once,

"How come my dad doesn't love me?" And my mom said, "Your dad loves you very much."

I saw my dad not doing the things that I saw other dads did for their kids. See, I did a lot of comparison back then. And their dads would hug them and come home and they'd feel love. They'd feel warmth in the family. We never really had that. And so I asked my mom how come my dad didn't love me, and the next morning I got some *money.* It was just on my cabinet. It was one hundred dollars or something like that. And that's the way my dad loved me, you know. He's always providing things for me, giving me money, but he could never say he loved me.

. . . Even when I go home to visit—even now—the worse times are the saying goodbye, because we don't know really how to say goodbye. He sticks out his hand, and you get a cold handshake, and [he] says, "Have a good year" or something like that. I know he loves me 'cause it's kind of like—his love is really based on—it's an unconditional kind of love, you know. I think that's different from a lot of my friends' parents.

The preceding response is fairly typical. For another respondent, what made the relationship with his father "special and close" was this simple realization:

. . . Just realizing how much my father really cares. He never shows emotion, and there's no overt show of affection ever. My father never hugged us. He shook our hands. It was very formal, which I don't think is healthy. No smiling at all, very little smiling.

What is lacking in verbal or physical forms of affection is compensated for by unparalleled self-sacrifice and devotion in other areas. For example, after the death of one woman's father, her mother saw her role in preparing meals to be more important than ever, seeing it as her "duty" that her daughter got through school by ensuring she was "eating well." Food preparation, in general, is not only a way in which to express hospitality, but was also commonly mentioned as both a symbolic and practical way in which close relationships were nurtured. Thus, a 27 year-old woman described her mother's "devotion" in terms of her "constantly feeding me."

She's constantly feeding me, preparing meals for a person who's certainly old enough to do it for herself. Basically, that's one thing Mom always did. At least that's the way I feel she showed her devotion.

Her boyfriend, too, was similarly described as devoted in this way.

Throughout the whole time I went to school, he always cooked for me on the weekends. He did the majority of cooking—breakfast, lunch, and dinner—every single weekend for a good three years, and he didn't complain about it because he wanted me to study.

Conversely, another woman remarked upon how the absence of such nurturance was what made certain past relationships problematic. What she "values most" about her present boyfriend is his very attention to such needs. While the instances may seem trivial, they point to some fundamental assumptions surrounding the expression of care.

I was taught that shelter and food were very important. It was a problem with other people. Some of them don't like to feed me. I went out with this guy and he fed me tea and sugar. Prior to him, I went out with another. He had a nice one-room cottage but when it came to food, that was a problem. He cooked spaghetti for dinner once, and while that was a good dinner, it was too little in quantity.

Being a good parent, in other words, means seeing that one's children lack little in the way of food, shelter, clothing, and other material needs. As already noted, perhaps most telling is the fact that parents formerly perceived as uncaring are suddenly seen in a new light once this other prism on their relationship surfaces. Thus, even that respondent who at one time "never thought" his father cared "'cause he was always running our lives" later acknowledged his father did many things, such as putting him through school, buying him a car, "making sure I have insurance," and spending "thousands and thousands of dollars" so that he could see the world and "give us things he never had." Not surprisingly, in those families that had to work hard to achieve even the basic amenities of life, there was little time to develop the kind of relationship modeled after dominant American notions of intimacy.

Both Chinese Amercan fathers and mothers were seen as expressing their love through tremendous self-sacrifice and self-denial. The following comments reflect the appreciation offspring have for such practical, material support.

(1) Sometimes my mom—she's so self-giving that my sister and I talk sometimes about how she never really had an independent existence of her own. And housewives nowadays talk about self-fulfillment, self-satisfaction. She never even thought about that kind of stuff for herself. It was always putting us first and just a tremendous amount of denial, self-sacrifice, you know. Then you think, what's wrong with her that she does this? I still think that sometimes today.

(2) I think I can say that the general sort of context in which they raised me was, they ended up putting us, the children, before their own interests and maybe their own needs sometimes. . . . Even now my mom calls up and says, "Do you need money for anything? We'll be glad to get you that money because. . . ." They don't come out and say it but what is said in the unsaying is that basically their whole life has been about seeing us get raised and putting us through school, and even now their life is still to some degree putting us before them in meeting our needs.

(3) My father will give me everything that he can, supports me financially in every way he can, and now even emotionally, and so does my mother, so that we can become successful. I think that I have to do the same for my children, to give them every advantage possible. And there's no question about it in my mind that if I had children, I'd try to send them to the best schools, get the best education, have the best opportunities possible.

Ambivalence about such sacrifices tend to be associated with the conflicts this presents in terms of individual autonomy and growth, especially in an American context where independence and self-sufficiency are competing values.

I think it's hurt me in some ways in growing. If I had a problem, I knew it wouldn't really be a problem, because I could always just go to dad and get money or something like that. He *always* bails me out. My rent or anything like that. Even now. And so in some ways, that's

> allowed me not to grow up because I don't have to go fending for myself. Like some of my friends, if they're in trouble, they'll have to get a second job or something. It's an issue in terms of growth.

CONCLUSION

As noted in the introduction, the initial goals of this research were to grasp the social context in which mental health problems are imputed. Because mental health referrals are typically made by close family or friends, the social support network is an important focus for researchers. The present chapter, however, focused on how intimacy and social support were ideally conceived and experienced among a sample of Chinese Americans and European Americans. A major underlying assumption was that whether one had individualistic or collectivistic assumptions helps explain what individuals value in their close relations. These differences in orientation have a patterned influence for why individuals idealize certain relations, what they perceive as problems, and the patterns of communication they seek to cultivate in intimate relations. Given the longstanding relationship that social scientists have noted between social support and health, these observations also have implications for mental health.

On the one hand, expectations or "needs" can be seen as a direct *extension* of cultural orientations which respectively emphasize individual and interdependency. On the other hand, individuals may have expectations or needs that *complement* these dominant value orientations (i.e., because they are generated by default).

Thus, among European Americans, a high value on individuality, autonomy, and self-direction meant that these considerations would be pronounced in intimacy and affect what individuals looked for in the other. A direct reflection of these values was an attraction to individuals who displayed self-sufficiency. Intimate relationships were ideally experienced as ones in which mutuality derived from the fact that both persons felt "equally strong." The imperative to self-actualize within an individualistic context, however, seemed to be incompatible with the notion of asymmetry in terms of individual "growth." Fear of dependency was as notable here among European Americans as it was absent among Chinese Americans. For the latter, autonomy meant freedom from overbearing authority. Among European Americans, individuality implied uniqueness and separateness from others. Consequently, an attendant longing for social connectedness, for "familyness" or a sense of community was generated. Likewise, the experience of relatively unfettered freedom created in its wake a certain need for structure, guidance, or direction. This reinforced the importance of "significant others" as role models. Important features of intimacy among European Americans reflected this dual tension. Where individuals talked about mental health problems, these problems were consistent with the dominant value orientation toward individualism, associated as they were with failures to achieve individual self-fulfillment or self-sufficiency. In sharp contrast, Chinese Americans were more likely to see some form of "self-centeredness" as the problem, looking critically upon individuals who appeared "frustrated" when they did not get their way or who failed to live up to social obligations or commitments.

We can similarly talk about intimacy within the Chinese American framework in terms of values or predilections that are direct reflections of a collective orientation. Whereas European Americans most valued those relations that supported individual self-actualization, Chinese Americans minimally expected from the other some form of devotion or self-sacrifice where practical, material, or "basic" needs were concerned. While there are aspects of Chinese culture that may encourage individualistic tendencies (Tong 1978), individualism has never been celebrated in and of itself, or if so, placed in the service of some larger cause. Born into a social network where relations are expected to be an important source of support throughout life, Chinese Americans are more likely to place greater value on adaptability, reciprocity, and social harmony, with fewer concerns about dependency than European Americans have. For the same reason, whereas intimacy for the latter presupposed intimate self-disclosure, even when this involved negative emotions, Chinese Americans were more reserved about such verbal revelation of one's emotions, especially when this involved conflict and the prospect of direct confrontation. Confiding under such circumstances depended more on the other's ability to detect a problem and then to steadfastly persist in trying to open the channels of communication. In general, conflict was viewed as delicate issue, best handled by being suppressed. While a great deal is invested in such relations, there were few role models for diplomatically articulating sensitive issues. Typically, misunderstandings will accumulate, and when expressed are, in their worst scenario, layered with history, blame, and raw emotions.

In the following passage, a 26-year-old man recalled how an argument between his father and grandmother escalated into a confrontation, which triggered the complete collapse of the relationship so that neither exchanged a single word with the other for ten years. At the same time, this problem was kept alive by the grandmother's unceasing denouncements of her son to friends and associates, a tendency that the respondent described as "abnormal . . . almost bordering on obsessive-compulsiveness."

> . . . For some reason, she just blew it out of proportion and stayed angry over it for—it must have been almost ten years—and she, she has lots and lots of friends—and she spread it out all over the place . . . [so that] my father became persona non grata for almost ten years. But the thing that made it abnormal almost bordering on obsessive-compulsiveness, is that she kept bringing it up over and over. To everybody. Day in and day out.

The irony, therefore, is that social relations are "taken for granted" because it is expected that they will endure. Yet their disruption or breakdown has serious or longstanding social as well as mental health implications. In general, the loss of social moorings and social support for Chinese Americans constituted a major source of inability to cope. In other situations where mental health problems were implicated, two cases ended in suicide and a third resulted in lifetime mental hospitalization. Although much more research needs to be conducted on the social circumstances under which mental illness was imputed, the existing mental health literature on Asian Americans supports the view that those mental health problems that eventually reach official attention tend to be far more serious than those reported for whites.

In sum, the preceding pages sought to illuminate and contextualize the subjective experiences associated with close relations given their direct implications for health, mental health, and referral to professional help. Individualistic or collective orientations offer one theoretical handle on these cultural configurations. Whatever the nature or hierarchy of values involved, strategies for intervention would be better informed by such understandings. The analysis also suggested the relevance of social history or social class issues for some of the cultural understandings already surfaced. (Thus, for example, to the extent that parent-child relations in some Chinese American families were unusually open and confiding, one might ask, to what extent is this the luxury of middle-class families or exposure to modern, urban, or Western cultural standards?) A study that systematically explores these issues at the outset could further build upon these cultural understandings as well as provide an alternate analytical handle. Finally, while intimate relationships were a strategic focus of this study, some of the findings here may have larger implications outside the intimate context—from school relations to work site.

ACKNOWLEDGMENT

I would like to express my thanks to both Dr. Allen Kanner and Professor Judy Yung for their thoughtful and detailed comments on earlier drafts of this chapter.

NOTES

1. Thus, in a study of interracial marriage, Chinese Americans who married white Americans indicated that they did so not only because they shared certain American or middle-class values, attitudes, and interests but because they sought to escape the repressive, overbearing, or domineering forms of relationships associated with their parents' traditions (Fong and Yung 1995/1996).

2. This emphasis on emotional self-control can be found in many Asian cultures. In one tragic instance, a Korean father was found guilty of murdering his daughter because the prosecution argued that his public reaction to the news of her death was so unemotional as to imply culpability. According to his wife, however, his grief was visible and apparent when the two of them privately shared their loss (Lin 1992).

3. In *Hearts of Sorrow,* a North Vietnamese woman interviewed by the author, James Farrow, explained how a conflict between a husband and wife would in the old country be mediated by the parents and relatives of both parties so that "face-saving" could occur all around. In the United States, however, geographical and residential dispersion reduced the possibilities of such familial intervention (p. 403).

REFERENCES

Cox, H. 1977. *Turning East: Why Americans Look to the Orient for Spirituality and What That Search Can Mean to the West.* New York: Simon and Schuster.

Fong, C. and J. Yung. 1996. "In Search of the Right Spouse: Interracial Marriage among Chinese and Japanese Americans." *Amerasia Journal* 21: 77-98.

House, J. S., K. Landis, and D. Umberson. 1994 "Social Relationships and Health." Pp. 83-92 in *The Sociology of Health and Illness: Critical Perspectives*, edited by P. Conrad and R. Kern. New York: St. Martin's Press.

Li, V. H. 1977. "Law and Penology: Systems of Reform and Correction." Pp. 140-150 in *Deviance and Social Control in Chinese Society*, edited by A. A. Wilson, S. L. Greenblatt, and R. W. Wilson. New York: Praeger.

Lin, J. 1992. "Was Jury Confused by Culture or Did He Kill his Daughter." *Philadelphia Inquirer*, April 28.

Spiegel, D. 1993. "Social Support: How Friends, Family, and Groups Can Help." Pp. 331-349 in *Mind Body Medicine*, edited by D. Goleman and J. Gurin. Yonkers, NY: Consumer Reports Books.

Tong, B. 1978. "Warriors and Victims: Chinese American Sensibility and Learning Styles." Pp. 70-93 in *Extracting Learning Styles from Social/Cultural Diversity*, edited by L. Morris, G. Sather, and S. Scull. Oklahoma City: University of Oklahoma, Southwest Teacher Corp Network.

Wuthnow, R. 1994. *Sharing the Journey.* New York: Free Press.

AIN'T IT A CRIME

BLACK WOMEN'S PERCEPTIONS OF INTIMACY WITH BLACK MEN INVOLVED IN CRIME

Sandra Lee Browning and R. Robin Miller

INTRODUCTION

The United States has the highest rate of incarceration in the world, with young black men disproportionately comprising the ranks within U.S. correctional institutions (Mauer 1990). For whites in this country, the incarceration rate is similar to that for whites from European nations—relatively low. But for blacks, the incarceration rate is similar to that in apartheid South Africa (Murphy 1990). The 1990 Sentencing Project Report found that one in four (23%) black men between the ages of 20 to 29 years were under some form of correctional supervision—jail, prison, probation, or parole (Mauer 1990). In their follow up study, "Young Black Americans and the Criminal Justice System: Five Years Later," Mauer and Huling (1995) found that within the span of five years, black male involvement in the criminal justice system rose to one in three (32.3%). Consequences, approximately 827,440 young black men in 1995 were under the control of the criminal justice system.

The high rate of black male involvement in the criminal justice system is not new. Scott Christianson (1981) in his article, "Our Black Prisons" argues that the state prison system has replaced slavery as a mechanism of social control. He notes that in the same year that the state of New York legislated the emancipation of the slaves (1796), it created the first state prison. Hawkins (1985), looking at data from North Carolina over a 100-year period, found that in 1870, the percentage of blacks in the prison population was 78 percent; by 1878 that proportion had increased to 90 percent, dropping to 51 percent in 1930. The black-white ratio in North Carolina was highest in 1880 with an incarceration ratio of 17 to 1, which dropped to 4.2 to 1 by 1980 (a ratio that has remained consistent throughout the nation, see Mauer 1990 and Mauer and Huling 1995).

In many cities, on any given day, half of the young black men are under the control of the criminal justice system (Miller 1996). The high rate of black men under the control of the criminal justice system has led to the alarming growth in the number and percentage of poor black female-headed households (U.S. Bureau of the Census 1996). In terms of marriage, the criminal justice system has effectively taken black men out of their communities and has consequently had a detrimental affect on not only the lives of these men, but on the lives of their wives, girlfriends, and significant others as well. This reprehensible trend has served to greatly diminish the already dwindling pool of black men available for marriage.

Marriage and the marital union are at the foundation of U.S. society. It is an expected norm for women to marry; and not just to marry, but to marry a man who will be able to support her and her children through employment at a legitimate job. Making a living through crime is not seen by society as an acceptable means of supporting a family. Setting aside the obvious moral question, the business of crime is often not steady or dependable. The heightened chances of black men being caught in the dragnet of the criminal justice system increases the odds against them sustaining themselves or their families through crime. Moreover, being involved in the criminal justice system tags one with a criminal record—a record that is with one for life, making it difficult to secure a legitimate job that can support oneself or a family.

With so many black men being under the control of the criminal justice system, the potential pool of black male marriage partners is drastically reduced. The black woman facing this problem is left with limited options. First, she can accept the possibility of remaining single, which would be viewed as abnormal. Second, she could date, and marry a man outside her race. Finally, she could date, and marry a black man involved in crime, thus playing and hoping against the odds that he will not be caught and sent to prison. Although a growing number of black women are opting to remain single, the messages that black women both receive and pass on to their children is that marriage is expected and preferable (Browning and Miller 1999). There is tremendous pressure to marry. With regard to dating and marrying a man outside one's race, the segregated history of blacks has discouraged race mixing. As a result blacks and whites tend to view interracial relationships as unacceptable. With the first two options possible but not probable, a black woman is faced with the reality that most of the men available to her will at some time have been involved in the criminal justice system.

SCOPE OF THE PROBLEM

A further examination of the problem illustrates the level of black male involvement in the criminal justice system. In 1992, Jerome G. Miller and the National Center on Institutions and Alternatives issued the report, "Hobbling a Generation: Young African American Men in Washington, D.C.'s Criminal Justice System." Focusing on the District of Columbia, Miller found that on average, 21,800 or 42 percent of the District's black men between the ages of 18 to 35 years were under the control of the criminal justice system. They were either in jail, prison, on probation or parole, or otherwise caught in the control of the system awaiting a disposition. As Miller explains (1992), the social programs of the 1960s and the establishing of a social safety net for the most disenfranchised groups in U.S. society have given way to a criminal justice dragnet; a dragnet that is cast to snare and catch within its web young black men. Specifically, Miller found that on average there were approximately 12,500 District residents in jail or prison. Of that total, roughly 7,800 were young black men between the ages of 18 to 35 years. As he noted, this represented about 15 percent of all the black men within that age group living in the District.

When probation and parole were examined, the control of the criminal justice system in the lives of young black men increased to 21 percent. On average, 6,000 black men between the ages of 18 to 35 years were on probation. There were another 4,000 on parole from the District and Federal prisons. This resulted in an overwhelming figure—some 11,000 young black men were under the control of the criminal justice system through either probation or parole.

The picture is brought further into focus when Miller includes those young black men awaiting criminal disposition: those who are out on bond, those who have outstanding felony warrants, and those who had misdemeanor warrants. When this category is included, approximately another 7,000 young black men, representing another 6 percent, were under the control of the criminal justice system.

As with the Sentencing Projects Reports, the National Center on Institutions and Alternatives conducted a five-year follow-up study. In this study, Eric Lotke (1997), found that within the five years since Miller's analysis of young black men and the District's criminal justice system, things have worsen and the dragnet has expanded. The 42 percent of young black men between the ages of 18 to 35 years who were under the control of the criminal justice system has increased to an astonishing 50 percent.

As Lotke noted in this study, the finding from the District of Columbia parallel those of the nation. Over a 15-year period (1980 to 1995) the number of blacks in prison grew from 146,900 to 541,900. The numbers of blacks during this period on probation and parole increased from 410,000 to 1,395,000 (Lotke 1997). This data supported the findings of the Sentencing Project, that one in three black males were under the supervision of the criminal justice system (Mauer and Huling 1995).

IMPACT OF THE WAR ON DRUGS

What are the causes of the over-representation of so many young black men under the control of the criminal justice system? Many social observers have attributed the burgeoning number of black men in the criminal justice system to the nation's War on Drugs. The political campaign against drugs began decades ago under former President Eisenhower; however, the strategies and programs that ultimately led to the new War on Drugs were initiated under former President Reagan. Under the Reagan administration, billions were cut from social programs that directly affected the health, education, housing, employment, and welfare of the nation's poorest citizens. In turn, the nation's attention and resources were directed toward a new War on Drugs. This new war increased the nation's allocation for fighting drugs from $65 million in 1969 to $1.65 billion in 1982 (Office of National Drug Control Policy 1999).

In 1989, former President Bush released a $7.8 billion federal antidrug plan. Fighting the War on Drugs, for Bush, meant increased criminal justice intervention and limited prevention, interdiction, education, and treatment. Bush stressed the need for police intervention in a 1989 televised speech to the nation, where he held up a bag of cocaine that he said was purchased by an undercover Drug Enforcement Officer in LaFayette Park just across from the White House. He argued that drugs were so prevalent in society that they had reached the White House door, and that drug dealers sold their wares without fear of immunity, even in front of the nation's capital. The drug dealer, a young black man as it was, later reported that he was lured to the park by the undercover agent, so that the arrest could occur across from the White House. When questioned about the events that led to the arrest in LaFayette Park, President Bush maintained that he saw nothing wrong with the staged event (Lusane 1991). After his national address, the image of black male drug dealers selling cocaine on the White House lawn was firmly implanted in the minds of the American people. With the increased public support for getting tough on drug dealers, Bush set about putting into place expanded criminal justice intervention.

Bush's Drug Czar William Bennett spelled out the administrations intentions when he announced that "a massive wave of arrests is a top priority for the war on drugs" (Lusane 1991, p. 69)—arrests that were to be achieved in part through warrantless searches and increased policing of poor and minority communities (Lusane 1991; Mann, 1993). In addition, mandatory minimums—sentences carrying a minimum number of days or years of incarceration were instituted. The mandatory minimums were originally intended, at least in theory to deter offenders. Although they have been used for a variety of offenses (e.g., committing an offense with a gun, driving while under the influence), mandatory minimums were also applied to various drug offenses—offenses that were now targeted for arrest by the president and his Drug Czar, William Bennett. The result of the increased police attention to the poor and minority populations and the reliance of mandatory minimums in the nation's prisons have accounted, in large part, for the more than tripling of the prison population since the 1980s (Lotke 1997). Since the enactment of mandatory minimum sentencing for drug offenses, just with in the federal prison system spending has increased by more than 1,350 percent, from $220 million in 1986 to approximately $3.19 billion in 1997 (Bureau of Justice Statistics 1997; Office of National Drug Control Policy 1997). Spending for the War on Drugs has continued to spiral out of control, and the current U.S. budget for the War has reached an all-time high of $17.1 billion.

RACE AND PRISON

Between 1990 and 1996, the number of blacks in federal prison for violent crimes and property crimes decreased by 726 percent. However, during the same time period, some 12,852 blacks were placed in federal prisons for drug violations (Bureau of Justice Statistics 1997). Blacks account for almost 37 percent of those arrested for drug violations. Overall, some 42 percent of those in federal prisons for drug violations, and almost 60 percent of those people in state prison are there for drug offenses (Bureau of Justice Statistics 1996; Bureau of Justice Statistics 1998; Substance Abuse and Mental Health Services Administration 1997). This, while only 11 percent of drug users in the United States are black.

Of those arrested and convicted of drug offenses, 54 percent of the blacks convicted are sentenced to prison, whereas only 34 percent of whites convicted of the same offenses are sentenced to prison. More specifically, 44 percent of blacks are sentenced to prison for possession, whereas only 29 percent of whites are sent to prison on this charge. Similarly, 60 percent of blacks are sentenced to prison for drug trafficking, whereas only 37 percent of whites were sentenced to prison for the same offense. (Bureau of Justice Statistics 1996).

Whether the issue is drug offenses, violent crimes, or property crimes, there are more young black men between the ages of 20 to 29 years under the control of the criminal justice system than in college (Mauer and Huling 1995; Haney and Zimbardo 1998). The absence of the black man in the community and the family has contributed to the rise of single female-headed families. It has also led to black women creating and adapting to alternative strategies to form marital relationships with the limited pool of black men who are available for dating, intimacy, and marriage.

CURRENT STUDY

The issue of control of the black man by the criminal justice system is central to our discussion of the black male/female relationship. The dragnet of control that the criminal justice system has over the lives of one in three black men has effectively limited the pool of eligible men to marry. This research investigates the perceptions black women have about the black man's involvement in crime, and his marriageability. As has been noted, the criminal justice system disproportionately affects the lives of black men, and by extension the lives of black women and their communities. A neglected aspect in the social science literature is how the black man's involvement in the criminal justice system affects his marriageability, and the perceptions black women hold about his involvement in crime. Are black women willing to date or marry men involved in crime and the criminal justice system? Is black male criminality viewed as an acceptable form of "employment" by black women? Has the disproportionate number of black men involved in the criminal justice system established a normative expectation? That is, have black women, faced with an increasing number of black men involved in the system, come to accept criminal involvement from the black men they become involved with, or is criminal involvement still viewed as a negative attribute in a mate or potential partner?

COLLECTION OF DATA

Data for this study were obtained from interviews with 286 African American women in Cincinnati, Ohio. Neighborhoods were identified using the U.S. Census of Population and Housing (1990) and the Cincinnati Neighborhood Profiles. A stratified sample by gender and income was selected. A random listing of 1,000 residential phone numbers representing the nine neighborhoods was obtained from the local phone company.

The interviewers asked to speak to the female head of the household. If consent was obtained, they then proceeded to ask a series of questions regarding marital status, age, and race. If the respondent was a white woman, she was thanked for her time and the survey was terminated. If the respondent was an African American women, the survey continued.

DATE A MAN INVOLVED IN CRIME: SURVEY QUESTIONS AND MEASUREMENT OF VARIABLES

The question of interest in this study is whether or not a woman would or has been involved with a man constrained within the criminal justice system. Questions 1 through 11 are each Likert-scaled ranging from 1 to 6, where higher numbers indicate stronger agreement with the specified statements. These questions are as follows:

1. I would consider dating, becoming involved with, or marrying a man who I know is involved in criminal activity.
2. Most of the men that I have been in a relationship with have had a criminal record.
3. I would not date/marry or become involved with a man with a criminal record.
4. For the men I date or am involved with, engaging in crime is their primary source of income.
5. If the man I am involved with has to sell drugs to support the family it is unfortunate, but acceptable to me.
6. If the man I am involved with has to sell drugs to support the family it is not acceptable and would keep me from dating/marrying or becoming involved with him.
7. If a man that I am involved with gets incarcerated I would continue my relationship with him.
8. If a man that I am involved with gets incarcerated I would allow him to have a strong parenting role in my child/children's lives.
9. If a man that I am involved with gets incarcerated I would not date other men.
10. I would date/marry or get involved with a man on probation.
11. I would date/marry or get involved with a man on parole.

Summing these items produces a metric in which higher numbers indicate a greater likelihood of having a relationship with a man in the criminal justice system. Questions 3 and 6, were reverse coded in order to keep the answers of the respondents consistent with the scale. When summing across all the questions we see a potential low value of 11, and a high value of 66. In looking at the mean for the entire sample across this scale, we will be able to determine the average response of women in terms of this scale. Higher average numbers would suggest greater likelihood that a would woman consider being in or remaining in a relationship with a man involved in crime or the criminal justice system.

All women were asked these questions. If a women was married, she was asked to answer the questions about the men she had dated before she got married. Completed data for all variables constituting this scale were obtained from 286 respondents in the sample. Further, by looking at the individual means on the questions that comprise the scale, we are able to characterize the respondents' attitudes toward developing such a relationship with greater clarity. Table 1 presents the actual range for this scale, which was 11 to 52, and the mean on this scale, which was 35.5. This suggest that on average the women in

Table 1. Descriptive Statistics for Date Crime Scale

Variable	*N*	*Minimum*	*Maximum*	*Mean*	*S.D.*
Date a man involvedmin crime	282	11.00	52.00	35.54	7.85

the survey have dated or would consider dating a man who is involved in crime and the criminal justice system.

The independent variables are economic status, and perceived pool of available men (whether there are enough men without a criminal record to marry). The control variables are age, number, and sex of children; level and type of external support; educational level; number of past marriages or divorces, or both; marital status of parents; and region of the country in which they were raised.

Another important dependent variable is the knowledge of family members or friends involved in the criminal justice system. The following question allows us to measure this variable:

Was any member of your family involved in the criminal justice system?

1. Father
2. Mother
3. Grandfather
4. Grandmother
5. Brother(s)
6. Sister(s)
7. Uncle(s)
8. Aunt(s)
9. Close family friend (s)
10. Other(s) (please specify)

DESCRIPTIVE STATISTICS

Table 2 provides descriptive statistics for the demographic variables that will constitute the bases of analysis. Of the 286 completed surveys, the youngest respondent was 18 and the oldest, 68. Most of the women (56%) were or had been in some type of marital relationship. Specifically, the majority (23%) were either married or currently living with someone (23.7%). Less than 10 percent (7.8%) were divorced. The largest single group of women (43.5%) reported that they were single and not living with a man. Of the women who reported being married, the majority (78.9%) were in their first marriage and the rest (21.1%) were in their second marriage. On average, the women were 25 years old at the time of their first marriage and had been married for nine

Table 2. Descriptive Statistics for Selected Variables

Variable	*N*	*Minimum*	*Maximum*	*Mean*	*S.D.*
Marital status	283	1	6	3.70	1.75
Number of times married	71	1	2	1.21	.41
Years married	71	1	42	9.32	9.52
Number of divorces	27	1	3	1.33	.55
Age	285	18	68	31.00	9.15
Age at first marriage	72	17	38	25.15	4.82
Age at first divorce	26	20	40	29.35	6.14
Race	283	1	1	1.00	.00
Have any children	273	0	1	.70	.46
Age at birth of first child	205	15	38	19.79	4.61
Number of children	205	1	8	2.17	1.00
Were all children born in wedlock	194	0	1	.31	.46
How many children born out of wedlock	111	1	4	2.10	.76
Occupational status	286	1	5	2.56	1.37
What she does for a living	222	1	4	3.00	1.23
Receive public assistance	279	0	1	.23	.42
Educational level	279	1	6	2.61	1.05
Income level	285	1	7	2.59	1.42
Total number of valid cases	286				

years. For the women who reported being divorced, the majority of women (65.2%) were divorced from their first marriage. Approximately one-third (30.4%) reported being divorced from their second marriage. The average age at the first divorce was 29 years.

More than 70 percent of the women (70.3%) reported having children. On average, the women had their first child at age 20. Most of these women reported having two children. A sizable number of these children (111) were born out of wedlock. A little more than 20 percent (20.7%) of the women had one child outside of marriage, slightly more than 52 percent (52.3%) had two children, and slightly less than one-quarter (23.4%) had three children outside of wedlock.

In terms of employment , more than 40 percent (42.3%) were employed in unskilled jobs. Less than 10 percent (8.4%) were employed in skilled positions. Slightly more than one-quarter of the women (26.9%) were in professional and semiprofessional jobs. Approximately 20 percent of the women in this sample identified themselves as unemployed or homemakers.

Slightly less than one-quarter (22.9%) of the women in the sample received public assistance. Of those women, less than 20 percent, respectively, received Aid to Dependent Children (18.2%) and food stamps (18.9%). Approximately another 20 percent (21.7%) were on Section 8 (housing subsidy) or lived in Metropolitan Housing (housing subsidy). The average length of time on public support was four years for Aid to Dependent Children, Section 8, and food stamps. Those on Metropolitan Housing reported being on this program an average of three years. The women were also asked how long they expected to receive public assistance. The average length of time that the women expected to receive public support was slightly over two years.

With regard to educational attainment, more than half of the women sampled graduated from high school (52%), with slightly over 40 percent reporting having some college (18.3%) or a graduate degree (22.3%). More than 50 percent (51.6%) reported that they were not trying to pursue additional education.

For income, excluding other sources of support, more than half of the women (56.5%) reported having incomes of less than $19,999 per year. About one-third (32.9%) had incomes between $20,000 and $39,999. Fully 68 percent did not believe that their incomes alone were sufficient to support the family.

Table 3. Regressions for Date Crime Scale for Selected Variables

	Beta	*Sig.*
Occupational status	–.196	.036
Number of times married		
Years married		
Number of divorces	–.284	.009
Age	–.240	.002
Age at first marriage		
Age at first divorce	.236	.039
Have any children		
Age at birth of first child	.206	.016
Number of children	.186	.041
Were all children born in wedlock		
Received public assistance	.207	.003
Educational level	–.402	
Income level		
Family crime involvement		
Enough good men available		
Adjusted R	.498	
Number of cases in the model	2.09	

Notes: Only those relationships whose significance levels were less that or equal to .10 are reported. ($p < .10$).

DATA ANALYSES

Regression Analyses

Table 3 presents the regressions for the date crime scale and selected variables. The R-square for the model was .50. Accordingly, occupational status, number of divorces, age at birth of first child, number of divorces, age, and educational level were all negatively significant in this model. This suggest that women who had a lower occupational status were more likely to date a man involved in crime. The fewer the number of divorces, the more likely they were to date a man involved in the criminal justice system. Finally, both age and educational level were negatively significant, while being on public assistance was positively significant. In addition, positively significant relationships were obtained for age at birth of first child and number of children. It would be expected that those women who had a larger number of children and those who had their children later in life would be more likely to date a man involved in crime.

Thus, it appears that a black woman's position in the social structure dictates the men her willingness to date and get involved with a man in the criminal justice system.

CONCLUSION

It is an expected norm within our society for women to be married. The black male/female relationship occurs within a structural environment of increasing social isolation, decreasing rates of black marriage, and an increasing level of black male involvement in the criminal justice system. This research found that women in more precarious circumstances (e.g., lower occupational status) were more likely to get into a relationship with a man involved in crime. A woman's place in the social structure (age, income, education) and her perceived pool of available men influences whether or not she gets involved with a man in the criminal justice system. If a women perceives that the only men available to her have engaged in crime or have been involved in the criminal justice system, those are the men available to her. No available evidence indicates that she prefers them.

Although women on welfare were more likely to date a man involved in crime or the criminal justice system than women not on welfare, on average all the women in the sample (on welfare and not on welfare) reported that they were willing to or had dated a man engaged in crime. Because of necessity women in the sample violated the dominant cultural norm—and dated a man with a criminal record.

Researchers have argued that there are a disproportionate number of black men involved in the criminal justice system. The recent rise in black male incarceration is due in part to the War on Drugs, and the continuing stereotype of the criminal black man. This has led to the system unjustly focusing it's law enforcement efforts on black men, which in turn has resulted in higher black male incarceration.

Black women are faced with an impossible situation. They must bridge the gap between the ideal cultural norm and the reality that many of the men available to them have a criminal record. Ain't it a crime?

Although black men under these circumstances are available to marry, their continued presence in the family is tenuous. Further detailed research into the relationship between black male incarceration and marriage/family dynamics is clearly indicated.

REFERENCES

Browning, S. L., and R. R. Miller. 1999. "Marital Messages: The Case of Black Women and Their Children." *Journal of Family Issues* 20(5): 633-647.

Bureau of Justice Statistics. 1996. *Sourcebook of Criminal Justice Statistics.* Washington, DC: U.S. Government Printing Office.

Bureau of Justice Statistics. 1997. *Sourcebook of Criminal Justice Statistics.* Washington, DC: U.S. Government Printing Office.

Bureau of Justice Statistics. 1998. *Prisoners in 1997.* Washington, DC: U.S. Government Printing Office.

Christianson, S. 1981. "Our Black Prisons." *Crime and Delinquency* 27(3): 364-375.

Cincinnati Neighborhood Profiles. 1980. *Interim Report.* Prepared by the Planning and Management Support System Division of City Planning Commission, Cincinnati, Ohio.

Haney, C., and P. Zimbardo. 1998. "The Past and Future of U.S. Prison Policy: Twenty-five Years After the Stanford Prison Experiment." *American Psychologist* 53: 7.

Hawkins, D. F. 1985. "Trends in Black-White Imprisonment: Changing Conceptions of Race or Changing Patterns of Social Control?" *Crime and Social Justice* 24: 187-209.

Lasane, C. 1991. *Pipe Dream Blues: Racism and the War on Drugs.* Boston, MA: South End Press.

Lotke, E. 1997. *Hobbling a Generation: Young African American Males in D.C.'s Criminal Justice System Five Years Later.* Alexandria, VA: National Center on Institutions and Alternatives.

Mann, C. R. 1993. *Unequal Justice: A Question of Color.* Bloomington, IN: Indiana University Press.

Mauer, M. 1990. *Young Black Men and the Criminal Justice System: A Growing National Problem.* Washington, DC: The Sentencing Project.

Mauer, M., and T. Huling. 1995. *Young Black Men and the Criminal Justice System Five Years Later.* Washington, DC: The Sentencing Project.

Miller, J. G. 1992. *Hobbling a Generation: Young African American Males in D.C.'s Criminal Justice System.* Alexandria, VA: National Center on Institutions and Alternatives.

Miller, J. G. 1996. *Search and Destroy: African-American Males in the Criminal Justice System.* Cambridge: Cambridge University Press.

Murphy, J. 1990. New York State Coalition for Criminal Justice. Testimony at Philadelphia Hearing of the National Commission on Crime and Justice.

Office of National Drug Control Policy. 1997. *Executive Office of the White House, National Drug Control Strategy. Budget Summary.* Washington, DC: U.S. Government Printing.

Office of National Drug Control Policy. 1999. *Executive Office of the White House, National Drug Control Strategy: Budget Summary,* Washington, DC: U.S. Government Printing Office.

Substance Abuse and Mental Health Services Administration. 1997. *National Household Survey on Drug Abuse: Population Estimates 1996.* Rockville, MD: Substance Abuse and Mental Health Services Administration.

U.S. Census of Population and Housing. 1990. *Population and Housing Characteristics for Census Tracts and Block Numbering Areas, Maps, Cincinnati-Hamilton.* Washington, DC: U.S. Government Printing Office.

U.S. Bureau of the Census. 1996. *Statistical Abstract of the United States: 1996,* 116th ed. Washington, DC: Government Printing Office

FAMILIAL VALUES, GENDER POLITICS, AND MEANINGS OF INFIDELITY
A CROSS-CULTURAL COMPARISON OF EXTRAMARITAL AFFAIRS

Jui-shan Chang

INTRODUCTION

Extramarital affairs exist cross-culturally, historically and in modern times (Kinsey, Pomeroy, and Martin 1948; Kinsey et al. 1953, p. 415; Fisher 1992; Greeley 1994). Most of the theories about gender roles, marriage, and family have been developed in regard to premarital and marital relationships or the wider kinship system. Compared to "conventional" activities of courtship, cohabitation between unmarried people, or relationships within marriage, extramarital relations can reveal more about romantic love, gender politics, and the relationship between marriage and family by virtue of their "unconventionality." Yet for all their ubiquity and capacity to be sociologically revealing, extramarital affairs have not been researched extensively. Nor have theories been developed to understand patterns of affairs and their trajectory; what less apparent functions extramarital affairs might be performing in modern times; and whether these functions vary across cultures. Of course, there is one obvious reason why extramarital affairs are understudied: it is very difficult to collect valid and reliable data owing to their nature of transgression.

There are, however, a few surveys and some empirical research in this area. It has been possible to ask respondents in surveys for attitudes and behavior about extramarital affairs. The focus of these studies, however, has mainly been on "objective" indicators of sexual behaviors and attitudes concerned with marital fidelity, rather than on subjective dimensions such as the meanings of the extramarital relationship and its implications for the quality of sexual experience and morality (Hunt 1969, 1974; Reiss, Anderson, and Sponaugle 1980; Saunders and Edwards 1984; Lawson 1988; Edwards and Booth 1994; Laumann et al. 1994; Greeley 1994; Bender and Leone 1995).[1] Yet even a cursory glance at magazines or newspapers reveals that much popular material is written regarding extramarital affairs. Whatever the frequency of extramarital affairs, it is clearly outnumbered by the frequency with which reports about extramartial affairs are read or discussed.

This study exploits this proliferation of material by using an unobtrusive approach that treats reports of extramarital affairs as "data," to examine how extramarital affairs are reported and depicted as real events in popular press, and how the problems of extramarital affairs are dealt with by people who lived with such events. Of course, such "revelations" from the reports in the media—even if they are *real* life stories—are not the same as conducting a survey with a random sample to find out people's attitudes and behaviors.

The purpose of this study is *not* to make a generalization to a population about people's extramarital attitudes and behaviors. Rather, this study aims to generate insights about the nature and the meanings of infidelity. For this purpose, the study of these real-life events reported in the media is more relevant than finding out the exact figures (if this were even possible) on such sensitive matters. Another important aspect of studying these media reports is that, even if individuals may not "live by" or "live through" such events, they can barely avoid "living with" such material, particularly if such material is taken as real.

Although extramarital affairs occur in a wide variety of cultures, the meanings of marriage and family may vary. The question that then arises is whether variations in familial and conjugal values are reflected in the depictions of transgressions to the norms that embody these values. For instance, it is sometimes suggested that there are deep-seated human needs that "break out" of the bonds of marriage, despite the varying normative orders. If these depictions of "breakouts" are similar across cultures, such a view of "basic needs" might find some support. Alternatively, it might be that the very values embodied in marital and familial norms generate different forms of transgression, in which case depictions of such transgressions would be more culturally specific. To this end, this paper deals with materials in two distinct cultural contexts: Chinese and the secular West.

This study proposes a cross-cultural approach to examine these issues by comparing Chinese and Western reports of extramarital affairs. What is the rationale for making a cross-cultural comparison in this study? Similarities across these different cultures can reveal whether there is a more fundamental and universal meaning of sexuality, marriage, and marital transgressions. The contrast of these dimensions and issues between different cultures can reveal how unique the meaning of these issues is in each specific culture. I want to argue that identifying the universal meaning as well as the culturally specific meaning is equally crucial for understanding the phenomenon of extramarital affairs. Such contrasts can enrich theories on love, sexuality, marriage, and family.

Furthermore, for the purpose of isolating and identifying the universal meaning or culturally specific meaning, or both, more compelling inferences can be made by comparing two very different cultures than by comparing two less different cultures. To this end, I wish to compare the reports of extramarital affairs between the Western context and a Chinese setting because they have very different value bases. In particular, these two settings have completely different traditions representing two contrasting ideals of individualism and familism. In this study, I have chosen Taiwan to represent a Chinese setting for such cross-cultural comparison with Western culture, because Taiwan can be seen as a baseline for the study of contemporary "orthodox" Chinese culture.[2]

The comparison between the Chinese and Western cultural contexts might appear as an artifact of some essentialist construction. Although there is diversity with respect to sexual behaviors and familial arrangements within both cultures on such dimensions as religion, ethnicity, class, political economy, generation, gender, or sexual orientation, the portrayals of marital transgression in particular are uniquely placed to reveal fundamental differences that go to the core of each culture. This is because the diversity of sexual attitudes and practices does not expand the limited number of ways in each culture in which responses or solutions to marital transgressions are dealt with. On the other hand, to refer in turn to a "core" of each culture might suggest a historicism that makes comparison impossible. Cross-cultural comparison remains viable as long as the base for comparison is explicit and there are culturally specific definitions of key comparative terms and concepts, such as love, family, sexuality, infidelity, and gender roles. The diversity of meanings between cultures should not hide the sheer fact that such terms operate within both cultures and have common reference points. The diversity or "variance" of meanings between cultures actually still hang upon some core central tendencies to the extent that in all cultures, men and women could fall in love, make and break commitments, enjoy or avoid sex, and engage in sexual activity beyond the bounds of what is acceptable in that society. In other words, a number of forces—normative, emotional, and situational—resolve themselves into patterns that appear across cultures, which makes cross-cultural comparision possible.

The research questions to be posed for this comparative study are as follows: Are extramarital affairs reported as having common trajectories in diverse cultures? Specifically, do these real events follow some "scripts" that define basic patterns of plots, moves, and resolution of conflicts? How do the reports of extramarital stories exemplify values of fidelity to people in modern times? More generally, what do the reports of infidelity reveal about the very nature of fidelity?

What do the depictions of extramarital affairs reveal more generally about any problematical aspects of manhood and womanhood within each culture? Love, sexuality, marriage, and family are the basic and universal dimensions of human beings, but to what extent are the meanings of these dimensions cross-culturally universal? To what extent are the meanings of these dimensions culturally specific or unique? Some of these questions can be answered directly by an analysis of the various reports under investigation. Others cannot be so directly answered by the data but they are suggestive of some cultural dynamics.

THEORETICAL RELEVANCE

The Cultural Centrality of Affairs

I want to argue that affairs (and sexual practices) can inform a core part of the culture—as a leading historian of sexuality, Randolph Trombach, emphasises that "sexual behaviour (perhaps more than religion) is the most highly symbolic activity of any society. To penetrate to the symbolic system implicit in any society's sexual behaviour is therefore to come closest to the heart of its uniqueness."[3] Following up Trombach's statement, Plummer highlights the point that "sexual stories are a major key to this system" (1995, p. 176).

The construction of sexuality does not only take one to the core of a culture; sexuality may not be contained by the core of that culture. Sex seems always to exceed the specifications of norms. Every culture has norms and regulations to protect its ideal form of sexuality from variation and excesses. This is why extramarital affairs reveal more about the forms of variation (actually deviation) as well as the nature of this "excess" aspect of sexuality (Murphy 1971). In other words, extramarital affairs can be seen as a "key" to where the system of marital expectations "leaks" and possibilities of transgression abound. Given the argument of the cultural centrality of sexual practices, in particular the extramarital affairs, I further want to argue that cross-cultural comparison of extramarital affairs can produce cultural shock in reverse, leading us to reexamine all the assumptions and meanings regarding sexuality and normative boundaries taken for granted in each culture.

Regarding the nature of transgression, the study of extramarital affairs can also inform the debates about the relationship between deviance and cultural values in general. Some argue that deviance mirrors or revitalizes the core cultural values (Shaw 1991). Deviance, therefore, strengthens the norms and clarifies the boundaries (Durkheim 1976). Others argue that deviance breaks the norms, which sometimes leads to social change (Parsons 1951). I also argue that sometimes breaking the "boundaries" is actually a part of the social change process. The cross-cultural comparison of marital infidelity in Chinese and Western contexts can isolate the culturally specific or universal linkage(s) between cultural values on marriage, family, and sexuality, (breaking) boundaries, and social change.

Cross-cultural Universality of Sexual Ethics and Gender Politics

The basic norms about marital fidelity and the mores of sexual propriety are not static and do not command a high degree of consensus in the West. Indeed any overview of the "history of sexuality" would have to highlight many areas of contention. Thus, an introduction to this core needs to address sexual ethics and more recent formulations regarding gender politics and sexualiy.

Most of the debates about sexual ethics post-World War II have been developed in the West without regard to the ethnocentric nature of their deliberation. Although there might have been a curiosity about how "other people do it"—as witnessed by the popularity of writings of anthropologists such as Malinowksi or Margaret Mead—the debates about sexual ethics have not sought to incorporate this cross-cultural diversity of practice. Rather, the Western ethical concerns about sex have been mainly focused on two fundamental viewpoints: romanticism and libertarianism. Sexual romantics link sex to bonds of affection, intimacy, and love. In this view, sex should not be approached casually or pursued as an autonomous erotic pleasure. Romanticism underscores normative exhortations to fidelity. Libertarians, by contrast, argue that sex has multiple roles: for pleasure, love, procreation, or just habit. The key ethical issues in such a polymorphous condition is that whatever is done has a basis in a choice process within the individuals (Seidman 1992).

The contrast between these two sexual ideologies can also highlight the debates regarding the issues of gender politics and sexuality within feminism: namely cultural feminism and its critique (Seidman 1992). Cultural feminists argue that men's sexuality is predominantly body-centered, loosely tied to emotions, more aggressive, controlling, exploitative, and promiscuous compared to that of women; whereas women's sexuality is more person-centered, tender, diffused, relational, defined by love, and inclined to monogamy. This kind of view has an affinity to sexual romanticism. Any "excess" finds its origins in men. The critiques of cultural feminism argue that the body-centered sexual pleasure and its pursuit as such can be applied to women and not just men. Romanticism is criticized for still imposing the patriarchal constraint on women and inhibiting their capacity to enjoy sex. In other words, romanticists' definition of the role and meaning of women's sexuality is still based on the definitions of sexuality and perspectives from men. In addition, the critiques of cultural feminism argue that women have a disadvantageous position in terms of having greater danger involved such as unwanted pregnancy, single motherhood (without support), and a double standard regarding sexual behavior.

To explore these arguments further, extramarital affairs (which are the focus of this study) can be seen as good "laboratories" to explore and examine the meanings and the plausibility of these debates about gender politics and sexuality. The focus on sex within intra- and extramarital contexts is particularly useful since one is "normal" and the other is "transgressive" yet not uncommon.

Past research into extramarital affairs in the Western contexts shows that extramarital affairs may be "functional" in terms of men and women fulfilling more flexible gender roles, which they may not be able to fulfill in a marital context. Lawson (1988) finds that, for men, they can feel constrained by their culturally defined dominant role or strong breadwinner role in marriage, but have emotions and needs that cannot be structurally met by the given marital and familial relations. In extramarital relations, men can play more flexible roles (by sometimes being a dependent boy to be looked after, or being emotionally vulnerable and receiving gentle nurture from a woman). For women, the flexibility in an extramarital relationship is more difficult to define but can be related to her sense of autonomy and control, which may be lacking in a marital context. Despite that, women may still desire a dominant male figure in an extramarital romance, as long as this autonomy is not constrained. In other words, such an argument implies that in the Western context, the familial structure and the conventional gender roles cause problems within marriage and push married people to seek a more fulfilling relationship outside marriage. In particular, under the trend of "feminization of love" (for men) and "masculinization of sex" (for women) in the West (Lawson 1988), there has been a convergence in rates of extramarital affairs for men and women (Richards and Elliott 1991).

There has also been a convergence in the reasons for women and men being involved in extramarital affairs: sexual freedom and short-term enjoyment for sexuality itself, variety, adventure, and the need to feel attractive and desirable (Hunt 1974; Pietropinto and Simenauer 1977; Seidman 1992). It has also been argued that, even if both men and women benefit from engaging in extramarital affairs, men are still in a more "advantageous" position than women in such extramarital contexts (Lawson 1988), in addition to the fact that men are in a more advantageous position in a marital relationship[4] (Duncombe and Marsden 1996; Elliott 1996; Holland et al. 1996). Are such arguments regarding the "problematic structure" in marriage, the convergence in the propensity to be involved in extramarital affairs between men an women, and "men always being advantageous in both marital and extramarital relations" valid? This study explores these arguments further by examining the way cases of extramarital affairs are presented in print. The advantage of this study is that these extramarital reports are typifications of the process by which a person gets involved in such an affair. These typifications can reveal gender politics in both marital and extramarital settings and the meanings of sexuality for both genders.

In these contemporary times, there is another purpose in examining the cross-cultural relevance of Western sexual ethics and the debates mentioned earlier. The comparison of extramarital affairs between a Western context and a Chinese setting can alert scholars and social commentators to the possibility that changing sexual mores in China need not follow a path similar to that in the West, nor yet simply extend existing Chinese values. I argue that in a Chinese context, the sexual constructions (as revealed in extramarital relations) in modern times have their own forms, which are not immediately congruent with the forms developed in the West. The shift of major feminist concerns from a critique of "women as objectified and exploited in sexual relationship" (Dworkin 1981; MacKinnon 1982) to an elaboration of possibilities regarding "women-centered pleasure" (Vance 1984; Jeffreys 1990) and "diversity and fluidity in women's sexuality" (Rich 1980; Stacy 1991; Robinson 1993) may not be able to inform the issues on gender politics and sexuality in the Chinese context. The crucial issues involved in the validity of these Western ideologies or debates are what the actual object of fidelity is within each specific society and what the meanings of love, sex, marriage, family, and manhood and womanhood are in each specific culture or society.

METHODOLOGY

Data

The reported *real* cases of extramarital affairs are the units of analysis. Forty-five Chinese cases of affairs come from reports occurring in Taiwanese newspapers from 1987 to 1994 (including affairs reported in advice columns, commentaries, features, etc.). Twenty-seven Western cases of affairs come from a series of interviews with women (who were either wives or third party in the events of extramarital affairs) reported by Leigh Cato in *Her Version* (1995).

For the Taiwanese data set, reports of extramarital affairs are taken from the newspaper clippings collected by the Women Awakening Foundation (*Fu-nu-hsin-chih*) in Taiwan. This foundation is an influential feminist organization founded in the early 1980s, which has been actively involved with numerous social reforms regarding Taiwanese women's issues. It lobbies to change legislation, publish documents and books, and mobilize women's movements. Its holdings of newspaper clippings (including news, columns, commentary, and articles) have been drawn from nine widely circulated newspapers under many subjects and indexes on women's issues.[5] These newspaper archives are open to all policy or academic users.

The coverage period of my Taiwanese newspaper data is from 1987 to the first half of 1994. My fieldwork was undertaken in Taiwan in June and July of 1994. The rationale for using 1987 as the starting line is that this is the year that the Marital Law was abandoned. This political liberalization has exposed even the taboo ideas of the past to Western and other globally dispersed ideas. The mass media have highlighted the many ways this further societal "liberalization" and openness have "speeded up" the modification of Chinese people's thinking about, and behavior toward, sexuality and marriage. I photocopied all of the clippings under the subject name of "extramarital affairs" from 1987 to June 1994. Since the unit of analysis in this study is each case of an affair being reported, the first step of "cleaning" the data was to screen all of the reports to see if each contained a reference to a "real affair" and a complete description of the whole process of such an event. Forty-five cases showing the complete process of affairs (from the beginnning to the end) were thus obtained.

The gender of the writers who reported such events is expected to be associated with their way of reporting about such events. In this data set there is not enough information to exactly identify the gender of the "reporters," because they (as well as the commentators or the writers of features) tend to use gender-neutral pen names. However, theoretically I would like to argue that most of the cases of affairs were likely to be reported by women. Regarding advice columns, most of the "writers" reporting such events (and problems) to the advice columns are women and, very often, the content itself reveals the gender (such as "my husband is having an affair . . ."). Most "advisers" on such so-called women's topics are also women. This same speculation can be applied to the authors of features or commentaries on such "soft" topics of marriage and family, who are also more likely to be women.

There are two advantages of using the newspaper clippings carried by this feminist organization (rather than by nonfeminist or other more "conventional" or male-dominant organizations) as the Taiwanese data source for this study. First, since women's issues are their focus, they can be expected to "clip" the relevant information more thoroughly. Second, the ideology of the group provides a particularly sensitive lens through which relevant clippings might be spotted and selected. Standard news clipping services or newsrooms may not have the necessary sensitivity to spot relevant clippings. In addition, news items that might be classified under some other topics in the standard news clipping may not then appear under the heading "extramarital affairs." Thus, the clippings of this organization are likely to "catch" most of the news and columns that fall under this heading. Their sensibilities are going to be particularly attuned to the rare odd cases that are especially "progressive" (in their terms) or especially patriarchal. In this way, there is likely to be a bias toward oversampling the extremes. But within this framework of isolating types of affairs, this kind of bias is an improvement of coverage from the point of view of the analytical inductions to be developed. The logic of the sampling is therefore theoretical rather than enumerative (Turner 1953; Barton and Lazarsfeld 1955; Glaser and Strauss 1967).

These same advantages justified for the aforementioned Taiwanese data can also be applied to the Western data set in this study. For the Western data, 27 real events of extramarital affairs in Leigh Cato's book *Her Version* (1995) were reported by women. These women, across all backgrounds, mostly on three continents (i.e., North America, Europe, and Australia), were interviewed by Cato over a period of ten years preceding the book's publication.[6] Cato is a woman who has, herself, married three times. She is concerned with these stories of ex-wives and the "other women" as they address the question, "What does it say about the status of women in our society, despite the advances of feminism in this century?" (p. 236). Also, Cato is fascinated by other people's lives, as she notes: "I have found that real life is richer and more complex than most fiction" (p. 1). Therefore, the reports in this book are real-life events presented by an author who is concerned with women's issues, which are comparable to the Taiwanese events of affairs (mainly reported and written by women) taken from the newspaper clippings carried by a feminist organization.

There is, however, a limitation in this Western data set: in all the reports the husband (or the man) was mainly the "initiator"—either the events were reported from the wives or from the third-party women. (There are six cases in which the other woman herself was also married.) The implication of this limitation is discussed further when I present the results of data analysis.

Coding, Data Analysis, and Generating Topologies

Through a comparison of affairs in each data set, a number of generic components or elements become visible. For instance, each "affair" has to refer to a married couple and a third party. The participants in the affair bring specific orientations and resources. After reading through all of the reports, I developed a coding scheme regarding extramarital affairs for each data set. This scheme codes the components of affairs, which include location; duration; demographic features of the first party, the second party, and the third party; how the affair happened; the first party's marital condition; the first party's "explanations" for the affair"; the third party's "explanations"; who was to blame where the spouse found out; and the outcome of the affair revealed in each report.

These components are relatively easy characteristics to notice and code. I first randomly chose ten reports to assess intercoder reliability. The other coder and I were consistent in our coding. After coding all the reports in terms of components, two major underlying variables regarding the characteristics of the participants and their situations could be discerned from the coded data. The combination of the categories of the two variables can determine the types of affairs. This set of different types of affairs—realized and unrealized—can be arranged into a theoretical typology. Some of the entries in this typology may not be "realized" in any reports, but this absence is as much a matter of investigation as those that do appear.[7] For each data set, a theoretical typology of four extramarital affairs that define plots, conditions, and outcomes can be generated. To assure the validity of these typologies generated, I plugged all the cases back into the typologies and they did fit into certain types of affairs in these typologies. Comparison and contrast can be made between the two typologies.

DIFFERENTIAL GENDER ALLUREMENTS TO INFIDELITY

In most cases in the data, affairs were introduced in terms of the kinds of persons who were involved. Participants can be characterized by their propensity to engage in affairs as depicted in the reports. There were specific situations and mechanisms involved in the whole process. Across all the reported affairs, two underlying variables characterizing the motive forces and situations can be discerned from each data set. One variable concerns the characteristics of the woman; the other variable concerns the man's condition. In this section, I shall compare and contrast these two variables underlying all the affairs between the two cultures.

The Woman Participant: Traditional versus Modern

This variable concerns the female participant in the affair. The woman was depicted in terms of her self-definitions ("modern" versus "traditional"). In a contemporary Taiwanese context, people perceive women with a "modern woman" ideology as highly educated, career oriented, financially independent, self-determined, and not necessarily focused on marriage as an ultimate goal. Meanwhile, individual women also use such terms to define themselves. In this paper, I refer to Taiwanese women with such an ideology as modern women, while I term the women without such an ideology traditional women.

In a Western context, two types of women, also characterized as being traditional and modern, were similarly depicted in the reports of affairs. Some aspects of the definition of this variable are similar to the Taiwanese ones, but some aspects are different. The modern women in the Western reports were depicted as having a career or some money and being more "calculating"—for instance, helping a man to set up a business or develop a career. Such women had autonomous minds and strong personalities. In contrast, the traditional women in the Western reports were seen as non-career oriented in general. They were often depicted as "dumb blondes" or even occasionally as "sluts" (particularly by the aggrieved wife). Such a woman offered good sex or comfort for a man who had become bored or marginalized by his wife—according to the reports from the third-party women.

The Man's Condition

This variable concerns the male participant in the affair, particularly his appeal to the woman. In the contemporary Taiwanese context, the conditions of the men who participated in the affairs can be defined according to the kind of support the male participant in the affair was depicted as being able to provide to the female participant. One kind of support was depicted as the man being rich or of high standing, someone who could provide material support for the woman. In this case, the lure of the man is seen in terms of his standing and command of resources (and respect). The other kind of support was depicted not as being materially powerful but as being spiritually supportive in terms of providing emotional support, mutual understanding, and sharing common interests in a career. Here, the lure is the man's "understanding" of the woman's needs or situation.

In Western reports, as previously mentioned, my data set mainly contains the reports with the husbands as the initiators of the affairs, so the "male participants" here would simply be referred to as married men. The key variable that marks the variety among the conditions of these men is the extent of their control over their own spouses. In some reports, the men were described as being in control of their domestic situation (or at least were quoted as if they were); their spouse did everything right to be a good wife and mother, but the husband took her for granted. Despite this, the man was depicted as being bored and in need of something more exciting. In contrast, the condition of having a status of "not controlling the spouse" was depicted as a marital relation in which the wife was a career woman, or was depicted as opinionated, demanding, or making the husband feel marginalized or tired. The thing missing in such a man's life was centrality or comfort.

In other words, in the Taiwanese context, a male participant in an affair was linked to the kind of resource he had and was able to offer, whereas in the Western context, the male participant's propensity of entering into an affair was triggered by something that was lacking in his marital relation.

TYPOLOGIES: PROSPECTS AND POSSIBILITIES IN MARITAL INFIDELITY

These two underlying variables discerned from all the reports deal with the two participants in affairs. Because the two variables define orientations of both genders who participated in affairs in the data, it becomes appropriate to cross these two variables to generate a set of types of affairs. In other words, these two variables define a typology of the possible modes of encounter. For each data set, a typology of four types of extramarital affairs can be isolated. However, the types are different, particularly given the different conditions that apply to men, for the two cultural settings.

Typology of the Chinese Affairs

A typology of extramarital affairs in the Chinese context which can be generated from these two underlying variables is presented in Figure 1. The diagram can be read as follows: Each gender is depicted in terms of a basic orientation in the affair. Men can provide either material or emotional support (vertical variable in diagram). Women can be defined in terms of traditional or modern self-definitions (horizontal variable in diagram). The result of this "encounter" are four types of affair.

Man depicted as: / Woman depicted as:	Traditional	Modern
Materially Supporting Third Party	*Savior*	*Manipulative Women*
Spiritually Supporting Third Party	*Unhappy Housewife*	*Soul Mates*

Figure 1. Typology of Depicted Affairs Key Players—The Chinese Case

From this typology, the unfolding of the various types of affairs is itself scripted—like a play or drama. In other words, the types of affairs isolated from all the Taiwanese reports can actually be seen as various "scripts." In each script there is a specific kind of plot, a key player who drives the action, and a specific conclusion to the process. Each type of script has an autonomy of its own. Once the affair starts along the path set for each script, it does not move to another. One cannot find or envisage an affair that reveals a "savior" role which then changes into a "soul-mate" role for instance.

Chinese Script 1: Traditional Masculine Savior

This first type, which I call the "savior," has the man depicted as materially powerful and the third-party woman as traditional. The form of the affair is depicted as one in which the man uses the explanation or excuse of "saving the woman." The reason why such an affair can work is that the woman's traditionalism is depicted as a lure (to the man) and a weakness (for the woman). The "powerful" man can be depicted as portraying himself as a savior. Conversely, the affair can be seen by the man as fitting a traditional setting in terms of enjoying the privilege of having a mistress in the postconcubine era.

The affair had the following plot. First, the man had power, money, or the opportunity to provide material support and to be a paternal figure for a subordinate (usually in a work context). The subordinate was depicted as young, pretty, and often a virgin. Second, the young woman had a poor family background and an innocent disposition. Third, there were "excuses" from this man about a weak marriage, a wife being bossy, or tough, or uncultured. Such accounts were shown as provoking compassion and enticing the young subordinate to "listen." Fourth, divorce was revealed to be a prospect, but no definite deadline was given. Fifth, the young woman felt "honored" by the attention and rapport and compassionate about the situation, which together provided the context for the women to "fall in love." The prospect of marriage gave the love a purpose and a justification.

The next part concerns the "exposure" of the affair. Usually the women provoked exposure: either the young woman began to question the deadline for marriage—perhaps owing to pregnancy—or the wife found out and demanded resolution. At this moment, there were three kinds of blame for this affair: First, the wife blamed the third party for stealing her husband. Second, the husband blamed the wife for not being able to share his pressure from work with her. Third, the third party blamed herself for being so stupid to trust the man's "promises."

The finale concerns outcomes. First, the wife fought and humiliated the third party or the two women fought. The wife awaited the return of husband. Second, the husband, in some contexts with the assistance of the wife, found a mechanism to get rid of this young person. Normally, the husband explained that his wife did not want a divorce, and gave the third party money and asked her to leave. Third, the third party felt heartbroken, cheated, humiliated, and regretful. Fourth, in some cases the wife was shown to develop mental problems (is depressed, neurotic, insecure, suspicious) after the event.[8] It is significant that in this final stage, the conflict was between the women rather than between the wife and the husband.

Chinese Script 2: Soul Mates

In the second type, termed "soul mates," the opposite characteristics of the actors are involved. The man is depicted as not using material power as a lure. Instead, this man is depicted as genuine, innocent, and honest about having a good marriage. The third-party woman is depicted as nontraditional, having a career, and not requiring marriage as a prospect.

The outline of the plot is as follows: First, in this script the man did not have power or money, but had the opportunity to be a work mate with a unmarried woman who was mature, intellectual, career oriented, and depicted as a "modern woman." Second, there was no "excuse" needed from the man in this affair. He told the woman that he had a fine marriage, divorce was not envisaged, the relationship with his wife was not exciting, but was harmonious. Nevertheless, he was interested in the work mate's concerns and aspirations, and was responsive (say, to her sense of humor.) Third, the woman was depicted as seeking romance and as enjoying the "spiritual" support from the man, did not want money, and did not care about marriage. She did not need any excuse or promise from the man. The affair occurred for whatever reasons—office politics created embattlements, or similar interests flourished, or an intensive project work brought them together. The affair took on a life of its own. She promised not to break up his marriage. Fourth, there was no future plan, but both enjoyed the romance and treasured the feeling of being a "free spirit"—with no feeling of responsibility or duty.

Then, there is the episode that reveals how the wife found out. In the process of resolving this situation, the question of blame would come to the fore. First, the wife blamed the third party. Second, the man might partly blame his wife—he could not share with her most of the intellectual interests and ideas regarding his career. He might also blame the modern woman for initiating the affair. Third, the modern woman did not have anyone to blame, and could be depicted as uninterested ("easy come easy go") or as lost ("what does this all mean to me?") or tragic ("the lot of women").

Finally, there is the episode that reveals the outcome. First, the wife humiliated or sued the third party. These stories were more likely to have the wife take on more aggressive retribution roles that in the first script. Sometimes, the third-party woman had to pay a "fine" to the wife, because the feminist had a career and money. Second, the man found a mechanism by which to quit the relationship. It was not very difficult to quit, because he did not promise her anything in the first place. Third, the third party felt heartbroken and empty, wondering what this romance meant. Fourth, the wife might be shown as sliding into mental problems (depression, neuroticism, insecurity, suspiciousness) since the event.

Chinese Script 3: Unhappy Housewife

The preceding two scripts had the husband engaging in the extramarital affair. This next script concerns the wife who is depicted as the key player who "caused" the affair. The cause was primarily lodged in her "situation" (of unhappiness) rather than in any "drives" she had toward promiscuity or an affair. This was a less common scenario in Taiwan, but its incidence had been increasing these recent years, and this had caused crises for some men.[9] These crises, in which the wife had an extramarital affair, were much more intolerable for men than the similar crises were for women: "wearing the green hat" is most humiliating for a Chinese man.

The plot goes something like this: There was an unhappy housewife. The husband took her for granted or mistreated her; she was not cared for, or satisfied by, her husband. She felt lonely. She did not have many opportunities to meet men by herself. She did meet men who were friends or work mates of her husband. The reports of this type of affair elaborate scenarios in which another man shows care and concern for her. She was shown to be attracted to him or even willing to give up this marriage for him. For her this was more a psychological need than a sexual need.

Then, there was an episode dealing with the husband's reaction when he found out. The husband was shocked and very angry. He felt he had lost face—the worst thing to happen to a Chinese man. He was completely emotionally out of control; he might beat up the man or his wife, or both. There was also another episode showing the wife as "brave" and blaming the husband for not treating her well in marriage. The finale was that the angry husband sued both the wife and the third party. The husband divorced the wife. The wife wanted to marry the third party, but in a tragic twist, these affairs then had the third-party man not wanting her and leaving her. Her life was ruined. She felt heartbroken and ended up with nothing but a bad reputation.[10]

Chinese Script 4: Missing Script

The question arises as to what a type would look like that has the man depicted as materially powerful and the woman depicted as modern. The type can be tied to a modern woman (regardless of marital status) who is shown as taking advantage of a powerful man to achieve more for herself and may be seen as being "manipulative." The gender politics revealed in this type seem very real, but this type is not revealed in any of my newspaper reports.

The plausible explanations for the absence of this type may be related to people's perceptions of modern women in Taiwan. Chinese people perceive and can imagine the possibility that a powerful man can be, or can be willing to be, taken advantage of by a poor woman, because such manipulation still reflects men's masculinity; for instance, as when a big man looks after a poor woman. Male writers in the newspaper sources may not like to think about the possibility of a scenario in which such a powerful man can be manipulated by a modern woman. Even if such women are perceived as intelligent, less feminine, capable, and sometimes threatening, they are still women. Female writers may not want to depict such a "negative" image for modern women nor "admit" such cases exist, because they tend to depict modern women (in particular, feminists) as conscientious, pro-justice, and supercapable.

Chinese Trends

Many reported affairs in my data set followed the savior type.[11] The relative frequency of each of the types of affairs in this theoretical typology can be expected to change. The main change in emphasis can be expected to move in a downward direction along the diagonal. The opportunities for personal relationships emerging from work between the sexes can be expected to grow. The opportunities for "romance" may be growing along with an increased significance of romance as an important aspect of relationships. By contrast, the implied traditionalism of young women believing the promises of powerful men can be expected to decline. Either way, one can expect greater interest to focus on the newer form of soul-mates type than on the traditional type (savior). However with this move there are subsidiary moves. The traditional women (as housewife or daughter) may succumb to the nonrich men who can provide spiritual support. Conversely, the modern women may be shown to make inroads into powerful or rich men by "playing them," though this type has yet to find much expression in these current materials.

Typology of the Western Affairs

The Western-reported affairs can also be framed within a typology that defines the possible modes of encounter. This typology is presented in Figure 2. The diagram can be read as follows. Two variables define this typology. For men, their conditions can be defined in terms of whether they can control the spouse in marriage (vertical variable in diagram). For women, they can be defined in terms of being traditional or modern (horizontal variable in diagram). The result of this encounter are four types of affairs that reveal various "push–pull" mechanisms in marital and extramarital relations.

In the Western context, the various types of affairs presented in Figure 2 do not accompany various corresponding "scripts." The trajectory of the affair is less constrained by the social structure and so lacks the structured variety of scripts that apply to the Chinese setting. Across all the Western reported affairs, no such script can be isolated. However, there is a persistent "theme"—sexually driven motive and consequence—that comes through, regardless of what types of extramarital affairs are involved.

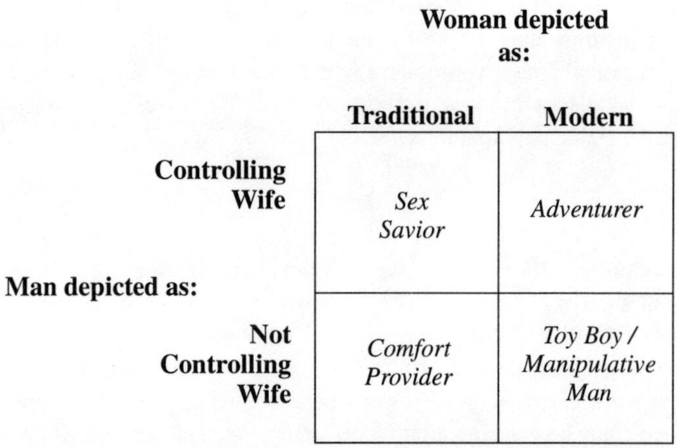

Figure 2. Typology of Depicted Affairs and Key Players—The Western Case

Western Type 1: Sex Saviors

The first type, of "sex saviors," had the man depicted as being able to control his spouse in a marital relation, and the third-party woman depicted as traditional. The condition of control over the spouse made that spouse appear boring and, as a result, made the man feel bored in marriage, or even in his life in general. This condition of "feeling bored" pushed this man away. The "pulling" force usually came from a third-party noncareer woman, who was often stereotyped as a "dumb blonde." She tried many ways to provide good sex to this man. The "ways" were depicted as wearing sexy lingerie, putting on makeup before he woke up, initiating sex in an unusual setting (for example, inside the phone booth), and so on. They were the best lovers for each other. In the reports, these women expressed pride at being able to "rescue" this unhappy man and bring excitement into his life. In this case, the man was the initiator of the affair, but the woman actually played an active role in the relationship in terms of acting as a "sex savior" for this man whose life lacks excitement.

Western Type 2: Comfort Provider

The second type, of "comfort provider," had a man who was less able to control his wife in marriage, and a third-party woman who was traditional. The wife was depicted as career-oriented, demanding, tough, or perhaps even a perfectionist. The man felt marginal or tired in his marriage. This marginal or tiresome condition pushed this man away. The "pull" force tended to come from a a young woman or a noncareer traditional woman. Here, the third-party woman could make the marginal man feel central and "in control" by being obliging. Or, she brought the "tired" man a lot of comfort by looking after him, happily doing all the domestic chores, and being available for sex anytime after he came back from work. Being with such a woman, this man could become dominant or feel relaxed, or both. In other words, this traditional woman was a comfort provider for this tired man. Compared to the sex savior, the depictions here have the third-party women being more passive and obliging.

Western Type 3: Toy Boy/Manipulation

The third type, of "toy boy/manipulation," had a man who also was depicted (in the accounts of the third party) as having had difficulty controlling his wife in marriage, but found himself with another woman who was also strong and modern. This other woman was depicted as autonomous, assertive, calculating, and having a career or money. In this type of "push–pull" situation, a man was pushed away from a demanding or powerful wife and pulled toward another strong and calculating woman. The encounter of these two types of people had generated two sorts of relationships revealed in reported affairs. In some cases, the man was depicted as being "weak" in terms of being indecisive and caught in between two strong women. The way of the third-party woman dealing with such an indecisive man was to be persistent and set up strategies to intervene in this man's marital life until his wife could not stand it and chose to give up her marriage. This man was controllable and could be like a "toy" to this strong woman. In other cases, the man was depicted as being manipulated by a powerful third-party woman's money. The woman would set up a business plan, or even give money to this good-looking big "dreamer" to assist his business ventures. At the same time, the woman was usually aware of, and alert to, the possibility that this man was using her. When this man's business did not take off, the man was suspected of being manipulative and the woman tried to "cut her losses."

Western Type 4: Adventurer

The fourth type, of "adventurer," has a man who has a controllable spouse and feels bored in marriage, and a modern woman who is autonomous and competent. The affair constructed between these two people could be very adventurous, such as a one-night stand at a conference, or a "self-contained" romance in terms of the woman not being interested in having a marital commitment and the man not intending to up his marriage. This type of affair is actually absent from all the Western reports in this study. The plausible explanation for this "missing" type may be related to the "selectivity" bias in the Western data set. All the reports included in the book *Her Version* came from the interviews with women who had been left and women who had been left for, so all the marriages between a wife and a "stray" husband must have been broken. The adventurous type of affair, which does not mean to break up any marriage, would not occur in this book, because such an event and the people involved would have always been hidden in the reality.

Western Trends

The majority of the affairs in the data set followed the traditional types of sex savior and comfort provider. The relative frequency of the four types of affairs in Figure 2 can also be expected to change. The irreversible trend toward gender egalitarianism would lead more women out of the traditional mode (regardless of being a wife or a third-party woman). More men will live in an egalitarian relationship in terms of "not controlling spouse." As a result, in an extramarital relationship, fewer women would like to be in a position of only playing a "provider" role—providing sex or comfort to men. More women would like to have a position of being able to assert, negotiate, and critically assess what they can get in such a relationship. Therefore, the newer form of toy boy/manipulation type can be expected to increase.

Major Western Theme: The Centrality of Sexuality to Personal Development

Regardless of the types of extramarital affairs, all the Western reported cases unfolded along the following lines. First, the man met this woman. Second, the woman felt that this man was the Mr. Right and the quality of the sex with this man validated this impression. Third, there were "excuses" from this man about a dead marriage, a wife being boring, not understanding, and not good in bed. Such accounts are shown as provoking compassion and rapport, and enticing the woman to "rescue" him. Fourth, the "best sex" or "always ending up in bed" was perceived by both parties as indicating "we were meant to be together" shortly after the affairs started. Fourth, divorce was to be a prospect.

The next part concerns the "exposure" and the consequence of the affair. Usually the wife found out and felt shocked. She felt that she had always trusted her husband. The marriage had been quite normal, even if not exactly exciting. Their sex life was okay, given they had been married for so long. The man in this phase first tried to deny the affair, then got angry. Usually after a few weeks, he moved out without explanations and stayed with the other woman. The wife was heartbroken and felt that she was not deserving of this treatment. She felt that she had done everything she could to be a good wife and mother. She could not help asking "what did I do wrong?" In some cases, the wife would try to warn the third-party woman of how bad this man really was, how untrustworthy, and so forth. At this moment, the third-party woman thought the wife behaved too nastily and that "doing this only pushes us closer." Normally she did not feel guilty about breaking up someone's family, because "if not me, it will be someone else." Instead, she felt proud of herself for being able to win this battle, bringing happiness to this man (because his wife failed to do so), and felt lucky to have this wonderful lover.

Regarding the final outcomes, they varied and depended on individual choices (or no choice), regardless of the kind of situation that caused the affair in the first place. There were two kinds of outcomes for the wives. On the one hand, the wife would hire the meanest lawyer in town and get the best settlement possible. After a while, she recovered from the anger and hurt. She felt the new life was actually better than before—having freedom, not having to compromise for anyone, being able to have lovers if she wanted or chose to, but not bothered to be married again. Some of the wives happily remarried—to a new husband who was so much nicer and solid, and so forth. On the other hand, in some reports the wife and children were abandoned and never received any maintenance from the husband. The wife had to struggle to make a living by herself—and there were no more men in her life. She could not afford to hire babysitters or go out to meet men. She did not want any man anymore, because men would "fool around."

There were two kinds of outcomes for the new couple. One was that the new couple were "happy ever after." In particular, given that the cost of the whole event was so high and the man had to pay a big settlement to his ex-wife and children, they really treasured this relationship and sex had always been and was still good after paying such a high price. The other outcome was messier. The new couple became "happy," but not "ever after." The man "repeated" the history again by leaving for another woman. At that moment, the new ex/old "other woman" really "felt for the ex-wife" and the ex-wife felt sorry for this other woman. These two women built up a bond because of this common "dick-driven" man (as some put it in Cato's book) in their lives. This other woman, after being hurt, either had to struggle to build a new life by herself, or found a new lover with a stronger emphasis on sexual fun in the future as a result of her disillusionment with having a permanent relationship.

Therefore, by contrast to the three Chinese scripts that were dramatized in the culture, and shared between individuals and among the society in general, no distinct script was revealed in the Western-reported affairs. The affairs in the West were depicted as more chaotic. However, there exists an underlying major theme about the centrality of sexuality for both men and women.

GENDER POLITICS AND INFIDELITY

Salvation and Resources

In both typologies, a savior role exists, but who the savior was varies between the two contexts. The contrast reflects the different gender roles in two cultures. In the Taiwanese scripts, a "resourceful" married man "rescued" a third-party woman by providing either material or spiritual support. In contrast, in Western-reported affairs, a third-party woman "rescued" a married man (who is either "bored" or "tired" in his marriage) by providing him good sex or comfort.

On the one hand, this contrast suggests that Western women's role seem to have been "progressive" from a passive and dependent (on men) role to that of a strong figure who could rescue a married man out of a miserable marriage. Western married men were depicted as "bored," "tired," or marginal. In contrast, Taiwanese men were depicted as strong and resourceful, while Taiwanese women were depicted as being "saved" or needing men's support.

On the other hand, this contrast reveals more progressive gender relations in Taiwan than in the West. Although in the scripts the "resources" a Taiwanese woman received from a man normally were the traditional material goods, a new type of resource emerged that was spiritual or emotional support from a poor man. This man (who could *only* supply spiritual or emotional support) would never have been "qualified" as a person owning concubines or extra wives, as many rich Chinese men did in the past. Today, the progressive modern Taiwanese women have "upgraded" the status of this kind of man as long as these men can provide spiritual, intellectual, or emotional company. In contrast, in reports the "resource" for a Western woman to "rescue" a married man mostly remained traditional—either being sexy (providing good sex for a bored man) or being obliging (providing "comfort" for a tired man).

Who Is to Blame

For marital infidelity, the Taiwanese wife tended to wait for the husband back home—everything was tolerable for a wife because she lived *for* the family, beyond an individualistic concern. Who is to blame? In Taiwanese scripts, men were never at fault. The only "fault" revealed in the reports may be the "nature" of men being more promiscuous. Therefore, a man's infidelity would not be as fatal to the marital relation or family system. Furthermore, if a women assumed that this "man's nature" was universal to all men, this assumption would only put women in a panic. They would feel vulnerable to the real possibility of their husbands engaging in extramarital affairs. The only object a woman in such circumstances could blame would be another woman rather than her husband. Therefore, all of the reports depicted women as perceiving other women as threatening—the feminists in particular were depicted as most threatening to the wives.

In contrast, in the Western reports, that man was the one being blamed by the wife. The wife could not accept the treacherous man again, because this was simply portrayed as intolerable for a Western woman from an individualistic perspective. In such circumstance, the wife would normally warn the other woman about the man's true nature. Sometimes the two women got together to attack the man for creating the mess or repeating the history. Some reports even depicted a situation in which both ex-wife and the other woman lived together to care for each other. Their emotional bond (or sisterhood) was established owing to this common man.

The Loser

In such a battle, who is the loser? In Taiwanese scripts, the ultimate principle to justify or assess the situation was familism. Thus, a marriage (actually a family) would normally not be broken by an extramarital affair. As a result, the third-party woman would be the loser. In contrast, a Western theme across all the stories suggests that "good sex" was a necessary and sufficient condition for such a big decision—a married man would leave his wife and children for a new (and good) sexual relationship with less regret. A woman, for the same sexually driven motive, could be depicted and refer to herself as taking another woman's husband without feeling guilty. Extramarital affairs were "justifiable" as long as this new couple was happy and the old marriage was (sexually) dead. In such context, the wife would be the loser.

In addition, the wife's role in the West was depicted as more vulnerable than the Chinese wife's role. When the wife was controllable, the man talked about being bored. When the wife was not controllable, the man was depicted as tired. Either kind of wife's role could push the husband away, even if she was a good wife and mother. The significant feature of this depiction is that the role of wife was seen as fundamentally threatened. In contrast, in my data the Chinese wife's role, as well as the position of being a wife, was depicted as more stable regardless of whether the husband had "resources" to offer to other women. Men normally tended not to leave their wives (actually their families) for another women, as long as their wives had played well the roles of good wife and mother. In other words, the Chinese familism secured the wife's role (although the wife might feel threatened by the other woman).

DISCUSSION

The two typologies developed from the reported extramarital affairs show similarities and differences between the Chinese and the Western contexts. Moreover these differences point to social change dynamics: the incidence or propensities of the structure and contents of certain types of the affairs. Given the irreversible trends toward gender equality in both structural and ideological senses, certain types of extramarital affairs can be theoretically expected to increase whereas other types of affairs can be expected to decrease. Other "missing" types that have not been occurring in my data are predicted to appear as a result of social change. Beyond these cross-cultural commonalities of pattern and linkage (both empirically and theoretically) between boundary breaking, values reinforcing, and social change regarding extramarital affairs, there are specific cultural meanings, mechanisms, and politics that merit attention. In particular, differential roles played by values of familism and individualism affect the meaning of love, sex, marriage, family, fidelity, and infidelity in each context. These can now be addressed.

Meanings of Romantic Love and Sexual Ethics

In both cultural contexts, romantic love features prominently. Within the Western context, romantic love is more central and controlling in all the reported affairs. Romantic involvement and sexual validation is the path to happiness, personal development, and a sense of identity. Within marriage, the expressive nature is emphasized and sexuality is necessary for intimacy and individual members' well-being. In this circumstance, sexuality is not only permissible, but obligatory (Edwards and Booth 1994, p. 233). Passion can fade, and consequently sexual life can also wane. This change is often taken as "personal"—being a bad situation or a personal failure. The women in this context see themselves as offering or having offered love. If it disappears, this is an occasion for loss, regret, self-doubt, resignation, or anger. Some move on to the next relationship—become alive again—while some become cynical about so-called "love" from men. Owing to the fact that the data in terms of my Western reports are limited only to women's narratives, the subjective feelings and meanings of romantic love of Western men cannot be directly discerned from the data. The only hint that can be indirectly derived from women's perceptions in these reports is that, for men, romance or "love" is mainly sexually driven.

The sexual ethics revealed in the common theme in my Western reports of extramarital affairs is more toward sexual libertarianism (rather than sexual romanticism). Sex is an act of self-expression, pleasure, or a sign of affection or love. Individuals ultimately have autonomy and are free to pursue their wants—particularly presented in the form and content of sexual pleasure, which often started in a casual relationship, or sometimes the man "repeats the history" in terms of constantly sleeping around. This may be a result of the selectivity bias of my data, which were based on reports in which men left their wives for other women. In other words, in this data set all the marriages were broken by extramarital affairs. These were "libertarian" types of behaviors in the first place. The spirit of sexual romanticism would be shown if there were reports about affair participants going back to their own marriages.

Within the Chinese context, of the straying married men, the ideal of romantic love may motivate the affair, but does not control it. Rather, a wider set of forces come into play. Romantic passion is seen as momentary and doomed to fade, and is not comparable to the importance of the familism ideal.

Sex is treated as a natural urge. For Chinese men, if the urge overflows its outlets, it means pressure was relieved and some "clean up" of consequences is required. The "extra" sexuality also brings Chinese men status symbols—"extra" wealth and power for materially resourceful men, and extra charm for "spiritually" resourceful men. This "deviance" is actually quite "normal," which mirrors the traditional value on Chinese male sexuality and one aspect of masculinity. For Chinese wives, the "objective" or culturally defined meaning of sexuality is more tied to familial responsibilities and gender roles such as procreation (the most important one) and pleasing the spouses. For women who are the first party or the third party in an extramarital affair, the meaning of sexuality is more tied to romantic love and emotional support than to physical satisfaction. In sum, reports of extramarital romance in Taiwan reveal an integration of the Chinese sociocultural legacy and the Western romantic love ideal. However, the concerns, consequences, and outcomes of extramarital affairs remain resolutely Chinese—familism remains the ultimate value for everyone in the circumstance in which a husband does stray, and the sexual ethnic is still only defined by the fidelity virtue of women.

Meanings of Infidelity

In the legacy of concubinage and prostitution, Chinese men's needs are given wider scope, but less significance. At worst, infidelity of Chinese men is a detour. He will return and is normally "welcomed" back home by his wife. The whole affair may have provoked both shame and guilt. Shame is brought upon the husband, the wife, and potentially their families. Shame provides a powerful incentive or requirement to simply bury the affair within, as it were, the garden of familial responsibilities. Therefore, the infidelity of men basically reinforces the conventional values and solidifies marriage boundaries. For Chinese women, infidelity is not a detour, but simply brings permanent damage—divorce. This dire consequence of female infidelity on marriage also reflects the persistence of, and has reinforced, the traditional values of gender roles, female virtues, and patriarchal familism. In other words, there are gender-specific meanings of infidelity in the Chinese context, which permeate the scripting of all the reports of extramarital affairs.

All the Western-reported affairs reveal that infidelity is less of a detour and more of an intersection. It is fraught with all the intensity of big decisions: which way to turn? The eruptions of extramarital affairs are treated with far more individualistic concerns and consequences. This is why no scripts exist among all the reported extramarital affairs in my Western data. Owing to the centrality of sexuality in affair participants' lives and the intolerance of the spouse from his or her individualistic concern, the prospects of return are less clear than the Chinese ones. However, the common prospect among all such individualistic events is that, regardless of who (husband or wife) is the first party or what the outcomes of such events are, affairs are less able to be buried in the Western context. Breaking the boundary often results in breaking up the marriage and family. In such an eventful process, the operation of shame is much reduced while the impact of guilt is attenuated, rationalized, and relocated onto many other objects. For example, in the statement, "She can't give him happiness, but I can!" Again, Western men's subjective feelings of guilt or shame are beyond my data, but in a few cases the wives or the third-party women mentioned that the men felt somewhat guilty initially, but their guilty feelings did not last very long. Overall, my data show the meanings of infidelity as less gender-specific in the West than in the Chinese context. This is consistent with, as well as related to, the argument that there is a trend of so-called masculinization of sex in the current of changing Western sexual mores which makes the gender gap smaller (Lawson 1988; Seidman 1992).

Constructions of Manhood/Womanhood and Gender Politics in Marital and Extramarital Relations

In the Chinese depictions, the force to push people into extramarital relations is contingent on whether a man has material or spiritual resources to be a savior for a woman who needs material or emotional support in her life in certain circumstance. The rule and format to end such an event is defined by the familism and the roles of being a husband/father and a wife/mother. As long as a married woman plays her wife and mother roles well the stray husband will not leave her and the children. He would quit the affair and come home to play his husband/father roles in the family. After all, individuals are not the concern except for their roles in the family. In other words, the cultural constructions of manhood and womanhood determined by the familial roles based on familism are clear, and define marital as well as extramarital relations. As a result of these primary familial roles and duties/responsibilities under the big picture of familism, there tends to be less room for playing, or being presented/represented by, gender politics in both marital and extramarital settings as compared to the Western context.

In contrast, in the Western-reported extramarital affairs, gender politics is a more significant feature and dynamic in both marital and extramarital relations. Based on the typology of affairs in this paper (when a married man is the initiator), the force that pushes a married man into a certain type of affair is determined by the gender politics in his marital relations in terms of whether he can control his wife. However, either kind of power structure in marriage presents some problems of being bored, tired, or marginal as an impetus for the man to seek a woman savior to provide him with some excitement, comfort, or centrality. The depictions of Western manhood and womanhood in marriage do present some structural problems for both married men and women. For a married man, regardless of whether he has power over his wife, something is missing in his marriage. For a woman, regardless of whether she is playing a traditional or modern role of being a wife or a mother, or both, or regardless of whether she is controlled by or controls her husband in a marital relation, either situation could push her husband away.

In such a context, neither men nor women have a more advantageous position. Married men are depicted as weak, bored, tired, marginal in marriage, and in need of rescue by a woman in an extramarital relation, while married women are depicted as playing an impossible role (always a loser) in the gender politics. In addition, in an extramarital relation, a man may be rescued by and depend on a traditional woman to fulfill his sexual and emotional needs. Or, a man could manipulate a powerful woman's resources while also constantly being watched by such a woman who is often calculating. Or, a man could be such a woman's toy. This kind of gender politics in the extramarital relations does not indicate a more advantageous position for men compared to the position of women involved.

Of course, there is a selectivity issue involved in this study inasmuch as all reports are about a married man who initiates the affair and leaves his wife and family for another woman. There are not enough cases in the data set to examine the circumstance, dynamics, and consequences when a married woman becomes involved in such an affair (rather than her husband). However, theoretically, I wish to argue that it is reasonable to imagine similar gender politics and the potential force to push a married woman into an extramarital relationship: either she controls her spouse in marriage, but feels bored, or she cannot control her spouse and feels tired and marginal. In other words, the power relation and politics in marital relations could create the same effect on both husbands and wives (given the cultural ideals of ultimately individualistic concerns and the centrality of sex for an individual's sense of self-worth and well being), which is a structural limitation in marriage and can be an underlying propensity for both genders to engage in extramarital relations.

The potential force or motive can be the same for both genders.[12] But the types of affairs (format, content, and gender politics within the extramarital relation) presented/represented when a married woman has an encounter with a man could be quite different from the four types of affairs presented before when a married man is the initiator, owing to the different gender roles in the culture. Therefore, this typology generated from reports of a married man being an initiator is partially reversible for the circumstance when a married woman is an initiator in terms of the linkage between a marital condition to the propensity for an extramarital affair.

Hence, the cross-cultural comparison of the reported extramarital affairs can not only reveal the boundary, norms, and values on marriage and family systems that is a core part of a specific culture, but also can show the relevance of the constructions of manhood and womanhood and the meaning of sexuality to marital infidelity in that culture. The Western-specific problematic manhood/womanhood construction and gender politics revealed in marital and extramarital relations can challenge the Western feminist view that "men are always in an advantageous position in any heterosexual relations." Moreover, in the Chinese culture, men and woman have given familial roles, and the emphasis of familism and familial structure remains significant in the contemporary Taiwanese context. The arguments about gender relations developed in the West might not be completely applicable in other cultural contexts unless we examine what the roles are and what the meanings of those roles are in a specific culture. The contentious ideological debates on sexual ethics (in terms of romanticism versus libertarianism) in the West also show little relevance to a context of Chinese sexuality. The cross-cultural comparison of reported marital infidelity has generated many insights for the important issues addressed in this study.

CONCLUSION

In talking or making exhortations about the importance of fidelity, the assumption is that it applies to the "other" in a marital relationship: one is to be faithful to the spouse. However, the preceding study and discussion suggest a deconstruction of fidelity is possible when viewed from the content of the reports of infidelity. In both cultural contexts, fidelity goes beyond the spouse: either fidelity extends to the depths of the self (in the West) or it extends to the depth of family responsibility (in Chinese culture). As a result, in the West, affairs threaten the marriage and yet release the self to desire new experiences. For both men and women an affair is an intersection point: whichever turn is taken is fraught with consequences. By contrast, in the Chinese context, affairs threaten individuals' esteem, which is bound to the family. Affairs ignite a momentary episode that is driven by the circumstance to revalidate the marriage, and then, with that validation completed, dissolve itself back into an intergenerational stream of familial expectations. The affair for men is more a detour. The patience and persuasion of the wronged wife and the wider kin bring the errant behavior back to the main road. However, for women who have initiated affairs, the road turns out to be, more likely, a dead end.

Therefore, for individuals who are "allowed" to have affairs or whose affairs are tolerated (i.e., both genders in the Western context and Chinese married men in my data), authenticity of self (West) versus responsibility of familial roles (Chinese) are the real objects of fidelity in each culture.

ACKNOWLEDGMENTS

This research was made possible by an Australian Research Council Grant. I wish to thank the Women Awakening Foundation (*Fu-nu-hsin-chih*) in Taipei for access to their extensive newspaper clippings. I also extend thanks to Richard Volpato for providing valuable insights about sexual mores from a comparative perspective and for commenting on this paper. An earlier version of this paper was presented at the Annual Cross-Cultural Research Conference, in Pittsburgh in February 1996.

NOTES

1. Salamon (1984) conducted interviews with the "kept women" in London, exploring the rationale and meanings of being a kept woman. However, kept women is only one category of extramarital relations involved.

2. Taiwan and the People's Republic of China can be seen as two "orthodox" Chinese societies (compared to Hong Kong, Singapore, or other overseas Chinese communities), but I chose Taiwan to represent a Chinese setting for this study. My reason is that, although Taiwan had been colonized by the Dutch for 40 years, and by Japan for 50 years, the Nationalist Government has been trying to preserve the Confucian values through cultural policies and the educational system since 1949. In contrast, in the People's Republic of China, the Confucian values had been under attack by the socialist ideology and political economy. In particular, the ten-year Cultural Revolution had been meant to destroy the Confucian Chineseness. Therefore, I argue that Taiwan can be used as a baseline for the study of social change and Chinese culture, as well as a baseline for the comparison of the "Chineseness" between Chinese societies.

3. This quotation about Trombach is obtained from Plummer (1995, p. 176).

4. Western feminists emphasize that masculinity in the West is defined in terms of aggression and sexual predatoriness, so the dominant sexual ideologies reflect and reinforce the sexual privileges and power of men. This "male-dominated heterosexuality" makes it more difficult for women to assert needs and desires (Elliott 1996) and construct heterosexual identity with the "male in the head" (Holland et al. 1996). Some radical feminists even insist that "heterosex *invariably* involves male domination." (This quotation about Dworkin is obtained from Duncombe and Marsden 1996, p. 223.)

5. These nine newspapers are *Chung-kuo-shih-pao, Lien-ho-pao, Li-pao, Tzu-li-tsao-pao, Tzu-li-wan-pao, Lien-ho-wan-pao, Chung-shih-wan-pao, Tzu-yu-shih-pao,* and *Min-sheng-apo*. All of these newspapers are privately owned and normally hold more liberal positions than the newspapers owned by the government or KMT party. *Chung-kuo-shih-pao* and *Lien-ho-pao* have the largest circulation among all of the newspapers in Taiwan (Tien 1989).

6. This is a statement based on my speculation. The author of this book does not inform readers exactly when these interviews were conducted. My speculation comes from this sentence, "The original concept of the book came over ten years ago when the marriage of one of my closest and dearest friends in Canada broke up. . ." (Cato 1995, p. 1).

7. This is the argument in the area of "sociology of the lack of knowledge" in which the exclusion of material is a matter of theoretical and empirical interest in the feminist research area (see Jane Haggis quoted in Reinharz 1992 p. 163).

8. A typical story with such a script can be seen in an article entitled "A Third Party Story" (*Tzu-li-wan-pao*, March 19, 1992).

9. News reported in *Min-sheng-pao*, August 4, 1989.

10. An example of such a script can be seen in a story in *Lien-ho-pao* (August 16, 1989).

11. I did not count the frequency for each type of affairs among the 47 cases included in the data analysis. This is because, as I mentioned before, the selection of these 47 cases from the Taiwanese newspaper clippings on extramarital affairs was based on one criteria—the case must contain a complete report of the whole process of the affair. Those reports that only mentioned part of the process were not selected for analysis. Therefore, methodologically it is not appropriate to conduct a frequency count for each type of affair out of these 47 cases selected as such. The main purpose of examining these 47 cases is to discern the *possibilities* of types of participants and resulting affairs.

12. Of course, even with the same potential force or motive, the real "force" may normally be stronger for married men than for women when they have young children. Young children would hold back women more than they do to men.

REFERENCES

Barton, A. H., and P. F. Lazarsfeld. 1955. "Some Functions of Qualitative Analysis in Social Research." *Frankfurte, Beitrage Zur Soziologie,* vol. 1.

Bender, D., and B. Leone (eds.) 1995. *Sexual Values: Opposing Viewpoints.* San Diego, CA: Greenhaven Press.

Cato, L. 1995. *Her Version.* Victoria: Australia Penguin Books Ltd.

Duncombe, J., and D. Marsden. 1996. "Whose Orgasm Is This Anyway? 'Sex Work' in Long-Term Heterosexual Couple Relationships." Pp. 220-238 in *Weeks*, edited by J. Holland and J. Holland. *Sexual Cultures: Communities, Values and Intimacy.* London: MacMillan Press, Ltd.

Durkheim, E. 1976. *The Elementary Forms of the Religious Life.* New York: The Free Press.

Dworkin, A. 1981. *Pornorgraphy: Men Possessing Women.* London: The Women's Press.

Edwards, J. N., and A. Booth. 1994. "Sexuality, Marriage, and Well-Being: The Middle Years." In *Sexuality across the Life Course.* Chicago: University of Chicago Press.

Elliott, F. 1996. *Gender, Family and Society.* MacMillan Press, Ltd.

Fisher, H. E. 1992. *Anatomy of Love: The Natural History of Monogamy, Adultery, and Divorce.* New York and London: W. W. Norton.

Glaser, B., and Strauss, A. 1967. *The Discovery of Grounded Theory: Strategies for Qualitative Research.* Chicago: Aldine.

Greeley, A. 1994. "Marital Infidelity." *Society* 31: 9-14.

Holland, J., C. Ramazanoglu, S. Sharpe, and R. Thomson. 1996. "Reputations: Journeying into Gendered Power Relations." Pp. 239-260 in *Weeks*, edited by J. Holland and J. Holland. *Sexual Cultures: Communities, Values and Intimacy.* London: MacMillan Press Ltd.

Hunt, M. 1969. *The Affairs: A Portrait of Extra-Marital Love in Contemporary America.* New York: World Publishing.

______. 1974. *Sexual Behavior in the 1970's.* New York: Dell.

Jeffreys, S. 1990. *Anticlimax: A Feminist Perspective on the Sexual Revolution.* London: The Women's Press.

Kinsey, A. C., W. B. Pomeroy, C. E. Martin, and P. H. Gebhard. 1953. *Sexual Behaviour in the Human Female*. Philadelphia: W. B. Saunders.

Kinsey, A. C., W. B. Pomeroy, and C. E. Martin. 1948. *Sexual Behaviour in the Human Male.* Philadelphia: W. B. Saunders.

Laumann, E. O., J. H. Gagnon, R. T. Michael, and S. Michaels. 1994. *The Social Organization of Sexuality: Sexual Practices in the United States.* Chicago: University of Chicago Press.

Lawson, A. 1988. *Adultery: An Analysis of Love and Betrayal.* New York: Basic Books.

MacKinnon, C. 1982. "Feminism, Marxism, Method and the State: An Agenda for Theory." *Signs* 7: 515-544.

Murphy, R. 1971. *The Dialectics of Social Life: Alarms and Excursions in Anthropological Theory.* London: Allen and Unwin.

Parsons, T. 1951. *The Social System.* London: Routledge.

Pietropinto, A., and J. Simenauer, 1977. *Beyond the Male Myth: What Women Want to Know about Men's Sexuality.* New York: Times Books.

Plummer, K. 1995. *Telling Sexual Stories.* London: Routeledge.

Reinharz, S. 1992. *Feminist Methods in Social Research.* New York: Oxford University Press.

Reiss, I., R. E. Anderson, and G. C. Sponaugle. 1980. "A Multivariate Model of the Determinants of Extramarital Sexual permissiveness." *Journal of Marriage and the Family* 42: 395-411.

Rich, A. 1980. "Compulsory Heterosexuality and Lesbian Existence." *Signs* 5: 631-660.

Richards, M., and J. Elliott. 1991. "Sex and Marriage in the 1960s and 1970s." In *Marriage, Domestic Life and Social Change: Writings for Jacqueline Burgoyne (1944–88)*, edited by D. Clark. London: Routledge.

Robinson, V. 1993. "Heterosexuality: Beginnings and Connections." Pp. 33-54 in *Heterosexuality: A Feminism: and Psychology Reader*, edited by S. Wilkinson and C. Kitzinger. London: Sage.

Salamon, E. 1984. *The Kept Woman: Mistresses in the '80.* London: Orbis Publishing.

Saunders, J. M., and J. N. Edwards. 1984. "Extramarital Sexuality: A Predictive Model of Permissive Attitudes." *Journal of Marriage and the Family* 46: 825–835.

Seidman, S. 1992. *Embattled Eros: Sexual Politics and Ethics in Contemporary America.* New York: Routledge.

Shaw, T. 1991. "Taiwan: Gangsters or Good Guys?" In *Deviance: Anthropological Perspectives*, edited M. Frelich, D. Reybeck and J. Savishinsky. New York: Bergin and Garvey.

Stacy, J. 1991. "Promoting Normality: Section 28 and the Regulation of Sexuality." Pp. 284-304 in *Off Center; Feminism and Cultural Studies*, edited by S. Franklin. London: Unwin Hyman.

Tien, H. M. (ed.). 1989. *The Great Tradition—Political and Social Change in the Republic of China.* Palo Alto, CA: Hoover Institute Press, Stanford University.

Turner, R. H. 1953. "The Quest for Universals in Sociological Research." *American Sociological Review* 18: 604-611.

Vance, C. 1984. *Pleasure and Danger: Exploring Female Sexuality.* London: Routledge and Kegan Paul.

PART III

COHABITATION AND ALTERNATIVE MARITAL FORMS

PART III

INTRODUCTION

We think of the marital relationship as only occurring within a religious, state-sanctioned union, when in truth there are a variety of "marital-like" relationships ranging from cohabitation to consensual unions that are neither state nor religiously sanctioned. "Nuptiality in Latin America: The View of a Sociologist and Family Demographer," by Susan De Vos, gives us a descriptive look at the increasing prevalence of "consensual unions"—those more fluid marriages without benefit of legal or religious sanction. While the frequency of cohabitation and nonmarital births has increased in industrial societies, consensual unions are different from these arrangements in many respects. This alternative marital form is critically important to our understanding of present-day marriage in that it is socioeconomically influenced, rather unstable, historically rooted, and increasingly common in Latin America.

Like De Vos, Lawrence Hong finds that long-held traditions are adapted in new and ingenuous ways in his "Marriage and Abortion in Japan: Adaptation and Alternative Strategies." Owing to such factors as the imbalance in sex ratio (i.e., many more males), delayed marriage by women, and high incidence of abortion, Japanese men have increasingly faced a marriage squeeze. In response, old customs have been adapted in new, alternative forms, such as *keekon joho sabisu* (marriage computer/information service), *pro nakodo* (professional go-between), and *Omiai deto* (look-meeting date). Marriage and marital forms are also adapted to bridge the gap between the rising incidence of abortion in Japan and the strong belief in fetal spirithood.

Wendy Carter addresses the issues of "Attitudes Toward Premarital Sex, Non-marital Childbearing, Cohabitation, and Marriage Among Blacks and Whites," examining whether race is important in determining attitudes toward various marriage and family issues, including cohabitation. Although many studies seeking to explain differences in behaviors of blacks and whites have employed culture of poverty–based explanations, this study found race to have little impact on attitudes toward these behaviors. In fact, age, gender, and marital status of respondent have greater effects on attitudes toward these behaviors than does race.

Suzanne Baker in "Marriages, Informal Unions, and Gender Ideologies in Revolutionary Rural Nicaragua" returns our attention to alternative marital forms in Central America by adding the influence of a profound ideological transformation following revolution. Part of this transformation was aimed at changing gender roles and identities. In addition, changes in socioeconomic forces provided a greater range of union choice. Ways of evaluating and justifying union choice were introduced.

"Cohabitation, Non-marital Births, and Traditionalism: Marriages in Rural Siberia From a Life Cycle Perspective," by Anna Mikheyeva, presents how Russia, as a newly industrializing nation both conforms to and deviates from patterns of the "second demographic transition," or falling births, then a rise in divorce, followed by a rise in cohabitation, and finally a rise in out-of-wedlock births. The rise in alternative marital forms, such as cohabitation and free unions, and the increase in extramarital fertility is interpreted in this context.

Finally, "Homosexual Relationships in the Netherlands," by Paul Vlaardingerbroek, extends our understanding of alternative marital forms to include those unions between homosexual individuals. Although industrial nations have recently given legal recognition of equal rights and benefits to men and women who are married and to unmarried couples, deriving workable solutions for homosexual couples with children has been much more problematic. Terms such as "registered partnership," joint parental authority, and adoption for homosexuals are reviewed giving a greater understanding of how "family" may be conceptualized in legal-structural terms.

NUPTIALITY IN LATIN AMERICA
THE VIEW OF A SOCIOLOGIST AND FAMILY DEMOGRAPHER

Susan De Vos

> Because changes in marriage patterns take place within the process of family formation, a major difficulty in assessing the future direction of marriage patterns in Latin America and the Caribbean is the slight knowledge of the dynamics of family formation in these regions. The origin of social acceptance of consensual and visiting unions and the undetermined conditions which have maintained such norms from the colonial era to the present have not been often addressed. . . . (United Nations 1990 p. 157)

INTRODUCTION

Even as we struggle to understand the role of marriage in contemporary American society, it is fruitful to consider union formation to our south in Latin America. Latin America shares some features with North America: It had an indigenous population that was overtaken by a European one; it experienced colonization over hundreds of years; parts depended on a slave and plantation economy while other parts had large urban labor forces; colonial status was overcome, but historical circumstances left a legacy of a mixture of races, ethnicities, and socioeconomic groups; and it is changing. As importantly however, Latin America differs significantly from the north. One difference is that North America was colonized primarily by the French and English whereas Latin America was colonized primarily by the Spanish and Portuguese of an Iberian Peninsula influenced by an intersection of Moorish and Catholic ideologies; thus, colonizers left different cultural legacies. Perhaps related to this is that one of Latin America's characteristics, despite the region's heterogeneity, is having an informal marital type known as "consensual union" in which a couple lives together with any children (but qualitatively different from the *barraganía* known in Iberia). To be sure, both cohabitation and non-marital childbearing are increasingly common in the United States and many other developed societies, but this is not the same as what has been a historically rooted but changing situation in Latin America.

The purpose of this paper is to try to understand, if only in a limited way, this system of two types of unions. After providing a short historical view of union formation in Latin America, I ty to understand the recent situation—the ambivalence toward union type, the timing of first union, and union dissolution. I also use census microfile data from a number of countries around 1980 to help show the tremendous diversity between and within countries, and the fact that union type has had a definite socioeconomic aspect that may have historical roots but which, in altered form, persists to this day. I also show that unions tend toward socioeconomic homogamy, marriages only slightly more so than consensual unions. It is important at the outset to point out, however, that this paper is written by a family demographer. What is family demography? And why should a family demographer be interested in a socioeconomic aspect of union type in Latin America? That topic deserves some discussion.

FAMILY DEMOGRAPHY

Family demography is the natural "marriage" between a sociological interest in the family and a demographic interest in population structure. The family is often considered *the* fundamental social unit, natural for sociological (and anthropological) study (see, e.g., Murdock 1949), whereas after age and sex, marital status may be one of the most important population characteristics there is (see Shyrock and Siegel 1973). If one considers the major population processes to be birth, migration, and death, then age, sex, and marital status tend to be the first characteristics by which crude measurement of these processes is refined. For instance, instead of looking at a crude birth rate (births per thousand in the population) one may wish to look at the crude birth rate standardized for a population's age distribution, or a birth rate per thousand women of childbearing age, or a birth rate among women at risk of bearing a child, often a marital-specific fertility rate. Students of migration have found it helpful to differentiate between married people and others, as well as by their age and sex, because married and unmarried people can have notably different migration rates. Finally, people have been intrigued to find that married men may have lower mortality than their unmarried counterparts.

Of population-related processes, it has seemed natural to study nuptiality, because marriage or union formation is associated with the production of children, but such demographic study developed independently from most sociological study of marriage and the family. For instance, marriage or exposure to the risk of pregnancy, conception, and gestation is considered an important "proximate determinant" of fertility (Davis and Blake 1956), whereas demographers usually ignore marriage's potential effect on other political, economic, social, or psychological factors. Demographers may learn about the European marital pattern of late marriage, small age difference between spouses, low marriage rates, but mainly in relation to Europe's historically low fertility (Hajnal 1965). It was usually family scholars, some of whom were also demographers, who pointed out what late marriage might mean for women's employment, ability to own property, or general social status. In general, until women demographers such as Judith Blake or Ruth Dixon (now Dixon-Mueller) came along, the relationship of the European marriage pattern, or any pattern for that matter, to gender relations was a topic outside most demographers' scope. The contribution of a later female age at marriage to fertility decline in East Asia has been traced (e.g., Cho and Retherford 1974; Mauldin and Berelson 1978), but again, it was a female demographer, Karen Mason, who explored the relation between women's status and fertility decline there. In a recent collection of articles on the fertility transition in Latin America, Rosero-Bixby (1996) investigates the degree to which changes in marriage patterns might help explain the general fertility decline, but how such changes in marriage patterns might be related to women's status was outside the scope of his study. Likewise, demographic interest in *union type* in Latin America is usually limited to the issue of its effect on fertility, asking, for instance, whether marriage or consensual unions lead to higher fertility (e.g., Chen, Wishic, and Scrimshaw 1974; Henriques 1979, 1989; Juarez 1989).[1]

A preoccupation with fertility as the reason to study nuptiality or union formation, along with the oft-unspoken assumption that a woman decides her own fertility or use of contraception, or both, resulted in many demographers concentrating on only half the married population—the women—thus further separating demographic study from studies that would naturally consider both elements in a marriage. For instance, demographers developed clever ways to measure fertility in the past or its "proximate determinants," including use of Im (an indicator of level of marriage) and Cm (the proportion of women married) (Bongaarts 1978; Coale and Treadway 1986), measures that concentrate on women. Beside a general bias in reproductive study toward the female (human or not), there is a reason for this limited concentration. It is sometimes called the "two-sex" problem (Keyfitz 1987). Mathematical and statistical models and methods can deal satisfactorily with only one unit at a time, usually the female. Attempts at developing two-sex models have not been satisfying. However, attempts are being made to bring males into the picture as well as to tackle the issue in other ways. For instance, instead of only collecting fertility information from women, fertility surveys now often contain a module directed at men. Demographic inquiry sometimes discusses a *couple's* decision-making process regarding the number of children in their family instead of using only the woman's preference (e.g. Thomson, McDonald, and Bumpass 1990). Family planning programs sometimes reach out to men despite the general focus on "maternal and child health." The 1997 Population of Association of America annual meeting was replete with papers addressing such issues as paternity and the male, family life course, as if demographers are finally discovering that men are important (see also Goldscheider and Kaufman 1996).

Some demographers have been less concerned with the demographic *processes* of birth, death, or migration than with the population's *composition:* the idea that a population's structure, including such factors as age, sex, marital status, race/ethnicity, households, education, and labor force participation, provides a backbone or skeleton upon which the rest of a social body is formed. It is here that the sociological imagination can take off, because sprouting from the demographic concern with population composition is the subfield of "family demography" that deals with "union formation and dissolution; family and household structure and change; and kinship" (Burch 1995, p. 85). Its strengths are "accurate measurement and description of what is" (Sweet 1977, p. 364) and it looks at men as well as women (Sweet and Bumpass 1987). Theory is not its strong point, although one might add that a life course or family development perspective seems to have provided good organizing principles (Glick 1955; Klein and White 1996; Uhlenberg 1978). And one of the stages in most life courses is the formation of a union. Indeed, Sweet and Bumpass seem to have organized their 1987 book on American families and households in part by marital status, while earlier Sweet (1977) had listed nine marital issues as topics in family demography: (1) age at first marriage—individual variation; (2) age at marriage—aggregate variation; (3) age at marriage and fertility; (4) age at marriage, age at first birth, and subsequent life changes; (5) premarital pregnancy; (6) marital disruption—aggregate trend; (7) marital disruption—differentials; (8) remarriage; and (9) marital disruption and fertility.

A sociologist's "marriage" between sociology and demography to produce family demography has advantages that formal adherence to one or another discipline could not provide. Demographers have a tradition of presenting "what is" coherently, but sociology has a tradition of thinking deeply about, and formulating numerous theories and hypotheses about, marriage that can help guide descriptive inquiries and interpret observed trends. In fact, the challenge has been to organize all the different perspectives sociology has to offer into a few overarching frameworks such as exchange, symbolic interaction, family development, conflict, or ecology (Klein and White 1996). Because a central sociological concern has been that of hierarchy and inequality, and because in the past actual "marriage" was most common among people of upper-class or European descent in Latin America, it seems natural to ask in this chapter whether there is a relationship between social hierarchy and union type in Latin America. In addition to the normal demographic inclination to look at individuals, we look at couples because of the guiding sociological idea of "homogamy," that likes tend to marry (or unite with) social likes. One sociological observation is that there tends to be a marriage "market" in which individuals or families exchange various attributes such as looks, family background, fertility, and earning potential in what ultimately appears to be a more or less fair exchange (culturally defined). Psychologists, too, speak of the benefits of shared values. Here, the focus is limited to the issue of socioeconomic status. To understand the recent situation, however, one must first delve into the past because the practice of having different types of union has historical roots.

A SHORT HISTORICAL VIEW OF UNION FORMATION IN LATIN AMERICA

Historically, Latin America was home to many different aboriginal groups who ranged from simple hunters, fishers, or gatherers to slash-and-burn farmers, small-scale irrigation farmers and pastoralists, to large-scale irrigation farmers. These pre-Columbian groups practiced an assortment of marital customs ranging from monogamy, male marriage to women from conquered tribes, to polygamy (Bernand and Gruzinski 1996; Steward and Faron 1959). Some women appear to have never married, instead serving as consorts; and the remarriage of widows also seems to have been handled in a variety of ways, including the levirate (a man marrying the widow of his dead brother). Some groups considered sibling marriage taboo, whereas others, at least among the elite, did not (e.g., Bernand and Gruzinski 1996). Regarding Mexico, McCaa (1994) suggests that a female's entry into marriage was fairly young, around 15, because mortality was high and it was important to maximize the woman's production of children.

Despite their pre-Columbian variety, conquered American Indians everywhere confronted two common situations. One was that the Church imposed its own view of marriage onto them, including its ideas about monogamy, concubinage, and incest. Women who might have been "second wives" under other conditions became mere concubines even when they had been baptized. Children by those consorts were "natural" and did not have the same rights as children by wives. The marriage of close relatives was banned, and even first cousins required special permission to marry (e.g., Bernand and Gruzinski 1996). The levirate was forbidden. However, sometimes the Church went further and outlawed the marriage of Europeans to Indians, even Catholic Indians, and often demanded a high price for a wedding. Thus, even when some Europeans wanted to formally marry Indians, they were prohibited from doing so by miscegenation laws, Church rule, or just the expense of a religious ceremony.

Second, the Europeans were predominantly men, perhaps six men to every woman, and many of these, even married ones, desired and had relations with aboriginal women (McCaa 1994; Bernand and Gruzinski 1996). We might think of true marriage as being between equals (more or less), but these relations were between conqueror and conquered. Today the population of many areas is predominantly *mestizo*, a mixture of white and Indian blood. For a variety of reasons, many of these people historically were not products of formal marriages, per se. And despite the fact that the current ratio of males to females is much more equal, the custom that men might have a marriage *and* extramarital relations persists. It is this persistence that I shall explore further in this chapter.

The story is not finished yet, however, because the situation was further complicated by the importation of African slaves to work on Spanish or Portuguese plantations. Africans had yet other marital customs (including polygamy and bride price) that were ignored by the plantation owners. Whereas slaves might have been encouraged to marry each other, they were also predominately male so there often was a striking sex imbalance (Love 1978). Also, it was not uncommon for a plantation owner to have a European family by legal marriage, and another family (or families) by less legal means (see also Nazzari 1996), or for a son of the owner to be initiated into sexual relations by a slave (see also Smith 1972). Thus, the populations of some areas have many people of mixed white and black blood called *mulatto,* while people of mixed Indian and African blood are sometimes referred to as *zambo*.

Jack Goody, among others, has speculated that a Western, Church-related ideology in Europe was historically different from that in places like North Africa, and the Middle East, or sub-Saharan Africa. Also, in Spain and Portugal European/Christian and Moorish/Muslim ideologies met. In his 1983 book on the development of marriage and the family in Europe, he elaborates on a scheme proposed by Guichard (1977 according to Goody) that distinguishes between Western and Eastern family structures in six ways: (1) system of descent, (2) conjugal pair, (3) kin groups, (4) matrimonial alliances, (5) position of women, and (6) the notion of honor. The Moorish system of descent was strictly patrilineal, whereas the European one was clearly bilineal. The Moorish conjugal pair had little strength whereas the European one was the "basic cell of social organisation. . . ." A Moorish kin group was "a group clearly defined in time and space in relation to an ancestor," whereas European kin groups were weak and existed only among the aristocracy. The Moors had strong "endogamous tendencies." The preferred marriage was "with the father's brother's daughter," whereas "To give a wife to another lineage (*linage*) is dishonorable and the wife-takers are superior to the wife-givers." In Europe, the tendency was towards exogamy (and prohibiting the marriage of first cousins). "Marriage alliances are valued; women circulate, bringing goods and honor; the wife-givers tend to be superior to the wife-takers." The Moors often observed a "Strict separation of the sexes and exclusion of women from the public sphere." In contrast, Europeans had "No rigid separation of the sexes; in the Germanic tradition, women appear to be able to play a public role and to have some political authority." The Moorish notion of honor was based on "being rather than having; feminine honor is passive, masculine honor, active." A European sense of honor was not supposed to be linked to gender or person: "Honor is tied to the possession of a title, a rank or riches; it can therefore be handed on and circulated in the social system." Thus, missionaries may have brought to Latin America, Catholic ideas about marriage, concubinage, and incest, but the Spanish and Portuguese colonial laity may have also been influenced by Moorish ideas, especially about gender roles and honor.

At the time of the European conquest of Latin America, the Iberian peninsula had expunged the Moors and had a European, feudalistic economic and social system. Spanish and Portuguese crowns often granted conquerors huge tracts of land and, perhaps, the labor of the land's Indian residents as well. Thus, while some people have likened the colonial situation in Latin America to the European feudal system, others convincingly disagree. First, they point out that European serfs were the same as their ladies and lords both racially and culturally. Indians, on the other hand, were different from their European masters and a status system developed involving skin color (Europeans and later Creoles being on top). Thus, a white person was often considered morally superior to a person of color simply by virtue of race. However, it was not unheard of for a relatively poor white to marry a rich Indian, *mestizo*, or *mulatto*, basically trading skin color for economic well-being. Second, many of the large land tracts existed far away from governmental control. Thus, owners were often "above the law," and did as they pleased, including taking sexual favors or concubines. Third, even white women in Latin America may have had lower social status than their European sisters, perhaps because of the presence of so many non-European women. European-based ideas about female rights of owning and inheriting property could do only so much in an environment of relative lawlessness in which few people had much property anyway.

Given the complicated situation of diverse peoples, homogamy, and interbreeding, one might assume that a formal Church marriage would be only one of a host of different kinds of marriage. But in colonial Latin America, there was only one kind of marriage recognized by the authorities, a marriage blessed by the Catholic Church. Later, in some countries, the institution of civil marriage separate from Church marriage developed, but it was patterned after the Church marriage in many ways. All other unions, of which there were often many, were lumped together into a "free union" category, and the resulting children were called "natural" or illegitimate.[2] In a thought-provoking article, Davis (1985) suggests that "marriage" is the only sexual union that commands public recognition, whereas common-law marriage lacks that recognition, and consensual union also lacks an assumption of permanence. This is confusing when trying to understand the situation in Latin America, however, because common-law marriage is an English concept not present in Latin American society, whereas customary marriage is a concept that is foreign to the societies of which Davis was thinking. Furthermore, if true marriage involves public recognition whereas other unions do not, who is the "public"—a European (and later Creole) administration? Perhaps instead of common-law marriage, we might speculate that there is a marriage type that could be called "customary," distinct from both the formal marriage of Church or state and consensual union. But how such marriage would be dealt with in relation to the other two types of union is unclear, especially as many unions that start out as consensual later become formalized through legal marriage.

NUPTIALITY IN LATIN AMERICA 1950 TO 1980

If a demographic view of marriage generally involves issues of formation and dissolution, one has to add and begin with a third issue, that of union type, since it has been shown that formal marriage is not the only kind of "marriage" in Latin America. Also included here must be the formalization of what had started as an informal union, sometimes even after the children are old enough to leave home. When viewing formation, the United Nations has generally found it helpful to consider both the timing and prevalence of first "marriage," where "marriage" includes all unions, not just formal ones (e.g., 1990). Dissolution also must consider the unofficial breakdown of informal unions as well as formal divorce or separation from legal marriage, especially as divorce was often illegal until recently, and still was in Chile and Paraguay in 1987 (Quale 1988).

Union Type

Although historic accounts of colonial "marriageways" (marriage-like unions in McCaa's view) allude to the commonality of non-marital sexual relations, their prevalence is problematic. Most contemporary Latin American censuses include a marital status of "consensual union" (not formally married) along with "married." This was not always the case as in the past when only "married" and "single" (and separated or widowed) were recognized. Indeed, the situation was so surreal that Hajnal (1965) omitted Latin America from his worldwide study of nuptiality with the observation that the Latin American data were "largely useless." When trying to trace marital structure from a series of Brazilian censuses, Greene reports (1991, p. 36): "Women in consensual unions were treated as single in 1940 and 1950, and as married in 1960 and 1970, appearing in the 'other' married category." This is probably typical. In 1970 census data for the Dominican Republic, I found that 62 percent of people 15 years of age and over were listed as single, but further probing disclosed many of these "single" people to be in consensual unions (see also Camisa 1978). The next (1981) census included consensual union as a marital status outright.

Although the true situation is probably somewhat different because of a social stigma sometimes attached to consensual union, census data reveal that in the recent past, between 5 percent (Chile 1982) and 54 percent (Honduras 1970) of all reported unions to women 15 years of age and over were informal. This is mapped out in Figure 1 based upon a grouping by Palloni and De Vos (unpublished) of under 15 percent, 15 to 25 percent, and over 25 percent.[3] Although the figures are for women, those for men are similar if not slightly lower, and reflect actually

Figure 1. Percent Adult Females in Union in Consensual Unions

maintaining a family with someone else. Palloni and De Vos observe: "We can only speculate at this point, but the differences *within* Latin America as of 1980 may have more to do with the prevalence of Indian and African-American populations and with the degree of ethnic intermarriage than with stages in an evolution of the patterns of union formation or the levels of socioeconomic development." The United Nations (1990, p. 157) similarly observes, after reporting some socioeconomic differentials in marriage: "However, preference for a type of union appears to interact with factors of a more cultural nature, such as ethnic group." Greene (1991) echoes this conclusion from her examination of Brazilian data noting, for instance, that there were marked regional differences in the proportion of unions that were consensual rather than formal.

There is also an apparent ambivalence about sexual relations in Latin America. On one extreme of a continuum would be *machismo*, a concept that includes male sexual *conquest* without marriage or responsibility for any offspring. At the other end, would be a faithful marriage that can only be broken through the death of a spouse. But even if married, men are still supposed to be very concerned with honor and virility, and may be "somewhat formal and restrained" with their children, even "stern and authoritarian" (Leonard and Louriero 1980, p. 211). The irony is to hold both familism and *machismo* as ideals simultaneously, although they are contradictory (Ingoldsby 1991). Marriage and *machismo* can and do coexist when there is the double standard that women must stay faithful in marriage, but men do not. (Remember Moorish ideas about honor and the position of women.) It is not uncommon for upper-class, married men to have lower-class *querida*s (sweethearts), sometimes in ongoing consensual relationships, perhaps because "often there is little emphasis upon the companionate aspects of the marriage" (Leonard and Louriero 1980; also Caravajal 1980). In contrast, their wives were supposed to find solace in their relationships with their children.

> Thus, the content of husband-wife conversation may hinge mostly on activities of the children, local gossip and the activities of friends and relatives. The husband is likely to belong to one or more clubs where he frequently goes in the evening to discuss finances, philosophy and politics with friends and associates. There are few women's organizations in which the wife can get involved. (Leonard and Louriero 1980)

Among the lower class, both men and women were supposed to be freer from the constraints and privileges of formal marriage as often they were not expected to contract them, and were, thus, not legally bound to the other. The catch is, however, that the resulting children usually were not legally bound to the father either. Sometimes the men in such unions are step-fathers to some of the children, but biological parents to others (see also Fonseca 1991; Ingoldsby 1995). Thus Greene (1991) argues that women traditionally preferred marriage to consensual union, in part because of child responsibility, even if the marital institution could hardly be considered ideal. Although in some places, the legal responsibilities of biological fathers may be clear, paternity may still be difficult to establish, and a man can "evaporate" if he so wishes, while mothers cannot. This may seem strange in a people who believe (profess) that children are prized possesions of the father's family, but it appears that in Latin America, as in the United States, children can be liabilities, rather than assets, and thus, remain with their mothers when there is a conjugal breakup. Some men may see children primarily as proof of their virility, rather than gifts to posterity (see also Desai 1992), and may see women primarily as sources of sexual gratification and the mothers of their children, rather than as equals emotionally or intellectually.

If responsibility for children has been a major reason women traditionally preferred formal marriage over consensual union with its risk of instability, then one wonders about the consequences of more widespread mechanisms of fertility control. That is, even in countries such as Brazil or Mexico that traditionally experienced noticeable increases in formal marriage at the expense of consensual union (e.g., Nyrop 1983; Ojeda de la Pena 1987), there appears now to be either stability or a rise in the tendency to form consensual unions (e.g., Quilodrán 1993; Greene and Rao 1995; also Zamudio and Rubiano 1991). It is also presently unclear how any overall figure may be owing to changes or stability among women in different social groups, by an increase among rather well-educated or well-to-do young women who are able to postpone childbearing (see also Parrado and Tienda 1997), or by less educated women whose bargaining power has been undermined by a "marriage squeeze" (Greene and Rao 1995). In any case, the marriage market will be changing, and the value placed on different assets will change along with it. Could it be that formal marriage per se will acquire more of a companionate quality, less of an institutional one (á la Burgess and Locke, 1945)?

The ambivalent attitude toward informal (consensual) union might best be noted through the fact that many unions that begin informally, eventually become formalized. Direct evidence on this matter has been difficult to come by as even the World Fertility Survey, which was so good in many respects, did not gather the kind of information on marital histories necessary for ascertaining the later legalization of a union that began consensually (except in Mexico). But there is *some* evidence. For instance, the proportion of women who are in consensual unions, tends to be much higher among younger women than older women. As an example, in Colombia (1985) over 41 percent of the women aged 20 to 24 years who were in a union were in a consensual union. By age 45 to 49, the figure was only 20 percent. Figures for Peru (1981) were 40 and 16 percent, respectively, and in Paraguay, 27 and 15 percent (1982) (United Nations 1990, table 19). Of all the World Fertility Surveys, the Mexican one (EMF) did gather information necessary to construct a life table that could show the termination of consensual unions because of their conversion into legal unions (for women). Pebley and Goldman (1986, p. 206) constructed such a table and found that "in the absence of separation or widowhood, two-thirds of consensual unions would eventually become legal marriages; . . ." In an earlier work, the same authors (Goldman and Pebley 1981) used 1960s data for women in rural Colombia, Costa Rica, Mexico, and Peru to estimate that "in the absence of separation and death of a partner almost half of consensual unions would eventually become marriages in Colombia, Costa Rica and Mexico, and about 65 percent in Peru" (p. 54). Commenting on the discrepancy between the earlier and later figures for Mexico, the authors note: "Such a difference could result from a number of factors including a change in the pattern of legalization subsequent to 1970, differential misreporting in the two surveys or differences in the sampling frames between PECFAL and EMF" (Pebley and Goldman 1986, p. 206). Of these possibilities, they go on to suggest that differences in question wording in the two surveys may have been a major cause.

A next natural question might be whether there are certain measurable characteristics of the woman or man that might make a union more or less prone to become legalized. Perhaps as the outgrowth of our notion about the "shotgun marriage," it was often speculated, for instance, that a pregnancy might induce union legalization. After careful analysis, Goldman and Pebley (1981, 1986)) did not find substantiation for the notion in either their rural samples of women in four countries in the 1960s (Colombia, Costa Rica, Mexico, Peru) or their national sample of women in Mexico in the mid-1970s. However, they did find that such factors as education, age, and whether the union was the woman's first union *did* have important influences on whether the union became formalized.

Timing and Prevalence of First Union

Although the range here is really from perhaps 18 to 23 years in 1950, and 20 to 24 years in 1980, the average mean age at first marriage for women in Latin America as a whole appears to have risen from not quite 21 years around 1950, to not quite 22 years around 1980. Likewise, despite a range for men, an average mean appears to have declined from almost 26 years around 1950, to around 25 years in 1980 (Figure 2). The age at first union for both males and females in Latin America appears "intermediate," since a European late age might be over 23 years for women and over 27 years for men, and an early age might be 20 years for women (early and intermediate age at first union for men does not appear to be an issue that has been addressed) (United Nations 1990). The narrowing age gap between men and women could be indicative of what demographers often refer to as a "marriage squeeze" or a mismatch between available men and available women. Greene and Rao (1995) suggest that a marriage squeeze in Brazil has resulted in an increase in consensual unions because men can then "circulate" better than if they were permanently matched. A marriage squeeze in favor of men may generally give them more leverage, while the opposite may start to be seen in future in such places as India, China, and Korea, where there appears to be a gross selection in favor of the survival of male infants at present. (One wonders how many of those prized males will be matched with a spouse later on.)

We should pause here to note that the figures are based on a measure, the singulate mean age at marriage (SMAM), that was developed to make use of census data on marital status *and* to deal with a differing incidence of non-marriage (Hajnal 1953). Of various ways to measure an average age at first marriage, SMAM has become quite popular, but it should be remembered that it will provide slightly different numbers from other measures, even ones based on census data instead of vital registration statistics (from registrations and such) (Shryock and Siegel 1973). In addition, it should be mentioned that census data provide a cross-sectional view

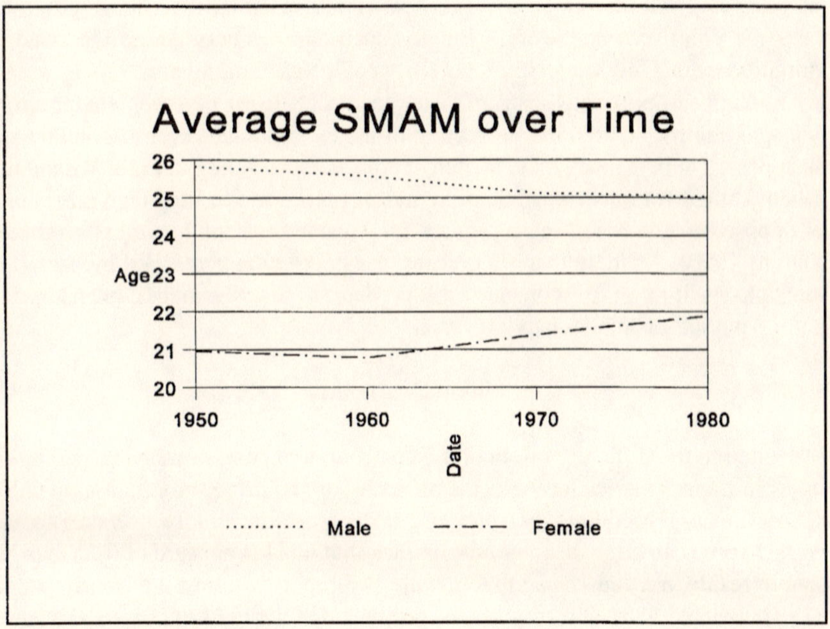

Average were based on SMAM for:

1950	*1950*	*1970*	*1980*
Bolivia	Argentina	Argentina	Argentina
Chile	Chile	Bolivia	Brazil
Colombia	Colombia	Brazil	Chile
Costa Rica	Costa Rica	Chile	Colombia
Cuba	Dominican Republic	Colombia	Costa Rica
Ecuador	Ecuador	Costa Rica	Cuba
El Salvador	El Salvador	Dominican Republic	Dominican Republic
Guatemala	Guatemala	Ecuador	Ecuador
Nicaragua	Honduras	El Salvador	Guatemala
Panama	Mexico	Guatemala	Mexico
Paraguay	Panama	Honduras	Panama
Venezuela	Paraguay	Nicaragua	Paraguay
	Peru	Panama	Peru
	Uruguay	Paraguay	Venezuela
	Venezuela	Peru	
		Uruguay	
		Venezuela	

Source: From Palloni and De Vos (unpublished) and United Nations.

Figure 2. Average Singulate Mean Age at Marriage (SMAM) in Latin America, 1950–1980

because different generations (or cohorts in demographic parlance) may enter into marriage at different average ages, yet census data refer to different age groups at the same point in time (period). Finally, such data cannot address the issue of mortality; even a modest level of mortality could affect the figures.

Although earlier figures are particularly suspect because of a problem reporting nonlegal unions, in 1980 the proportion of males or females ever married by age 50 appeared to be about 91 and 90 percent on average (Palloni and De Vos unpublished; United Nations 1990). Such figures suggest a moderately high prevalence as opposed to the much lower figures for Western Europe historically (United Nations 1990). As more than 95 percent of a population is married by age 50 in some places, it would be more accurate, perhaps, to describe marital prevalence in Latin America as "moderately universal."

Union Dissolution

Related to the difficulty of obtaining an accurate picture of people's real union status in Latin America has been the difficulty of acquiring information on union dissolution, as it is commonly believed that it can only be dissolved when a spouse dies. Divorce can be so dishonorable, in fact, that it is not unheard of for a married man to reside in a consensual union while still reporting himself as married (see, e.g., Caravajal 1980). One might also speculate that the rigidity of laws and views regarding formal marriage could make the option of consensual union appear attractive for both men and women, despite its unsanctioned quality. Divorce per se was illegal until recently in many countries except for the gravest of reasons, and still is illegal in Chile. Even legal separation (*diquete*) was difficult to obtain. It should come as no surprise, therefore, that the reported divorce rate in Latin America is very low when statistics on it are even available: "The rates of divorce . . . are about one half to one fifth those of Europe because registered dissolutions in Latin America occur only to the *legally* married, who form a smaller percentage of the total adult population" (Goode 1993, p. 191).

The picture is very different, however, if one considers union dissolution broadly to include consensual unions because they are much less stable than formal marriages. For instance, using World Fertility Survey data, Quilodrán (1985) estimated that the likelihood of dissolution after five years was perhaps six times higher for consensual unions than for formal marriages during the 1970s in various Latin American countries. Perhaps a third of the consensual unions tallied were not still operative after five years among women aged 15 to 49 years. Thus, instead of having a *low* dissolution rate, the region could be considered to have a rather *high* dissolution rate. Indeed, some people have described Latin American union formation as "serial monogamy" in which people have successive partners who may behave as if they are married for a limited period of time. (This is probably related to the fact that a relatively large proportion of children live in female-headed households in Latin America, but that is a topic outside the scope of the present study.) Divorce has become increasingly acceptable socially (and legally), however, perhaps reducing the appeal of a consensual union.

A SOCIOECONOMIC DIMENSION TO UNION TYPE IN LATIN AMERICA

One often reads about the socioeconomic correlates of consensual union or how union type was connected to "class" (a related but different concept to that of status). But when I was challenged to provide substantiation for that claim, I was amazed to find little hard evidence. And what I *did* find tended to be limited to women who had participated in fertility surveys. For instance, in a recent paper that used Demographic and Health Survey data from three Latin American countries (Dominican Republic 1986, Peru 1988, Brazil 1986) Quilodrán (1992) presented a table showing separately that women who lived in urban areas, and had seven or more years of school, were more likely to be married than in a consensual union, both among 20-to 24-year-olds and amond 40- to 44-year-olds. Her figures also suggested that among the younger group, women who had worked outside the home before their union were *less* likely to be formally married. Although one can easily consider the limitations of her analysis (such as limited trivariate analysis, perhaps because of sample size), I was in the awkward position of finding it much better than nothing but "common knowledge" that nobody bothered to substantiate. Indeed, I looked at her references in the hope of finding other studies similarly documenting socioeconomic differentials, but I came up empty-handed.

Greene's (1991) use of 1984 data for one Latin American country, Brazil, limited to women, was a big step in the right direction. She ran multivariate analyses at the national level among different age groups, including variables for race and region, as well as years of education, urban/rural residence, and presence of earned income (see her table 5.4). Because the race and region variables were significant, even after including the other variables, she had evidence to support the claim that there seemed to be a cultural factor behind union type not explained by education, urban/rural residence, or income. One is also left to ask whether, otherwise, was general "common knowledge" so clear that one did not need documentation of "fact"? As the sparse evidence all pertains to women, there is no information about men, even in the societies for which there *is* some evidence. A failure to document what many considered to be facts does not seem particularly scientific to me, even if there is now interest in showing the emergence of a "new" as opposed to traditional kind of consensual union (e.g., Parrado and Tienda 1997). And while demographers may have traditionally focused on women, a family demographer includes men as well as women in his or her purview.

The main sociological question is how to think about "social structure" when the formation of consensual unions seems to be more cultural than socioeconomic. Take, for example, education, an indicator often used to signify socioeconomic sta-

Table 1. Logit Results of Regressing Union Type (married = 1/union = 0) onto Education and Urban Residence Controlling for Years of Age Among Females Aged 20 to 24 and Males Aged 23 to 27 (standard error in parethesis)

	Argentina 1981		*Brazil 1980*		*Chile 1982*		*Colombia 1985*		*Costa Rica 1984*	
	F	*M*	*F*	*M*	*F*	*M*	*F*	*M*	*F*	*M*
Region 1	—	—	—	—	—	—	—	—	—	—
Region 2	−.572**	−.586**	.729**	.587**	.065	−.072	−.328**	−.287**	.429**	.430**
	(.08)	(0.8)	(.06)	(.05)	(.10)	(.11)	(.04)	(.04)	(.09)	(.10)
Region 3	−.286**	−.430**	.799**	.745**	−.025	−.153	.036	.104*	−1.141**	−1.325**
	(.09)	(.09)	(.06)	(.06)	(.08)	(.08)	(.04)	(.04)	(.09)	(.09)
Region 4	.824**	.727**	1.364**	1.134**			−.341**	−.241**		
	(.14)	(.15)	(.08)	(.08)			(.04)	(.05)		
Region 5	−.335**	−.495**	.588**	.606**			−.805**	−.901**		
	(.12)	(.13)	(.07)	(.06)			(.07)	(.08)		
Age	.140**	.087**	.084**	.064**	.146**	.074**	.090**	.101**	.035	.029
	(.02)	(.02)	(.01)	(.01)	(.03)	(.03)	(.006)	(.007)	(.02)	(.03)
Urb	.005	.090	−.629**	−.692**	−.255**	−.219*	−.501**	−.386**	.125	.137
	(.07)	(.08)	(.05)	(.04)	(.10)	(.10)	(.02)	(.03)	(.08)	(.08)
Ed1	—	—	—	—	—	—	—	—	—	—
Ed2	.788**	.917**	.465**	.730**	.642**	.791**	.411**	.282**	.579**	.689**
	(.12)	(.12)	(.05)	(.05)	(.18)	(.20)	(.03)	(.04)	(.17)	(.18)
Ed3	1.686**	1.644**	.624**	.969**	1.249**	1.322**	.742**	.508**	1.323**	1.268**
	(.14)	(.13)	(.06)	(.06)	(.18)	(.20)	(.04)	(.04)	(.17)	(.17)
Ed4	2.340**	2.466**	.801**	1.003**	2.010**	2.056**	1.090**	.839**	1.474**	1.483**
	(.13)	(.13)	(.06)	(.06)	(.18)	(.20)	(.04)	(.04)	(.18)	(.18)
Ed5	2.949**	2.743**	1.588**	2.055**	2.247**	2.353**	1.918**	1.287**	2.558**	2.068**
	(.16)	(.19)	(.12)	(.12)	(.26)	(.25)	(.04)	(.04)	(.20)	(.20)
Ed6	2.614**	2.790**	2.133**	2.044**	3.281**	2.773**	2.888**	2.088**	3.218**	2.907**
	(.23)	(.19)	(.28)	(.37)	(.42)	(.28)	(.09)	(.06)	(.30)	(.26)
% in CU	17.5	17.1	16.1	16.5	8.9	8.4	39.8	40.0	22.3	21.7

(continued)

Table 1 (Continued)

	Dominican Republic 1981		*Ecuador 1982*		*Panama 1980*		*Paraguay*	
	F	*M*	*F*	*M*	*F*	*M*	*F*	*M*
Region 1	—	—	—	—	—	—	—	—
Region 2	.149*	.179*	−2.743**	−2.737**	.228**	.267**	.182*	.247**
	(.07)	(.08)	(.06)	(.06)	(.09)	(.09)	(.09)	(.09)
Region 3	−.473**	−.397**	−.873**	−.938**	.315**	.386**	.610**	.522**
	(.08)	(.09)	(.07)	(.07)	(.07)	(.08)	(.07)	(.07)
Region 4	.133	.070	-2.721**	−2.665**				
	(.08)	(.08)	(.08)	(.09)				
Age	.084**	.127**	.096**	.067**	.112**	.136**	.036	.074**
	(.02)	(.02)	(.01)	(.01	(.02)	(.02)	(.02)	(.02)
Urb	.475**	.567**	.118*	.186**	.400**	.483**	−.253**	−.283**
	(.06)	(.06)	(.05)	(.06)	(.07)	(.07)	(.07)	(.07)
Ed1	—	—	—	—	—	—	—	—
Ed2	.423**	.388**	.520**	.824**	.732**	.296	.733**	.293*
	(.08)	(.09)	(.06)	(.08)	(.15)	(.18)	(.10)	(.12)
Ed3	.592**	.515**	.914**	1.214**	1.217**	.951**	.816**	.426**
	(.11)	(.13)	(.06)	(.08)	(.14)	(.16)	(.11)	(.13)
Ed4	1.483**	1.167**	1.275**	1.217**	1.845**	1.634**	2.207**	1.197**
	(.08)	(.09)	(.07)	(.09)	(.14)	(.17)	(.25)	(..22)
Ed5	3.229**	2.115**	2.523**	2.002**	2.940**	2.511**	2.963**	1.772**
	(.13)	(.12)	(.14)	(.12)	(.15)	(.17)	(.25)	(.21)
Ed6	4.134**	3.261**	2.670**	2.578**	3.825**	3.177**	3.576**	2.643**
	(.22)	(.16)	(.16)	(.13)	(.18)	(.18)	(.52)	(.31)
% in CU	76.1	70.8	31.2	30.8	60.4	57.8	26.7	29.6

Notes: F, Female; M, male; * $p < .05$; ** $p < .01$.

Urban = 1 if yes; 0 if no.

Education is: illiterate/no education, < primary school, primary graduate, middle school or some high school, high school graduate more (such as some or all of college).

% CU is the percentage of all unions (marriages and consensual unions) being consensual.

tus. In Argentina in 1981, of women in unions, aged 20 to 24 years, with less than a primary school education, 35 percent were in a consensual union, and 65 percent were married. Among comparable women with more than a primary education, 7 percent were in a consensual union, while 93 percent were married. How does one interpret this? Can one say that there is something structural as opposed to cultural about an educational difference? Does not education affect both ideas and one's position in society? Or take urban/rural residence for another example. In Argentina in 1981, again, among women in unions, 20 to 24, who were urban residents, 16 percent were in a consensual union while 84 percent were married. Among their rural counterparts, 27 percent were in a consensual union, and 73 percent were married. How does one interpret this? What is socioeconomic about living in an urban or rural area? Since people with the same culture occupy different positions on a social hierarchy, how would you interpret an education difference among people in the same area who presumably share a common culture? I would interpret it in terms of a socioeconomic effect. Call that "structural" if you wish.

In this exercise, let us use education as our indicator of socioeconomic status, but stay more agnostic about urban/rural residence. (Sociologists also like to use characteristics such as occupation [see Portes 1985], but I find that bad for studying women, even when a society is primarily industrial.[4] Adequate income data usually exist in very few countries.) We furthermore hypothesize that people with more education tend to marry rather than be in a consensual union—that there is a positive relationship. We could simply look at figures for other Latin American countries such as the ones presented for Argentina, but better yet, we can assess a *net* effect of education on union type that persisted after controlling for factors that might muddy that true differential. For instance, we know that the probability of being married rather than in a consensual union increases with age for both men and women, and therefore want to control for age. One way is to limit the analysis to a five-year age group, as Greene did in her 1991 study. I found this to be only partly satisfactory, however, especially among young people, because even one year makes a substantial difference. Hence, I limited my purview to young people, and furthermore controlled for years of age in multivariate analyses of these groups. That is, since age at first union for women tended to be somewhere in the 21- to 22-year area, I limited my view of women to 20 to 24 years of age. Since men's age at first union was somewhere around 25, I limited my view of men to 23 to 27 years of age.[5] Within these groups, I added a control for years of age.

A second factor, after age, is urban/rural residence. Although Quilodrán (1992) and Greene (1991) both found consensual union more common among rural than urban residents, the issue is complicated and may differ from place to place. It seemed important to include urban/rural residence in our model, however, if we really want to assess the *net* effect of education on union type. Finally, every country, no matter how large or small, has regions that can sometimes be very different. The situation in Ecuador might be the most dramatic because less than

10 percent of the women aged 20 to 24, or men aged 23 to 27, who were in a union in 1982 were in a consensual union if they lived in the mountains, but over half were in a consensual union if they lived on the coast.[6] (When we additionally ran separate models within regions instead of whole countries, we obtained similar results for the effect of education but different ones [sometimes] for the effect of urban/rural residence.)

Results from these multivariate analyses, separately for males (23 to 27) and females (20 to 24) are shown in Table 1. Throughout Latin America, there is a definite positive relationship between education and union type among young people that persists after controlling for region, urban/rural residence, and age, and that is as true for women as for men. Women and men tend to be married instead of in consensual unions the more education they have received. It also seems the case that, except in Argentina, there is more of a gradient by education among women than men.[7] But since results generally seem so similar for men and women, does this mean that men and women of similar education tend to marry or enter a consensual union with each other? It appears so.

The issue of homogamy is a complicated one, and discussion of the methods one might use to study it also are complicated. Additionally, the examination of the issue here cannot be definitive even if we wanted because, among other things, my data do not contain information that could link all spouses. Instead, I have information on "relationship to head" and can only link heads to their spouses, or vice versa. Because many young people do not live in their own households even when they have formed a union (another topic), such links can only be performed on a subgroup of all the people, generally 75 percent or so of them selected because of their "relationship to head." What bias this might entail in representing the entire sample has to be an open question. The proportion of the entire sample included in the analysis is listed in Table 2. Here, I compute correlations between heads and spouses of heads for in-union women, 20 to 24 years old, by type of union for this subgroup.

In general, and as one might expect, it appears from the figures in Table 2 that among young women, aged 20 to 24, who were in a union and co-headed a household, those unions tended to be socially (educationally) homogamous whether formal marriages or consensual unions. However, they tended to be somewhat more so in formal marriage than in a consensual union. That is, whereas the correlation between the man's and woman's education ranged from .33 (Paraguay) to .61 (Panama) among those in a consensual union, it ranged from .53 (Chile) to .71 (Ecuador) for those in a formal marriage. The difference seemed greatest in Paraguay, where the education correlation for those in a consensual union was .33 compared with .59 for those in a formal marriage, and smallest in Panama where the correlation for those in a consensual union was .61 compared with .66 for those in a formal marriage.

Table 2. Proportion of Total Sample Included in the Analysis (co-heads) and Correlation Between Woman's and Spouse's Education* by Union Type: Nine Latin American Countries Circa 1980

	In Consensual Union		*Married*	
	%	r^2	%	r^2
Argentina 1981	75	.50	71	.65
Brazil 1980	88	.51	91	.60
Chile 1982	79	.40	75	.53
Colombia 1985	71	.56	75	.70
Costa Rica 1984	89	.46	91	.63
Dominican Republic 1981	87	.43	86	.59
Ecuador 1982	81	.56	81	.71
Panama 1980	79	.61	82	.66
Paraguay 1982	83	.33	84	.59

Notes: * Education is a six-category variable where 1 = illitreate/no school; 2 = less than all of primary school; 3 = primary graduate; 4 = some high school or middle school; 5 = high school graduate; and 6 = more than high school.
All correlations were significant at $p < .001$.

Source: From in-house calculations of census microfiles of the *Changing Household in Latin America* at the University of Wisconsin-Madison's Center for Demography and Ecology.

CONCLUSION

Deeply entwined in Latin America's social system is a nuptiality pattern in which there have been rather stable formal marriages, at least for public record, and informally, rather unstable consensual unions. Although most marriages today are civil rather than religious, the idea of a Catholic-like marriage that is very difficult to untie on the one hand, and everything else that can dissolve "easily" because it is not legal on the other, has obvious historic roots. Ideologically, these two types of union are very different, and predictably have a strong socioeconomic as well as cultural dimension to them. Marriage is often preferred over consensual unions and is positively related to educational attainment, but the two types are both common and well entrenched. The coexistence of the two types suggests an ambivalent attitude toward union formation, union dissolution, and gender relations that has never been resolved in a coherent manner. For instance, many unions that start consensually end up becoming formalized, even if the children are already grown. If a marriage goes sour, the husband (usually) may form a consensual union with someone else while officially staying "married" instead of divorcing and remarrying. There has been a coexistence of contradictory *machismo* and marriage ideologies because men can act "macho," but women are supposed to be faithful. Whereas the idea of the dialectical nature of union formation is hardly unique to Latin America, what *is* unique is how this has evolved there.

Traditional demographic interest in nuptiality may have been limited primarily to nuptiality's role in fertility, but family demographers have found union formation patterns to be of interest in their own right because they consider marriage-like unions to be a fundamental population characteristic, of as much demographic interest as occupational or ethnic composition. However, they and other social scientists have generally eschewed studying nuptiality in Latin America, in part because of the difficulty in obtaining reasonable-looking data. Until recently, censuses and other government agencies have often tried to act as if consensual unions did not exist, although such unions are an important type of union in most of Latin America. One result is that great strides have been made in contrasting a historic European family system with an Asian one (e.g., Hajnal 1982; Goody 1996), but family systems in Latin America remain big enigmas.

The situation of the recent past should not obscure the fact that it is changing, as the social meaning of union formation seems to be doing everywhere. But trying to understand this change as part of an overall social change (see Ogburn 1932) is hampered by a poor understanding of what family formation traditionally was. Social scientists have generally neglected study of marriage and the family in Latin America in favor of studying the region's political processes, economic development, or overall population change. In an exceptional piece that was focused primarily on ideology, Todd (1985) emphasized the importance of marriage and inheritance systems in helping to form family types. He assigned to Latin America two family systems, both associated with neo-local residence, but having different marital and inheritance customs. Unfortunately, the empirical substantiation of his assertion is difficult to understand and is therefore suspect, but it does seem correct in at least proposing that most households contain only one married couple (see also De Vos 1995). Yet one has to conclude (tentatively) that the Latin American nuptiality situation is both distinctive and of probable value when trying to develop an overarching theory of marriage that can help us understand the wide variety of forms marital union takes around the world.

ACKNOWLEDGMENT

I am grateful to Elizabeth Arias for comments on an earlier draft, to senior programmer Julia Gray for preparing data for computer analysis, and to the University of Wisconsin's Center for Demography and Ecology's *Changing Household in Latin America* for providing the necessary data. The Center's facilities funded by NICHD center grant HD05876 are also gratefully acknowledged.

NOTES

1. In the end it seems that results vary between places, perhaps because the different union types have different implications for the exposure to the risk of pregnancy in different places.

2. Sometimes distinction was made according to the father's marital status, by whether he was already formally married to someone else.

3. Less than 15 percent: Argentina (1980), Brazil (1980), Chile (1982), Mexico (1980), Uruguay (1975); 15 to 25 percent: Bolivia (1950), Costa Rica (1984), Paraguay (1982), Peru (1981); greater than 25 percent: Colombia (1985), Cuba (1981), Dominican Republic (1981), Ecuador (1982), El Salvador (1971), Guatemala (1981), Honduras (1974), Nicaragua (1971), Panama (1980), Venezuela (1981).

4. Females in Latin America tend to receive education along with males whereas they do *not* tend to have an "economic activity" other than housework if they are in a union. Thus, the social demographer can examine the socioeconomic dimension to union type by looking for a relationship between education and union type. Whether one is more concerned with class than status is another issue.

5. Just as researchers in the past have lamented the fact that it can be impossible to determine whether a woman's present marriage originally began as a consensual union, one has to lament the fact that most censuses cannot determine whether someone's current union is the first or is of some higher order. One way people have dealt with the limitation is to limit their analysis to women of a young age in the expectation that by doing so, they are dealing mainly with first unions. We follow that course.

6. The regions were as follows: ***Argentina:*** Central (depts. Buenos Aires, Cordoba, Santa Fe, Capital Federal); Northeast (depts. Corrientes, Chaco, Entre Rios, Formosa, Misiones); Northwest (depts. Catamarca, Jujuy, La Rioja, Salta, Santiago del Estero, Tucuman); West Central (depts. La Pampa, Mendoza, San Juan, San Luis); Patagonia (depts. Chubut, Neuquen, Rio Negro, Santa Cruz, Tierra del Fuego). ***Brazil:*** North ("states" of Rondonia, Acre, Amazonas, Roraima, Pará, Amapá); Northeast (states of Maranhao, Piaui, Ceara, Rio Grande do Norte, Paraiba, Pernambuco, Alagoas, Sergipe, Bahia, F. Noronha); Southeast (states of Minas Gerais, Espirito Santo, Rio de Janeiro, Guanabara, Sao Paulo); South (states of Parana, Santa Catarina, Rio Grande do Sul); Central (states of Mato Grosso Sul, Mato Grosso, Gois, Distrito Federal). ***Chile:*** Santiago (regions of Tarapaca, Antofagasta, Aysen, Magallenas, Santiago); Wide Center (regions of Coquimba, Maule, Bio-Bio, Araucania, Los Lagos); North and South (Atacama, O'Higgins, Valparaiso). ***Colombia:*** Pacific (depts. of Choco, Chauca, Nariño); Atlantic (depts. of Atlantico, La Guajira, Cesar, Magdalena, Bolivar, Sucre, Cordoba, Norte de Santander, San Andres, Casanare); Highlands (depts. of Boyoca, Cundinamarca, Valle, Meta, Antioquia, Quindio, Tolima, Huila); Bogotá (Bogotá D.E. dept.); National Territories ("depts." Caqueta, Arauca, Putumayo, Amazonas, Guania, Guaviare, Vaupes, Vichada). ***Costa Rica:*** San José (prov. of San José); Other valley (provs. Alajuela, Cartago, Heredia); Limon and outer (provs. Guanacaste, Puntarenas, Limon). ***Dominican Republic:*** Capitol (prov. of Distrito National); North (prov.s of Dajabon, Espaillat, La Vega, Monte Cristi, Puerto Plata, Salcedo, Santiago, Santiago Rodriguez, Valverde); Southwest (provs. of Azua, Bahoruco, Barahona, Elias Pina, Independencia, Pedernales, Peravia, San Juan); Southeast (prov.s of Duarte, El Seybo, La Altagracia, Semana, San Cristobal, San Pedro de Marcoris, Sanchez Ramirez, Maria Trinidad Sanchez). ***Ecuador:*** Sierra (prov. of Carchi, Imbabura, Pichincha, Cotopaxi, Tungurahua, Bolivar, Chimborazo, Canar, Azuay, Loja); Coast (prov. of Esmeraldas, Manabi, Los Rios, Guayas, El Oro); East (prov. of Napo, Pastaza, Morona Santiago, Zamora Chinchipe); Galapagos (prov. of Galapagos). ***Panama:*** South (prov. of Colon, Darien, Panama); Center (prov. of Cocle, Herrera, Los Santos); North (prov. of Boca del Toro, Chiriqui, Veraguas). Paraguay: Capitol (dept. Asuncion, Cordillera, Paraguari); North & West (dept. Concepcion, San Pedro, Amabay, Pte. Hayes, Boqueron, Olimpo); East (dept. Guaira, Caaguaza, Caazapa, Itapua, Misiones, Alto Parana, Neembucu).

7. This was not assessed statistically.

REFERENCES

Bernand, C., and S. Gruzinski. 1996. "Children of the Apocalypse: The Family in Meso-America and the Andes." Pp. 161-215 in *A History of the Family Volume II: The Impact of Modernity,* edited by A. Burguière, C. Klapisch-Zuber, M. Segalen, and F. Zonabend. Cambridge, MA: Belknap Press/Harvard University Press.

Bongaarts, J. 1978. "A Framework for Analyzing the Proximate Determinants of Fertility." *Population and Development Review* 4: 105-129.

Burch, T. K. 1995. "Theories of Household Formation: Progress and Challenges." Pp.85-108 in *Household Demography and Household Modeling,* edited by E. van Imhoff, A. Kuijsten, P. Hooimeijer, and L. van Wissen. New York: Plenum.

Burgess, E., and H. J. Locke. 1945. *The Family from Institution to Companionate.* New York: America.

Camisa, Z. 1978. "La Nupcialidad de las Mujeras Solteras en la América Latina." *Notas de Poblacíon* 18: 9-75.

Caravajal, M. J. 1980. "The Costa Rican Family." Pp. 335-387 in *The Family in Latin America,* edited by M. S. Das and C. J. Jesser. New Delhi: Vikas.

Chen, R. H., S.M. Wishic, and S. Scrimshaw. 1974. "Effects of Unstable Sexual Unions on Fertility in Guayaquil, Ecuador." *Social Biology* 21: 353-359.

Cho, L. J., and R. D. Retherford. 1974. "Comparative Analysis of Recent Fertility Trends in East Asia." Pp. 163-181 in *International Population Conference 1973,* vol. 2. Liége: International Union for the Scientific Study of Population.

Coale, A. J., and R. Treadway. 1986. "A Summary of Changing Distribution of Overall Fertility, Marital Fertility, and the Proportion Married in the Provinces of Europe." Pp.31-181 in *The Decline of Fertility in Europe,* edited by A. J. Coale and S. C. Watkins. Princeton, NJ: Princeton University Press.

Davis, K. 1985. "The Meaning and Significance of Marriage in Contemporary Society." Pp. 1-21 in *Contemporary Marriage,* edited by K. Davis. New York: Russell Sage Foundation.

Davis, K., and J. Blake. 1956. "Social Structure and Fertility: An Analytic Framework." *Economic Development and Cultural Change* 4: 211-235.

Desai, S. 1992. "Children at Risk: The Role of Family Structure in Latin America and West Africa." *Population and Development Review* 18: 689-717.

De Vos, S. 1995. *Household Composition in Latin America.* New York: Plenum.

Fonseca, C. 1991. "Spouses, Siblings and Sex-Linked Bonding: A Look at Kinship Organization in a Brazilian Slum." Pp. 133-160 in *Family, Household, and Gender Relations in Latin America,* edited by E. Jelin. London: Kegan Paul and UNESCO.

Glick, P. C. 1955. "The Life Cycle of the Family." *Marriage and Family Living* 17(1): 3-9.

Goldman, N., and A. Pebley. 1981. "Legalization of Consensual Unions in Latin America." *Social Biology* 28: 49-61.

Goldscheider, F. K., and G. Kaufman. 1996. "Fertility and Commitment: Bringing Men Back." Pp. 87-99 in *Fertility in the United States: New Patterns, New Theories,* edited by J. B. Casterline, R. D. Lee, and K. A. Foote. New York: The Population Council.

Goode, W. J. 1993. *World Changes in Divorce Patterns.* New Haven: Yale University Press.

Goody, J. 1983. *The Development of the Family and Marriage in Europe.* Cambridge: Cambridge University Press.

______. 1996. "Comparing Family Systems in Europe and Asia: Are There Different Sets of Rules?" *Population and Development Review* 22(1): 1-20.

Greene, M. E. 1991. "The Importance of Being Married: Marriage Choice and Its Consequences in Brazil." Ph.D. Dissertation in Demography, University of Pennsylvania.

Greene, M. E., and V. Rao. 1995. "The Marriage Squeeze and the Rise in Informal Marriage in Brazil." *Social Biology* 42(1-2): 65-82.

Guichard, P. 1977. As cited in Goody, 1983. *Structures sociales 'orientales' et 'occidentales' dans l'Espagne Musulman*. Paris.

Hajnal, J. 1953. "Age at Marriage and Proportions Marrying." *Population Studies* 7(2): 111-136.

______. 1965. "European Marriage Patterns in Perspective." Pp.101-143 in *Population in History: Essays in Historical Demography,* edited by D. V. Glass and D. E. C. Evers. London: Edward Arnold.

______. 1982. "Two Kinds of Preindustrial Formation System." *Population and Development Review* 8: 449-494.

Henriques, M. H. F. T. 1979. "Legal and Consensual Unions: Their Fertility Implications in Latin America." Pp.271-286 in *Nuptiality and Fertility,* edited by L. T. Ruzicka. Liége, Belgium: Ordina Editions.

______. 1989. "Brazil: Changes in Nuptiality and Their Fertility Implications." Pp.163-174 in *International Population Conference New Delhi,* vol. 3. Liége, Belgium: International Union for the Scientific Study of Population.

Ingoldsby, B. B. 1991. "The Latin American Family: Familism vs. Machismo." *Journal of Comparative Family Studies* 22: 57-62.

______. 1995. "Poverty and Patriarchy in Latin America." Pp. 335-351 in *Families in Multicultural Perspective,* edited by B. B. Ingoldsby and S. Smith. New York: Guilford.

Juarez, F. 1989. "Marriage Patterns, Family Formation and Demographic Change." Pp. 189-202 in *International Population Conference New Delhi,* vol. 3. Liége, Belgium: International Union for the Scientific Study of Population.

Keyfitz, N. 1987. "Form and Substance in Family Demography." Pp.3-16 in *Family Demography,* edited by J. Bongaarts, T. Burch, and K. Wachter. Oxford: Clarendon.

Klein, D. M., and J. M. White. 1996. *Family Theories: An Introduction*. Thousand Oaks, CA: Sage.

Leonard, O. E., and E. A. Louriero. 1980. "The Family in Bolivia." Pp. 199-234 in *The Family in Latin America,* edited by M. Singh Das and C. J. Jesser. New Delhi: Vikas.

Love, E. F. 1978. "Marriage Patterns of Persons of African Descent in a Colonial Mexico City Parish." *Hispanic American Historical Review* 58(1): 79-91.

Mauldin, P., and B. Berelson. 1978. "Conditions of Fertility Decline in Developing Countries, 1965–1975." *Studies in Family Planning* 9: 90-147.

McCaa, R. 1994. "Marriageways in Mexico and Spain, 1500–1900." *Continuity and Change* 9(1): 11-43.

Murdock, G. 1949. *Social Structure.* New York: Macmillan.

Nazzari, M. 1996. "Concubinage in Colonial Brazil: The Inequalities of Race, Class and Gender." *Journal of Family History* 21(2): 107-124.

Nyrop, R. F. 1983. *Brazil, a Country Study,* 4th ed., DA pam 550-20. Washington, DC: U.S. Government Printing Office.

Ogburn, W. 1932. *Social Change.* New York: Viking.

Ojeda de la Pena, N. 1987. "Family Life Cycle and Social Classes in Mexico." Ph.D. Dissertation, the University of Texas at Austin.

Palloni, A., and S. De Vos. "Change in Families and Households in Latin America, 1950–1980." University of Wisconsin. Unpublished manuscript.

Parrado, E. A., and M. Tienda. 1997. "Women's Roles and Family Formation in Venezuela: New Forms of Consensual Unions?" *Social Biology* 44(1-2): 1-24.

Pebley, A., and N. Goldman. 1986. "Legalization of Consensual Unions in Mexico." *Social Biology* 33(3-4): 199-213.

Portes, A. 1985. "Latin American Class Structures: Their Composition and Change during the Last Decades." *Latin America Research Review* 20(3): 7-39.

Quale, G. R. 1988. *A History of Marriage Systems.* Westport, CT: Greenwood.

Quilodrán, J. 1985. "Modalitès de la Formation et Evolution des Unions en Amerique Latine." Pp.269-283 in *International Population Conference,* vol.3. Liége: International Union for the Scientific Study of Population.

______. 1992. "La Vida Conyugal en America Latina: Contrastes y Semajanzas." Pp.245-264 in *The Peopling of the Americas Veracruz 1992,*vol.3. Liége: International Union for the Scientific Study of Population.

______. 1993. "Cambios y Permanencias de la Nupcialidad en México." *Revista Mexicana de Sociologia* 15(1): 17-40.

Rosero-Bixby, L. 1996. "Nuptiality Trends and Fertility Transition in Latin America." Pp. 135-150 in *The Fertility Transition in Latin America,* edited by J. M. Guzmán, S. Singh, G. Rodríguez, and E. A. Pantelides. Oxford: Clarendon.

Shyrock, H. S., J. S. Siegel, et al. 1973. *The Methods and Materials of Demography.* Washington, DC: Government Printing Office.

Smith, T. L. 1972. "Marriage and the Family in Brazil." Pp.459-483 in *Brazil: People and Institutions.* Baton Rouge: Louisiana State University Press.

Steward, J., and L. C. Faron. 1959. *Native Peoples of South America.* New York: McGraw-Hill.

Sweet, J. A. 1977. "Demography and the Family." *Annual Review of Sociology* 3: 363-405.

Sweet, J. A., and L. L. Bumpass. 1987. *American Families and Households.* New York: Russell Sage Foundation.

Thomson, M., E. McDonald, and L. Bumpass. 1990. "Fertility Desires and Fertility: Hers, His and Theirs." *Demography* 27(4): 579-588.

Todd, E. 1985. *The Explanation of Ideology: Family Structures and Social Systems.* Oxford: Blackwell.

Uhlenberg, P. 1978. "Changing Configurations of the Life Course." Pp. 65-97 in *Transitions The Family and the Life Course in Historical Perspective,* edited by T. Hareven. New York: Academic Press.

United Nations. 1990. *Patterns of First Marriage: Timing and Prevalence.* New York: United Nations. (ST/ESA/SER.R/111).

Zamudio, L., and N. Rubiano. 1991. *La Nupcilidad en Colombia.* Bogotá: Universidad Esternado de Colombia.

MARRIAGE AND ABORTION IN JAPAN
ADAPTATION AND ALTERNATIVE STRATEGIES

Lawrence K. Hong

INTRODUCTION

This paper examines both the changing functions of marriage in Japan and the cultural responses to the changes. In the West, industrialization has been the driving force in the modernization of marriage and the family. In Japan, the situation is more complex because of the multiplicity of factors involved. As one of the world's largest and most advanced economies, Japan has been slow in modernizing its social institutions. For example, the custom of *miai kekkon* (arranged marriage) persists, albeit it in the minority and in modified forms (Population Reference Bureau 1989, p. 5; Applbaum 1995; De Mente 1996, p. 319). "Glass ceilings" for women in white collar occupations are conspicuously low. These employees, known as Office Lady (OL), seldom advance beyond the clerical level. In 1990, only about 8 percent of the working women were in administrative or managerial positions (Japan External Trade Organization 1992, p. 20). Ancient fertility rites glorifying the power of the phallus, though in decline nationwide, are still in vogue in various localities, including a suburb adjacent to Tokyo (Bornoff 1991, pp. 83-84; De Mente 1996, p. 141).

In spite of the obstacle of traditions, Japanese marriage and family are changing (Kumagai 1995). Industrialization has played a role in the change, just as it has in other modern societies. However, there are also other forces at work in Japan, some of which are demographic and some, ideological. No doubt, those factors do not operate in a vacuum and are probably correlated with others, such as industrialization and internationalization. It is not my intention here to sort out the causal sequence or to map out their interrelationships. Rather, my aim is to examine how the two most important functions of marriage and the family in Japan—namely the status-giving and the reproductive functions—are being influenced. Furthermore, an equally important focus is to describe the ramifications and resolutions that have emerged. Information for this paper is based on published sources and field observations in Japan.

CHALLENGE TO THE STATUS-GIVING FUNCTION

One of the most important functions of marriage in Japan is conferring adulthood on individuals. While age is an important consideration for adulthood, and 20 is the age of majority, marriage can override it (Hendry 1981, p. 27; Applbaum 1995, p. 38). By getting married, a person attains adulthood regardless of age. Conversely, by not getting married, a person is considered irresponsible and not *ichinin-mae* (real adult), even though he or she may be many years over the age of majority. Because responsibility and maturity are perceived to be associated with marital status, it is especially important for *sarariman* (male salaried white-collar workers) to be married. It has been reported that discrimination against bachelors in career advancement is common in the workplace (Fukami 1989; Iwao 1993; De Mente 1996). Female employees, as mentioned earlier, are discriminated against regardless of marital status, and most leave work for home upon marriage.

The importance of marital status in the corporate culture also can be seen in the role it plays in courtship. Many large companies provide in-house matchmaking services for their employees. Some of the most prestigious corporate go-betweens are Sanwa Bank's "Green Bridal Department," The Mitsubishi Group's "Diamond Family Club," and Daiichi-Kangyo Bank's "Heart Family Club" (Yoshida 1990). Others, such as The Nissan Motor Company in Kanagawa, contract the services of outside marriage brokers to help their employees find a mate ("The Changing Face of Marriage" 1990).

The status-giving function of marriage, however, has been seriously challenged by the new generation of women who are not eager to get married. This is reflected in the age at first marriage among women, which rose from 23.6 years in 1950 to 26.9 years in 1990 (Tsuya 1994, p. 96). Breaking the 25 years-of-age barrier is a watershed event for Japanese women because traditionally this marked the "deadline for women seeking marriage" (Applbaum 1995; De Mente 1996, p. 320). Traditionally, the ideal age *(tekireiki)* is lower, usually in the 22- to 23-years-of-age range (Hendry 1981, p. 118). Women are delaying marriage for a variety of reasons, one of them being educational aspiration. In 1993, more than 43 percent of female high school graduates went to college as compared to less than 6 percent in 1960 (Kokusei-sha Corporation 1994, p. 553). Apparently, more and more women are choosing education over marriage in their early adulthood, displaying little interest in the status that marriage may offer.

Financial independence is another factor. Industrialization and the prosperity of the past decades have vastly increased employment opportunities for women outside the home. Even though women have always been an important part of Japan's labor force, they were concentrated primarily in the agrarian and domestic sectors before industrialization and in the poorly paid textile industry in the early days of industrialization (late nineteenth century). After World War II, the labor participation rate for women continued to be high due to the high rate of widowhood resulting from the casualty of war. However, most work available to women remained in the domestic and the low-paying segments of the economy. It was not until the 1960s, the onset of the great economic expansion, that large numbers of women began to move into other manufacturing and white-collar occupations that offer better pay and working conditions. Between the 1950s and the 1990s, the proportion of "family workers" in the female labor force dropped from approximately 60 percent to 16.7 percent (Kodansha International 1995, p. 196). With better economic options available, marriage has become less attractive for women. It follows that employment is competing with matrimony as a symbol of adulthood.

For men, the status-giving function of marriage has been undermined by the reality of the "marriage squeeze." Since the mid-1980s, a shortage of marriage partners for men has become a social problem in Japan (Iwao 1993; Applbaum 1995). A 1991 survey found that "40 percent of the men said they were having a tough time finding a wife" (Watanabe 1992).

The most important factor contributing to the marriage squeeze is the imbalance in the sex ratio. In 1985, statistics revealed that single men exceeded single women by 2.01 million in the prime age group for marriage, 20 to 34 (Yoshida 1990). The gaps remain almost the same in 1989 and 1991, with single men out numbering women by 2.28 and 2.7 million, respectively (Kokusei-sha Corporation 1990; "By the Numbers" 1991; Tsuya 1994). This imbalance has caught national attention, prompting the media to declare that Japan has entered the "Era of Too Many Men."

Adding to the difficulties in finding a spouse for men is the *chonan* (eldest son) problem. Traditionally, the eldest son is expected to take care of their parents in old age by sharing the same place of residence with them. This creates a stem-family, which has always been a very common living arrangement in Japan. In a 1988 study, Martin and Tsuya (1994) reported that of the men who were in coresidence with their parents, 59 percent indicated that it was their duty as eldest sons to do so.

Many young women in the 1990s, however, do not see it as their duty to take care of their in-laws, and are reluctant to marry men who take on the *chonan* responsibility. Thus, many men who assume this time-honored duty are at a disadvantage in finding a mate. In short, whether it is by choice, as in the case of women, or by circumstances, as in the case of men, the capacity of marriage as a status-giving institution has been compromised.

RESPONSES TO THE MARRIAGE SQUEEZE

Nevertheless, it is still important for men to get married for career reasons if not for anything else. For example, De Mente (1996, p. 32) reported that: "A man who remains unmarried after he is 30 is less likely to get promoted in his job." In response to the marriage squeeze, many men turn to other countries to solve their problem. A nation known for its ethnic and racial homogeneity, Japan registered a record number of international marriages in 1993 ("The Changing Color of Marriage" 1994). In that year, 20,092 men married non-Japanese wives (compared to only 6,565 women who married non-Japanese men). The largest numbers are Filipino women (31%), followed by Korean (25%), and then Chinese (23%). Others attempt to improve their chances in finding a spouse by enrolling in *Hanamuko Gakko* (Bridegrooms Schools), an educational innovation and profit-making opportunity created by the marriage market. These schools teach men the art of courtship, including conversational skills, proper grooming, and dressing well (Fukami 1989; "A School for Bridegrooms" 1994).

The most dramatic response to the marriage squeeze is the emergence of the computer matchmaking industry. Formally known as *Kekkon Joho Sabisu* (Marriage Information Service), the first one opened for business in 1978, and the numbers soon grew to nearly 5,000 (Ito, Murray, and Loveless 1991). At one point, OMMG, one of the best-known companies in the field, had branches in 67 cities and a clientele of more than 62,500 men and women. By paying an annual or semi-annual membership fee, individuals are provided with a list of computer matches based on compatibility analyses of self-reported personal information submitted on questionnaires. If two matched members are mutually interested, the service makes arrangements for both parties to meet. The providers of these services claimed that about 25 to 35 percent of their clients were successful in getting married (Ito et al. 1991).

The popularity of computerized matchmaking in Japan is not surprising because it is simply a high-tech version of *Omiai,* the time-honored method of mate selection that first came into popular practice in the Edo period, AD 1600 to 1868 (Kodansha International 1995, p. 194). In fact, *Omiai* itself is receiving renewed attention as a means of finding a spouse.

Omiai, which literally means "look meeting," is a formal meeting in which the prospective partners and their parents size each other up for the purpose of deciding whether the couple should marry. Because the decision of whether or not to marry usually is made shortly after the meeting, marriages that go through this custom are often called *miai kekkon* (look-meeting marriage), better known as an arranged marriage.

Although there has never been a standard procedure for arranged marriage, it generally involves the following sequence of events (Hendry 1981). When a son or daughter reaches the ideal age of marriage *(tekireiki),* parents approach a friend or a relative to serve as go-between *(nakodo)* in finding a potential spouse for their child. When a candidate has been identified, the go-between arranges for an exchange of resumes *(rirekisho)* that contain personal and family data, including photographs of the parties involved. In the old days, before photography, a "hidden look" *(kagemi)* would be arranged by the go-between so that the candidates could get a glimpse of each other without the knowledge of the other party (Hendry 1981, p. 123). If everything looks promising, a date is set for that all important meeting, *Omiai.* In describing such a meeting in recent times, Bornoff (1991) wrote:

> Both in immaculate tailored suits, the young man and woman in question stare unwaveringly at the tablecloth, hardly daring to exchange a glance—let alone a word. The girl eats practically nothing. The parents talk over their heads. The boy's father reels off his son's academic achievements and his prospects; the girl's father will extol her virtuous nature, schooling, hobbies, housekeeping abilities and fondness of children. At some point, the two will be expected to say something, generally a tremulous and extremely modest version of the already very reticent paternal summary. (p. 50)

After the meeting, both sides are expected to inform the *nakodo* as to their decisions. Both the candidates themselves and, in the past, their parents might veto a marriage for any reason. On the other hand, if the decisions are favorable, the go-between will work with the families in making plans for the wedding *(kekkonshiki).* This method of mate selection was fast dwindling after World War II (Blood 1967), but there has been a renewed interest in it in recent years as a result of the marriage squeeze. Knight (1995) reported that in some rural areas, "male celibacy" has become such a problem that the local governments are promoting *Omiai* by offering "go-between rewards." Today, approximately 25 to 30 percent of all marriages are *miai kekkon* (Population Reference Bureau 1989, p. 5; Applbaum 1995, p. 37; De Mente 1996, p. 319).

It must be pointed out that there is a crucial difference between the new *Omiai* and the old. Today, *Omiai* is used not to make decisions on marriage but, rather, to make decisions about dating. Couples who are interested in each other after the *Omiai* go on dates for an extended period of time before deciding on marriage. Because the Japanese term for dating is *deto,* a more appropriate term for the current practice is *Omiai deto*.

Another modern variation of *Omiai* is the use of paid go-betweens (Applbaum 1995). Instead of asking friends and relatives to volunteer as *nakodo,* it has become acceptable to hire professional matchmakers or *pro nakodo*. These professionals belong to go-between associations where they exchange information on their clients and identify potential matches for them for a fee. When a match is found, the *pro nakodos* for each side discuss the prospects with their candidates, and, as in the former practice, if the prospects are mutually interested, a formal meeting is held.

Unlike the traditional *Omiai,* however, the meetings arranged by the *pro nakodo* are frequently attended by the couples alone, without the presence of the parents. The go-betweens accompany their candidates to a public meeting place, usually a restaurant, and leave after a brief introduction. Applbaum (1995), who has conducted an in-depth study of this form of *Omiai,* reported that about 20 percent of the couples were successful in finding someone they like after the initial meeting.

It appears that in times of marriage squeeze, the advantage of using *Omiai* for mate selection is once again being appreciated either in the high-tech form or in a modified version of the old-fashioned way. Men who are eager to find a mate will utilize as many methods as possible, including *Omiai,* to meet the opposite sex. Women, in spite of their favorable position in the marriage market, are receptive to *Omiai* because they also want to increase their chances of meeting the right man.

CHALLENGE TO THE REPRODUCTIVE FUNCTION

Another traditional expectation of marriage, both in Japan and in many other societies, is to view marriage as a vehicle for reproducing offspring to continue the family line and to replenish societal membership. Legacies of a pervasive fertility cult carried over from its agrarian past continue to thread through the cultural fabric of contemporary Japan. Although the numbers of shrines and festivals devoting to phallic deities have dwindled, the remaining ones, numbering about 40, are jolting reminders of the centrality of fertility in Japan's Shinto belief system (Bornoff 1991, pp. 83-84; De Mente 1996, p. 141). In describing the annual *Jibeta Matsuri* (Fertility Festival) held at the *Kanamara-Sama Jinja* (Metal Phallus Deity Shrine) near Tokyo, Bornoff (1991) wrote: "Exhorting fertility, local parents and grandparents pointedly place toddlers of either sex atop two monster phalli set on wooden frames like cannons" (p. 86). "Children . . . lick at phallic lollipops or munch banana topped with a circular blob of pink chocolate. There are phallic cakes, phallic signs and pictures, phallic toys and phallic *object d'art*" (p. 91).

Coupled with fertility devotion is another tradition that holds marriage as the sole legitimate means for procreation. Although young couples may initially expect romance and companionship in their marriages, most of them eventually drift back into the traditional emphasis on childbearing and child caring, especially for the wives (Kodansha International 1995, pp. 190-191). The norm confining procreation to marriage is so strong that it was not until 1995 that the Tokyo Court of Appeals struck down the term "illegitimate child" *(shiseiji)* for births outside of wedlock in the resident registration form *(juminhyo)*. In a step backward, however, the Supreme Court, in the same year, upheld the constitutionality of another law, Article 900, paragraph 4, that discriminates against illegitimate children. The legislation stipulates that the right of succession for illegitimate children is to be one half that of legitimate children (see www.law.tohoku.ac.jp/new-law-e.html).

Against this background of pro-childbearing values is the anomaly of widespread abortions among married women. Japan has one of the highest abortion rates in the world, with over 60 percent of women having an abortion by age 40, of which about the same percentage are married women (Coleman 1991, p. 5; Gelb 1996, p. 119). According to the Ministry of Health and Welfare statistics, the most likely persons who have an abortion are married women, 30 to 39 years of age with one or two children. In contrast, there are very few teenage abortions; less than 7 percent in 1993 (Gelb 1996, p. 124).

The high abortion rate together with other demographic factors, such as late marriage, has contributed to a steep decline in fertility. The legislation that has the most direct bearing on the liberalization of abortion is the Eugenic Protection Act of 1948 (revised in 1949 and 1952), which allows abortion "when there is risk that pregnancy or birth will cause substantial harm to the health of the mother due to physical or economic reasons" (Coleman 1991, p. 20). In 1952, further revisions were made granting the *shitei ishi* (designated physician who is certified to perform abortion) the sole authority in deciding whether or not a pregnancy qualified for *chuzetsu* (interruption). Before the change, requests for abortion required the review of a committee (Coleman 1991, p. 20).

The effect of legalized abortion on fertility was immediately evident. In 1947, a year before the enactment of the law, the total fertility rate (TFR) was 4.5 children per woman (15 to 49 years old). It dropped to 2.0 by 1957, leveling off between 2.0 and 2.7 until 1974, and then resumed its downward movement, reaching 1.5 in 1990 (Tsuya 1994, p. 94). With a TFR of 1.5 children, Japan's fertility rates are below the standard replacement level of 2.1. Hence, Japanese couples are not reproducing enough offspring to replace themselves, a fact that highlights the diminishing reproductive function of marriage.

RESPONSES TO THE CULTURAL DILEMMA

While low birth rate may be incongruent with the value placed on the reproductive function, the method of achieving it presents a cultural dilemma of a much higher order. For a variety of reasons, oral contraceptives have remained illegal for birth-control in Japan. Many women and women's groups are unconvinced that the pill is safe and have not pushed for its legalization (Gelb 1996, p. 128). Furthermore, the pill is strongly opposed by the condom industry for obvious reasons, and by the Japan Medical Association, whose *shitei ishi* members are dependent on the "pregnancy interruption" practice (Tamkins 1996). Hence, the availability of oral contraceptives continues to be restricted, and legal only for specific medical applications, making abortion the singular means of birth control when prophylactics and other less reliable methods of contraception failed or were not being used. (According to recent press reports, legislation legalizing the pill has been passed in 1999.)

Hereupon lies one of the most glaring contradictions in Japanese society as well as a most unique solution to the dilemma. Buddhism, the counterpart of Shintoism in Japan's dual religious system, is adamantly opposed to abortion. Among all the Buddhist teachings, the first and foremost is "I will not willingly take the life of a living thing," and this includes the life of a fetus (LaFleur 1992, p. 11). Japanese Buddhism advances the status of the fetus further by attributing a soul or spirit to it. Since the seventeenth century or maybe even earlier, there has existed a belief that the soul of an aborted fetus, known as *mizuko* (water child), would go to *sai no kawara* (children's limbo) waiting to be born again (Coleman 1991, p. 60). The medical profession, which is keenly aware of this popular sentiment on "fetal spirithood" (Oaks 1995), has adopted the term *ninshin-chuzetsu* (pregnancy interruption) for abortion in order to avoid the common term *datai,* which has a more explicit connotation of "killing the fetus."

For women who have had an abortion and are troubled by it, the problem may not be summarily dismissed by linguistic play; they may need to do something more concrete, as evident in the millions who have turned to Buddhism itself to seek consolation. Often accompanied by their spouses, these women go to the temples or to the special cemeteries for *mizuko* to pray in front of the statues of Jizo, a *Bodhisattva* (Buddha-to-be) who is known as a patron of children, born and unborn, including the miscarried, the stillborn, and, especially, the aborted. The repentant parents bring flowers, toys, and snacks to the children whom they believe they have denied a chance at life, and ask for their forgiveness. Many also leave signed written confessions and apologies on the altars or bulletin boards.

Since the 1970s, with the country becoming more affluent and the people having more abortions, these memorial services *(mizuko kuyo)* and Jizo figurines have become a major source of revenues for many temples and cemeteries (Desmond 1996). In Hase-dera Temple behind the famous "Great Buddha of Kamakura," there is literally an entire hill of Jizo statues. The temple began installing the statutes for parents in 1974 and, within a decade, the numbers have grown to 50,000 (Coleman 1991, p. 60; LaFleur 1992, p. 4). Each image, usually about 2 feet in height, sometimes wearing a red apron, is meant to represent both the *mizuko,* the lost child, and Jizo, the *Bodhisattva* who takes care of him or her in the other world.

The rise of Jizo worship needs an explanation because it shows how religious ideas were adapted to meet human needs. The Jizo tradition, before it became a *mizuko kuyo,* has its origin in India, and entered Japan most probably between AD 600 and 700 by way of China, where Jizo is known as Ti-Tsang. By the nineteenth century, the Jizo had become "the most Japanese of all Japanese divinities," even though it has received very little attention from Japanese Buddhist scholars (LaFleur 1992, pp. 44-45). Jizo's popularity is to be found on the street corners and on the hillsides, where its icons, in various sizes and styles, are chiseled in stones.

In Buddhist theology, Jizo (called *Ksitigarbha* in Sanskrit) is best known for his great compassion and his celebrated vow, "Unless all the souls in hell have been saved, I vow not to enter Buddhahood" (Motizuki 1989, p. 17; Total Buddhism Publisher 1995, p. 11). Largely because of this vow, Jizo was viewed as savior of individuals in suffering. In subsequent Buddhist teachings, his altruism has been expanded to include helping women escape the injustices of a patriarchal society (Total Buddhism Publisher 1995, p. 24). From this vantage point it is easy to understand why Japanese women are drawn to Jizo and became, as early as the Edo era, the "principal creators and shapers of the Jizo cult" (LaFleur 1992, pp. 50-51). Responding to the desires of women, Jizo eventually became a guardian of children, especially the unborn. LaFleur (1992, p. 47) has noted that, corresponding with Jizo's rise as a patron of children, his rendition, which was originally adult in form, has become increasingly childlike (e.g., see Fujiwara and Sugimura 1996). But while Jizo may have shrunk in size, he has grown immeasurably in stature in the hearts of the masses.

CONCLUSION

In examining the normative expectations of marriage in contemporary Japan—namely the status-giving and the procreative functions—a pattern emerged. In both cases, the main thrusts of the responses to social change have brought forth traditional customs with a broad base of support in society. This is not to say the responses are static and trite. On the contrary, they are creative applications of time-tested practices that address the exigencies in a novel way.

Faced with the problem of a marriage drought for men, *Omiai* has resurfaced in a variety of forms—computer matchmaking, *pro nakodo,* and *Omiai deto*—helping men to attain the status that only marriage can confer. When confronted with the seemingly insurmountable moral conflict of two diametrically opposed positions—the belief in the sanctity of fetal life and the acceptance of abortion as a means of birth control—Jizo rituals provide a mechanism to bridge the contradictions and ease the guilt.

In fact, the Japanese way of handling the dilemma has attracted international attention (Desmond, 1996), because it may offer an alternative direction in the abortion debate. Paraphrasing the noted feminist Naomi Wolf, Desmond (1996) raised the question of whether "a Westernized form of the *(mizuko)* memorial service might soothe consciences troubled by abortion guilt." Adding to her intriguing thought, I wonder what would happen if pro-choice and pro-life advocates both were to share the Japanese view on the sanctity of the fetus. In the West, the discourse on abortion has been bogged down over the issue of whether or not the fetus represents human life. Perhaps if we were all to accept that it does, we would then be able to move on to the next level of the abortion debate and find a common ground for justifying the interruption of pregnancy.

REFERENCES

Applbaum, K. 1995. "Marriage with the Proper Stranger: Arranged Marriage in Metropolitan Japan." *Ethnology* 34(1): 37-52.

Blood, R. O. 1967. *Love Match and Arranged Marriage: A Tokyo-Detroit Comparison*. New York: The Free Press.

Bornoff, N. 1991. *Pink Samurai: Love, Marriage, Sex in Contemporary Japan*. New York: Simon and Schuster.

"By the Numbers." 1991. *Japan Times Weekly International Edition,* January 28-February 3, 21.

"The Changing Color of Marriage." 1994. *Japan Times Weekly International Edition,* October 3-9, 20.

"The Changing Face of Marriage." 1990. *Japan Times Weekly International Edition,* September 24-30, 10.

Coleman, S. 1991. *Family Planning in Japanese Society.* Princeton: Princeton University Press.

De Mente, B. L. 1996. *Japan Encyclopedia*. Lincolnwood, IL: Passport Books.

Desmond, J. F. 1996. "Apologizing to the Babies." On-line. Available: www.leaderu.com/ftissues/ft9610/desmond.

Fujiwara, T., and T. Sugimura. 1996. *Warabe Jizo*. Tokyo: Suzuki. (Japanese)

Fukami, A. 1989. "Teaching the Art of Courtship." *Los Angeles Times,* September 14, 22.

Gelb, J. 1996. "Abortion and Reproductive Choice: Policy and Politics in Japan." Pp. 119-137 in *Abortion Politics: Public Policy in Cross-cultural Perspective,* edited by M. Githers and D.M. Stetson. London: Routledge.

Hendry, J. 1981. *Marriage in Changing Japan*. Tokyo: Charles E. Tuttle.

Ito, T. K., V. Murray, and R. U. Loveless. 1991. "Computerized Matchmaking." *Mangajin* 11: 16-20.

Iwao, S. 1993. *The Japanese Woman*. New York: The Free Press.

Japan External Trade Organization. 1992. *U.S. and Japan in Figure II.* Tokyo: Japan External Trade Organization.

Knight, J. 1995. "Municipal Matchmaking in Rural Japan." *Anthropology Today* 11(2): 9-17.

Kodansha International. 1995. *Japan: Profile of a Nation*. Tokyo: Kodansha International.

Kokusei-sha Corporation. 1990. *Nippon: A Charted Survey of Japan 1990/91*. Tokyo: Kokusei-sha Corporation.

______. 1994. *Nippon kokusei zue 94/95 [A Charted Survey of Japan 94/95]*. Tokyo: Kokusei-sha Corporation. (Japanese)

Kumagai, F. 1995. "Families in Japan: Beliefs and Realities." *Journal of Comparative Family Studies* 26(1): 135-163.

LaFleur, W. R. 1992. *Liquid Life: Abortion and Buddhism in Japan*. Princeton: Princeton University Press.

Martin, L. G., and N. O. Tsuya. 1994. "Middle-aged Japanese and Their Parents: Coresidence and Contact" Pp. 1153-178 in *Tradition and Change in the Asian Family,* edited by L. Cho and M. Yada. Honolulu: East-West Center

Motizuki, S. 1989. *Jizo Bosatsu.* Tokyo: Gakusei-sha. (Japanese)

Oaks, L. 1994. "Fetal Spirithood and Fetal Personhood: The Cultural Construction of Abortion in Japan." *Women's Studies International Forum,* 17(5): 511-523.

Population Reference Bureau. 1989. "Japan." *Population Today* 17(2): 5.

"A School for Bridegrooms." 1994. *Mangajin,* March 33, 11.

Tamkins, T. 1996. "Japan Should Legalize Birth-control Pills, Expert Says." *Medical Tribune News Service,* February 14

Total Buddhism Publisher. 1995. *Canon of Ti-Tsang Bodhisattva*. Taipei: Total Buddhism Publisher. (Chinese)

Tsuya, N. O. 1994. "Changing Attitudes Toward Marriage and the Family in Japan." Pp. 91-119 in *Tradition and Change in the Asian Family,* edited by L. Cho and M. Yada. Honolulu: East-West Center.

Watanabe, T. 1992. "In Japan, a Goat Man or No Man." *Los Angeles Times,* January 6, A1.

Yoshida, R. 1990. "Getting Married the Corporate Way. *Japan Quarterly* April-June: 171-175.

ATTITUDES TOWARD PREMARITAL SEX, NON-MARITAL CHILDBEARING, COHABITATION, AND MARRIAGE AMONG BLACKS AND WHITES

Wendy Y. Carter

INTRODUCTION

Differences in marital and family behavior between blacks and whites are important in American society. The "transition to adulthood" continues to change, and it is important to study the social acceptance of alternative sequences into this transition, especially if different social groups follow different avenues. If marriage is becoming less central to the process, it is still normative in the United States. I am particularly concerned with comparisons of society's attitudes by race on issues of premarital sex, non-marital childbearing, cohabitation, and marriage.

The sociological literature on attitudes follows two distinct theoretical lines of thought: structural or cultural. Structuralists theorize that persons who share the same relationship to the social structure and institutions often share similar attitudes because of their shared life experiences. In contrast, cultural theorists focus on the influence of group attitudes as an intervening mechanism on social outcomes. Cultural theory presumes that attitudes influence behavior.

Previous sociological attempts to explain differences in behaviors between blacks and whites have often centered on a culture-based theory that focuses on group differences in attitudes transmitted through socialization. Although many have speculated on the preexisting attitudes, few have actually studied the influence of prevailing social norms and individual preferences of blacks and whites empirically. This research examines the extent to which attitudes on marriage and family behavior are consistent with actual behaviors within black and white populations in the United States.

Much of the traditional research on racial differences in marriage and family behavior painstakingly highlights divergent behaviors and infers cultural differences in attitudes, normative beliefs, and values on the basis of these differences. For example, Bernard (1966) and Furstenberg (1976) have suggested that the higher rates of illegitimacy among blacks relate to a lower stigma associated with non-marital childbearing.

In contrast, Staples (1985, p. 1006) writes that the "unconventional family arrangements and lifestyles easily can mislead outsiders to assume that blacks are strongly in accord with newly emerging alternative family lifestyles." He suggests that the "dominant force" that accounts for differences between black and white family structures can be found in the structural conditions of the black population. Moreover, Kluegel and Smith (1986) maintain that attitudes are based on one's position in the social structure and are separate from one's behavior. If Kluegel and Smith are correct, then blacks and whites with similar socioeconomic and demographic characteristics should have similar attitudes on many issues in spite of the behavioral differences between the two groups.

I am particularly concerned with comparisons of attitudes by race on issues of premarital sex, non-marital childbearing, cohabitation, and marriage. I address each issue separately. To integrate behavior into my discussion of attitudes, I begin by providing an overview of the recent literature on each issue, followed by a synopsis of the current research findings examining racial differences in these four specific behaviors.

The marked racial differences in marriage and family behavior provide the context in which these behaviors may be differently accepted in society. This paper first considers differences in the structural composition of the two groups, and then conducts a comparative analysis of the attitudes of blacks and whites on premarital sex, cohabitation, non-marital childbearing, and the tolerance of singleness in relation to marriage as a preferred adult status using the National Survey of Families and Households.

This paper is organized as follows. In the first section, I briefly review the literature on each topic previously discussed. Next, I provides the data source, the independent variables considered in the analysis, and statistical methods used. Then I discusses the bivariate and multivariate results. The last section provides the conclusion and implications of this research.

LITERATURE REVIEW

Premarital Sex

Traditional research on sexual permissiveness links sexual attitudes with behaviors without reference to a particular causal order of this association. The debate in this field centers on either attitudes alone or on the ability of behavior to predict attitudes. Many sociologists (Duberman 1974; Singh 1980; Robertson 1987; Walsh 1989; Caplow 1991) note that sexual attitudes have become more permissive over time. Walsh (1989) finds that the trend in premarital sexual permissiveness is concomitant with increased sexual behavior. Others (Duberman 1974; Robertson 1987; Thornton 1989; Caplow 1991) cite the widespread availability of oral contraceptives as the main antecedent for changes in behavior and attitudes because it separates procreation from sexual satisfaction. Thornton (1989) suggests that increasing acceptance of premarital sex and cohabitation has lessened the importance of marriage as a "sanctioning institution for sex" (p. 889). Rindfuss and Parnell (1989) suggest that the sexual revolution has meant a fundamental change regarding sexual intercourse for many unmarried individuals.

Much of the empirical research on premarital sex is somewhat antiquated and focuses primarily on adolescents and young adults. Moreover, few studies examine racial differences in premarital sexuality. Those that do, find substantial racial differences in the prevalence and timing in premarital sexual behavior (Zelnick and Kanter 1980). Zelnick and Kanter find an increase in both the rate for females and the percentage of adolescents and young adults engaging in premarital sex. Since 1970, the disparity between black and white sexually active teens narrowed because of a substantial increase among white teens. The rates by ages for males and females converged as well. However, comparisons by race still show that premarital sexual intercourse is earlier and considerably higher among blacks than whites. By the late 1980s, most 18-year-old females had experienced coitus, as had one of four by age 15. A majority of males had experienced sex by age 17. Although blacks might experience sex at an earlier age than whites, by age 19 these differences all but disappear (Moore and Stief 1991).

More recent studies contribute to our understanding of premarital sexual attitudes among blacks and whites in terms of premarital sexual permissiveness (Singh 1980; Brown 1985; Smith 1992). Smith (1992) contends that a consequence of differential experiences in: (1) age of first intercourse, (2) teenage sexual activity, (3) non-marital childbearing, and (4) marital instability help to "predict" greater sexual permissiveness among blacks. Despite the differences in experience, Singh (1980) notes that the difference in premarital sexual permissiveness is less pronounced between blacks and whites than between men and women. In addition, studies consistently find that church attendance is an important factor in influencing black female sexual permissiveness (Singh 1980; Brown 1985). Their results also show that sexual standards are not uniform among all blacks.

Similar to Singh, this study will examine the variation in attitudes between black and white adults considering the marked differences in premarital sexual behavior. Although Smith assumes that behavior is able to predict attitudes, this study questions this assumption and instead regards behaviors as the social setting in which these attitudes exist. Because of the risk associated with premarital sexual intercourse (e.g., acquired immunodeficiency syndrome [AIDS], pregnancy, sexually transmitted diseases), it is important to understand the extent to which current patterns of premarital sexual behavior are in keeping with the attitudes and values of the society in which people live (Moore and Stief 1991, p. 362).

Non-marital Childbearing

Research on non-marital childbearing documents that it is more prevalent and more acceptable today than in prior decades (Farley and Bianchi 1987; Bumpass and McLanahan 1989, Mare and Winship 1991). The question of the underlying causes for the substantial increase in non-marital fertility has been unresolved by the current debate in the literature.

Some of the major arguments center on the decline in marriage and the reduced stigma associated with evidence of non-marital sex. Bumpass and McLanahan (1989) write that several decades ago, a non-marital pregnancy would have been "stigmatizing evidence of sexual intercourse outside marriage" (p. 279). Whereas non-marital cohabitation is considered a relatively new phenomenon in relation to the decline of marriage, its impact on non-marital childbearing has received little attention in the literature.

Two studies (Tanfer and Horn 1985; Bachrach 1987) address cohabitation as evidence of a non-marital sexual relationship by focusing on the frequency of sexual intercourse and the increased risk of non-marital pregnancy among cohabiting couples. Bachrach (1987) finds that cohabiting couples have sex more frequently than non-cohabiting couples. She suggests that the frequency of intercourse is influenced by the living arrangement, not by the couple's marital status (p. 635). Therefore, she predicts that the dramatic rise in cohabitation will increase the proportion of cohabiting couples with children. Her evidence for this prediction is the sharp increase in non-marital births (32%) to young cohabiting women (aged 20 to 29 years), versus 13 percent for single women who have never cohabited (Tanfer and Horn, 1985).

Other explanations suggest that the increase in the number of non-marital births could possibly be attributed to the economic independence of women, the availability of income transfer payments, or social acceptance of this practice (Farley and Bianchi 1987).

Whatever the causes, studies consistently find considerable racial differences in the rates and prevalence of non-marital childbearing among blacks and whites (Wilson and Neckerman 1986; Bumpass and McLanahan 1989; Mare and Winship 1991). While the rate of non-marital childbearing has been decreasing among blacks and increasing among whites, the racial gap is still quite large (Espenshade 1985; Bumpass and McLanahan 1989).

The major arguments used to explain these differences incorporate not only the economic barriers to marriage for blacks, but also the changing demographic composition of the population as a consequence of delayed marriage. Comparisons by race show that the decline in marriage has been more pronounced for blacks than for whites (Espenshade 1985). Mare and Winship (1991) find that "the main cause for the tripling in the number of black families headed by women since 1940" has been the increasing number of women at risk of having a child out of wedlock (p. 175).

Bumpass and McLanahan (1989) looked at the marked differences in non-marital fertility between blacks and whites and wondered "whether there is a different orientation toward marriage and childbearing in the black community" (p. 280). They find similar rates of premarital births by race depending on whether the individual comes from a high-risk or low-risk background. Non-marital childbearing is often closely related to class and is disproportionately prevalent among women with less education (Hogan and Kitagawa 1985; Bennett, Bloom, and Craig 1989; Rindfuss and Parnell 1989). Correspondingly, Wilson and Neckerman (1986) suggest that poverty is a major contributor to the formation of black single-parent families in that "male joblessness" could be the most important factor in determining this family structure.

Moreover, Bumpass and McLanahan (1989) suggest that the weakened taboo against "illegitimacy" and the high rate of divorce contribute to the social acceptance of non-marital fertility. Mare and Winship (1991) make a comparable argument by concluding that "separation of childbearing from marriage" is viewed not only as feasible, but acceptable (p. 195). Although these conclusions might be true, they provide little in the way of analytical insight into the relationship between attitudes and behavior. Beyond that, these conclusions are nothing more than speculation on the presumed coupling of behaviors and attitudes.

Pagnini and Rindfuss (1993) provide a first look at the an increasing acceptance of non-marital fertility. They find that a majority of black women agree that "there is no reason why a single women shouldn't have children and raise them if she wants to" (p. 334). This study extends and augments the literature on this issue by exploring the underlying relationship between a high level of acceptance among blacks and the compositional differences between blacks and whites using a logit regression model. A multivariate analysis should provide further insight into the nature and strength of this relationship.

Cohabitation

Cohabitation between unmarried persons is often discussed in relation to the decline of marriage. Recent studies (Thornton 1989; Bumpass, Sweet, and Cherlin 1991) suggest that cohabitation among unmarried couples is a relatively new phenomenon with considerable impact on the decreasing marriage rates. Thornton (1989) shows that while marriage rates have been decreasing rapidly, cohabitation has emerged as a new alternative for young couples. Bumpass and Sweet (1989) find that by their late 20s or early 30s "nearly half of all young American adults has cohabited at some point in their lives" (p. 619). The duration of cohabiting unions is often relatively short and often ends by separation (Bumpass et al. 1991; Schoen and Owens 1992). In addition, Bumpass, Sweet, and Cherlin (1991) find that education is negatively correlated with cohabitation in that the least educated in society show the highest rates of cohabitation.

Comparisons by race show that while the differential is not large, cohabitation is not only more frequent but also lasts longer for blacks than whites (Schoen and Owens 1992). However,

Bumpass and Sweet (1989) find that when education and family background are taken into account, the rate of cohabitation among never-married persons is 26 percent lower among blacks than among non-Hispanic whites. Given the marked delay in marriage, which results in a higher prevalence in cohabitation among blacks, we might expect their attitudes to be more approving, regardless of whether cohabitation is a cause or consequence of prior behavior.

If cohabitation is only a phase, it is likely that these views are temporarily correlated with their behavior. Perhaps a more important comparison is not between cohabitors and non-cohabitors, but between married couples who previously cohabited and those who did not. Cohabitation might be a temporary alternative to marriage for some, but it is not a deterrent to marriage for most. That most people eventually marry might suggest that their temporary behavior does not fully explain their overall attitudes toward marriage, cohabitation, or being single.

Two important studies by Thornton (1989) and Sweet and Bumpass (1990) specifically deal with attitudes regarding non-marital cohabitation. Both studies were consistent in finding that cohabitation without marriage is acceptable to a majority of young people. Moreover, Thornton suggests that approval of non-marital cohabitation is increasing. Sweet and Bumpass find "that groups that are likely to be facing the greatest economic hardships (blacks and those with 12 years of education or less) are most likely to approve of cohabitation" under certain circumstances (p. 23).

Despite these findings, the lack of clarity about factors that account for racial differences in attitudes regarding cohabitation still remains. The present study not only accounts for differences in attitudes by race but simultaneously takes into account other factors such as age, marital status, income, and education. I suspect that the prevalence of non-marital cohabitation among blacks has less to do with a cultural acceptance of non-marital cohabitation and more to do with structural and demographic differences in the black and white populations. Whether cohabitation is a cause or a consequence of delayed marriage, measuring society's attitudes toward this alternative lifestyle might provide some further insight into marriage, the acceptance of non-marriage, marital disruption, and cohabitation. If blacks have lower rates of marriage and higher rates of marital disruption, one would expect that blacks should have higher rates of approval of cohabitation than whites.

Marriage

The literature on marriage documents a marked decline in marriage rates (Becker 1981; Espenshade 1985; Schoen 1985; Goldscheider and Waite 1986; Thornton 1988; Sweet and Bumpass 1990; Mare and Winship 1991). Specifically, trends in marriage behavior and racial differences in marriage have been well documented. Studies on marriage consistently reveal (1) a trend toward later marriage, (2) a decline in the proportion ever marrying, (3) a smaller fraction of total lifetime spent in the married state, and (4) a shorter duration of marriage (Epenshade 1985; Goldscheider and Waite 1986; Bumpass and Sweet 1989).

The bulk of research on contemporary variation in black and white marriage patterns concentrates on the divergent pattern which has emerged since 1950 (Espenshade 1985). These studies typically find that the decline in marriage has been more pronounced and has occurred earlier for blacks than for whites (Espenshade 1985; Bennett et al. 1989; Bumpass and Sweet 1989).

Traditional research that accounts for the marked differences in marriage between blacks and whites focuses on the high levels of non-marriage among blacks (Staples 1985; Wilson and Neckerman 1986; Farley and Bianchi 1987). The high prevalence of non-marriage is treated as a "problem" to be explored in terms of examining the obstacles to marriage. Most studies (Spanier and Glick 1980; Schoen and Kluegel 1988; Bennett et al. 1989; Mare and Winship 1991) that involve comparisons by race examine the demographic compositional factors which contribute to non-marriage. This body of research examines the differences in demographic composition of both populations and encompasses a broad view of marriage that involves (1) the availability of suitable partners, and (2) the propensity to marry. These studies consistently find that the size of the marriageable pool for black women is quite small and negatively associated with education (Bennett et al. 1989; Mare and Winship 1991). Consequently, many black women are faced with the choices to marry older men, men who have been previously married, men with less education, or to forego marriage (Spanier and Glick 1980). Lichter, McLaughlin, Kephart, and Landry (1993) find that "mate availability in local marriage markets is a significant factor contributing to delayed marriage, and perhaps non-marriage among black women" (p. 797).

Most of the empirical work on the propensity to marry is motivated by Becker's (1973) theory of the gains to marriage. Becker suggests that black women have less to gain from marriage than white women because of the relatively smaller gap between black male and female labor market earnings. By concentrating on the decision to marry, Bulcroft and Bulcroft's (1993) analysis show that black women are less likely to want to marry a man with fewer resources or who has had previous marital/family experience (p. 352).

Research on the desire to marry has received very little attention in the literature. The current attitudinal research on marriage considers *singleness* in contrast to marriage and finds that most young people and their parents have lessened their disapproval of persons choosing singleness as an adult status (Glick and Norton 1977; Thornton and Freedman 1982; Thornton 1989; Sweet and Bumpass 1990). One study that considers the desirability of marriage for blacks and whites finds that the lack of consensus on the benefits and costs of marriage among black men and women is largely responsible for the prevalence of non-marriage among blacks (South 1993).

Although these findings vary across different time periods and have contributed to our understanding of differences in the way men and women view marriage, less is known about preferences of singleness or marriage as an adult status. Whereas the behavioral differences might be due to the unequal access to the resources necessary for forming and maintaining an independent household, these differences provide little insight into the desirability of marriage for either group. While South focuses on the desirability of marriage for never-married persons, this study examines racial differences in attitudes toward marriage in comparison to singleness irrespective of current marital status. While the findings in these studies have shown blacks have a lower marriage rate than whites, these findings have yet to show whether the paths were taken by choice or were involuntarily imposed (Bennett et al. 1989).

DATA AND METHODS

The data for these analyses are from the National Survey of Families and Households (NSFH), a national survey of 13,017 adults living in the continental United States between March 1987 and May 1988. The main cross section sample consists of 9,643 households, in addition to an oversample of blacks, Puerto Ricans, Mexican-Americans, single parents, families with stepchildren, recently married couples, and cohabitors. In each household a randomly selected adult was interviewed. The response rate for the main sample was .77 and the total response rate was .74. The data for this paper are restricted to black and white respondents aged 20 to 59 years, a total of 9,014 persons of whom 1,910 are black and 7,104 are white.

The independent variables considered are race, gender, marital status, household income, and education. Household income is measured here relative to the U.S. government poverty standard. This needs standard takes into account the size of the household and number of related children under 18. The poverty line is computed by adjusting the published poverty thresholds in 1984 to the 1986 dollars (NSFH code book p. R-45). The four dichotomous dependent variables are measures of attitudes toward (1) premarital sex, (2) non-marital childbearing, (3) cohabitation, and (4) being married versus staying single.

Five questions were examined to analyze attitudes toward cohabitation, marriage, premarital sex, and non-marital childbearing. All of the respondents were asked to what extent they agreed with the following four statements:

Q1. It's better for a person to get married than to go through life being single.
Q2. It is all right for unmarried 18-year-olds to have sexual relations if they have strong affections for each other.
Q3a. It is all right for an unmarried couple to live together even if they have no interest in considering marriage.
Q3b. It is all right for an unmarried couple to live together as long as they have plans to marry.

The response was based on a five-point, Likert-type scale ranging from strongly agree to strongly disagree. The response variables were transformed into the dichotomy—agree or disagree. Refusals, "don't know," and "no answer" responses were deleted. To get an overall sense of an attitude toward cohabitation irrespective of whether or not marriage is intended, questions 3a and 3b were combined. Responses less than four were coded agree; if either 3a or 3b were equal to 4 or 5, the response was coded disagree. It is quite possible that some agreed with cohabitation only in special circumstances: however, that issue is not addressed here. The "marriage question" (Q1), asking whether it is better to get married or go through life single, was coded differently. Because I was interested in the non-normative attitudes toward marriage, responses less than 3 were coded agree, otherwise disagree.

The scale for the last question is quite different from the previous set of questions. Respondents were asked to answer this question based on a seven-point, Likert-type scale ranging from strongly approve to strongly disapprove. The respondents were asked to respond to the following:

Q4. Here are some questions about your views on some other topics. Please circle the number that best represents how much you approve or disapprove of the behaviors described.

Women Who Have a Child Without Getting Married

Response categories (1, 2, 3) were coded approve. "Don't know," refusals, and "no answers" were eliminated from the analysis. The descriptive results in the following analysis are based on weighted estimates. The multivariate analysis is based on unweighted, grouped data.

The logistic regression model was used to predict the dichotomous dependent variable. The response variable predicts the log odds of agreement or approval.

Table 1. Percent of NSFH Respondents in Selected Social Categories by Race and Sex (20 to 59 year olds in 1988)

	Men		*Women*					
	Black	*White*	*Black*	*White*	*Black*	*White*	*Men*	*Women*
N	506	3641	631	3863	1138	7505	4148	4495
Education								
<12	30%	10%	24%	12%	27%	11%	13%	14%
12	37	35	40	44	39	40	36	43
13-15	22	26	25	24	24	25	26	24
16+	11	28	10	20	10	24	26	19
Total	100	100	100	100	100	100	100	100
Age in Years								
20-29	35	30	34	29	34	29	30	30
30-39	29	31	31	30	30	31	31	30
40-49	20	23	18	22	19	22	22	21
50-59	16	16	17	19	17	18	16	19
Total	100	100	100	100	100	100	100	100
Marital Status								
Never married	35	24	35	15	35	20	26	18
Currently married	48	66	35	69	41	68	64	64
Previously married	17	10	29	16	24	13	10	18
Total	100	100	100	100	100	100	100	100
Income/Poverty Line								
<.75	19	9	36	13	29	11	10	15
.75-1.49	10	8	15	9	13	8	8	9
1.5-2.49	21	14	18	15	19	15	15	16
2.5x	50	69	32	64	40	66	67	60
Total	100	100	100	100	100	100	100	100

Compositional Differences by Race

To the extent that attitudes vary between blacks and whites because of their differences in composition on other variables, it is important to note the marked differences in composition between the two groups.

Table 1 shows that 27 percent of blacks have less than 12 years of schooling, compared to 11 percent of whites. At the other extreme, 24 percent of whites have over 16 years of schooling compared to 10 percent of blacks. The pattern is the same for the distribution of income. While 29 percent of the blacks in the sample have needs adjusted incomes less than 75 percent of the poverty line, just 11 percent of whites are in the same position. The majority of whites (66%) have incomes 2.5 times the poverty line compared to 40 percent of blacks.

More striking are the compositional differences between blacks and whites by marital status. Compared to blacks, whites are more likely to be currently married (68% versus. 41%) and are less likely to be previously married (13% versus 24%) or never married (20% versus 35%), respectively.

FINDINGS

Premarital Sex

Table 2 presents the proportions approving of premarital sex. A majority of both blacks (58%) and whites (53%) approve of premarital sex. A greater per-

Table 2. Percent Agreeing that "It Is All Right for Unmarried 18-Year-Olds to Have Sexual Relations if They Have Strong Affections for Each Other": NSFH 1987–1988

	Men		*Women*					
	Black	*White*	*Black*	*White*	*Black*	*White*	*Men*	*Women*
Total	59%	60%	57%	47%	58%	53%	60%	49%
Education								
<12	60	55	49	45	54	49	56	46
12	56	60	59	47	58	53	60	49
13-15	68	65	60	52	63	58	65	53
16+	55	57	60	44	58	51	57	45
Age in Years								
20-29	70	78	71	67	71	72	77	68
30-39	58	63	56	49	57	56	63	50
40-49	55	49	45	34	49	42	50	36
50-59	44	36	43	29	44	32	37	31
Marital Status								
Never married	71	80	72	68	72	76	79	69
Currently married	48	51	46	41	47	46	51	41
Previously married	69	67	51	56	57	67	67	55
Income/Poverty Line								
<.75	48	56	61	44	57	48	54	49
.75-1.49	45	61	56	49	52	55	59	51
1.5-2.49	58	60	50	47	54	53	60	47
2.5x	55	57	48	44	51	51	57	45

centage of men (60%) approve than women (49%). As well, a greater percentage of those less than 40 years old approve of premarital sex than do persons aged 40 and older.

Clearly, the difference between blacks and whites is substantially less than the difference between men and women. When marital status is accounted for, whites who are not currently married approve of premarital sex more than blacks in the same category. Those who are not currently married approve of premarital sex more than those who are currently married. Table 2 shows that 51 percent of men and 41 percent of women who are currently married approve of premarital sex. Nearly 70 percent of all individuals who have never married approve of premarital sex; 79 percent of men and 69 percent of women approve. Among never-married

Table 3. Parameter Estimates for the Log Odds of Approval of Premarital Sex (Model E)

Parameter	*Gross*	*Net*	*S.E.*	*T-ratio*	*Odds*
Constant		0.103	0.241	0.428	1.108
(Black)					
White	0.658	0.749	0.233	3.220	2.115
(Men)					
Women	0.042	–0.125	0.179	–0.697	0.883
White Women	–0.768	–0.611	0.201	–3.039	0.543
White 30-39	–0.488	–0.488	0.243	–2.008	0.614
White 40-49	–0.588	–0.634	0.284	–2.234	0.530
White 50-59	–0.812	–0.812	0.306	–2.651	0.444
(Income <.75)					
.75-1.49	–0.105	–0.114	0.156	–0.728	0.893
1.5-2.49	–0.084	–0.014	0.144	–0.094	0.987
2.5x	–0.120	0.157	0.124	1.267	1.170
(Age 20-29)					
Age 30-39	–0.174	–0.113	0.214	–0.528	0.893
Age 40-49	–0.529	–0.457	0.255	–1.789	0.633
Age 50-59	–0.762	–0.731	0.276	–2.655	0.481
(<High school)					
High school	0.138	0.106	0.124	0.855	1.112
Some college	0.294	0.071	0.137	0.520	1.074
College +	0.129	–0.071	0.146	–0.484	0.932
(Currently married)					
Previously married	0.419	0.670	0.105	6.354	1.954
Never married	1.150	0.892	0.129	6.936	2.440

persons the race difference in approval is smaller than the sex difference: 80 percent of white males approve of premarital sex. Smith (1992) argues that the never married and the previously married are more permissive because it is in their self-interest to be so. An analogous argument made by Kluegel and Smith (1986) is that attitudes are able to predict relevant behaviors when the attitudes matter to the individual. Because those who are currently married have little interest in the issue of premarital sex, it is not surprising that respondents in this category show some of the lowest levels of approval.

Table 3 suggests that the odds of approval of premarital sex are affected by race, marital status, and age (Appendix 1 shows the fit statistics for several models). The t-ratios indicate that the coefficients for the race and sex, and the race and age interaction terms, marital status and age, are statistically significant. Having been previously married and being married increase the odds of approval.

Figures 1 and 2 facilitate the discussion of the results of this model from Table 3. Whereas Smith (1992) argues that blacks are more permissive than non-blacks because of their lower age at first intercourse and the prevalence of non-marital childbearing among blacks, the results show that, irrespective or race and gender, younger people are more likely to approve of premarital sex than are older people.

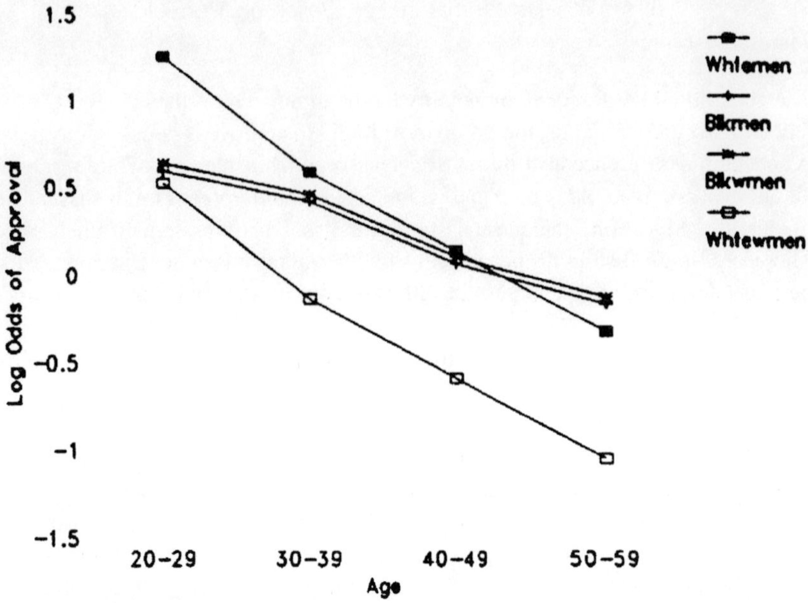

Figure 1. Gross Effects of Race, Gender, and Age on the Log Odds of Approval of Premarital Sex

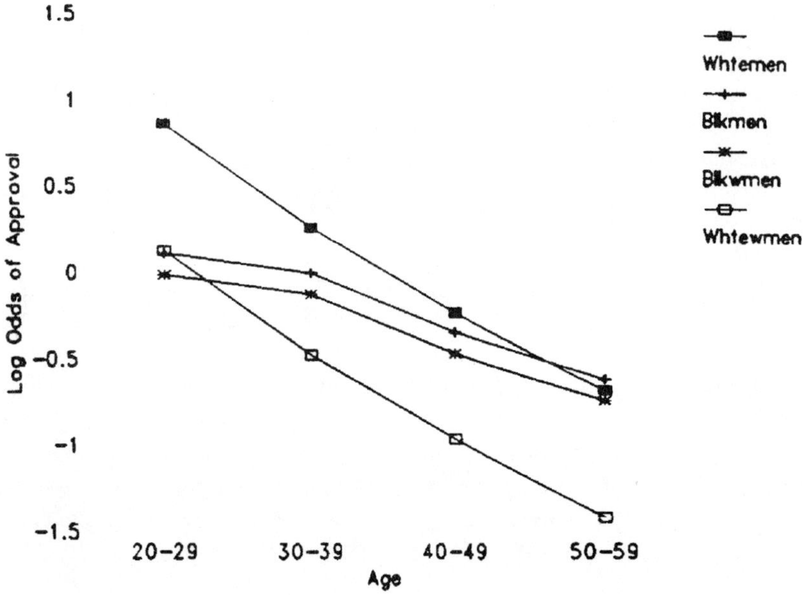

Figure 2. Net Effects of Race, Gender, and Age on the Log Odds of Approval of Premarital Sex (Based on Model K)

The generational differences are notably larger among whites than blacks. Consistent with the earlier results, the group most likely to approve is young white males. A substantial difference in attitudes exists between white men and white women at all ages. In contrast, black men and women share similar views on this issue. Net of income, education, and marital status the gross patterns remain unchanged. Therefore, these factors do not account for the racial differences in approval of premarital sex despite the large race differences in these variables.

Non-marital Childbearing

Cutright and Smith (1988) suggest that racial differences in sexual activity, contraceptive use, abortion, and legitimation account for the higher black than white non-marital fertility rates. A major component of this is the greater propensity among whites to legitimate premarital pregnancies. The high propensity for whites to legitimate premarital pregnancies might suggest that whites disapprove more of non-marital childbearing. However, the lower tendency of legitimization among blacks does not imply that blacks approve of non-marital childbearing. The results in Table 4 show that neither blacks (26%) nor whites (16%) approve of a woman having a child without getting married. Despite the current high levels of

Table 4. Percent Approving "Women Who Have a Child Without Getting Married": NSFH 1987–1988

	Men		*Women*					
	Black	*White*	*Black*	*White*	*Black*	*White*	*Men*	*Women*
Total	20%	15%	31%	16%	26%	16%	15%	18%
Education								
<12	18	19	29	15	23	17	19	19
12	23	13	34	15	30	14	14	17
13-15	17	16	28	18	23	17	16	20
16+	23	14	30	18	27	15	15	18
Age in Years								
20-29	24	17	40	23	33	20	18	26
30-39	19	17	31	17	26	17	17	19
40-49	16	13	25	12	21	13	14	14
50-59	16	9	18	9	17	9	10	10
Marital Status								
Never married	25	17	43	30	35	22	18	34
Currently married	14	13	24	12	19	12	13	13
Previously married	24	22	26	22	26	22	22	23
Income/Poverty Line								
<.75	33	12	39	23	37	19	16	27
.75-1.49	27	18	28	21	27	19	19	22
1.5-2.49	22	12	27	18	25	15	13	19
2.5x	17	15	25	13	21	14	15	14

non-marital fertility among blacks, less than 35 percent of blacks approve of non-marital childbearing. While black women at all levels approve more than white women, the greatest difference is among the never married women: 43 percent among blacks compared to 30 percent among white women. Whereas 35 percent of blacks compared to 22 percent of whites who have never married approve of non-marital childbearing, the percentage difference between men (18%) and women (34%) is slightly higher.

Since the data show that both groups tend to disapprove of non-marital childbearing, a cultural theory that illegitimacy is generally acceptable among blacks is not supported. While the simple bivariate analysis does show that blacks are more approving than whites, the subsequent multivariate analysis should help to ascertain the influence of compositional factors on the approval of non-marital childbearing among blacks and whites.

After experimenting with several model specifications in Appendix 2, the baseline additive model (Model A) was chosen predicting the odds of approval of

Table 5. Parameter Estimates for the Log Odds of Approval of Non-marital Childbearing (Model A)

Parameter	*Gross*	*Net*	*S.E.*	*T-ratio*	*Odds*
Constant		–0.791	0.205	–3.86	0.45
(Black)					
White	–0.673	–0.438	0.115	–3.81	0.65
(Men)					
Women	0.157	0.051	0.104	0.49	1.05
(Income <.75)					
.75-1.49	–0.400	–0.292	0.182	–1.60	0.75
1.5-2.49	–0.476	–0.256	0.167	–1.53	0.77
2.5x	–0.611	–0.192	0.144	–1.33	0.33
(Age 20-29)					
Age 30-39	–0.315	–0.221	0.123	–1.80	0.80
Age 40-49	–0.659	–0.535	0.154	–3.47	0.59
Age 50-59	–0.854	–0.814	0.177	–4.60	0.44
(<High school)					
High school	–0.260	–0.189	0.149	–1.27	0.83
Some college	–0.156	–0.175	0.164	–1.07	0.84
College +	–0.134	–0.020	0.175	–0.11	0.98
(Currently married)					
Previously married	0.570	0.527	0.127	4.15	1.69
Never married	1.066	0.744	0.134	5.55	2.10

non-marital childbearing. Evidently race, age, and marital status have a great impact on attitudes toward non-marital childbearing. The t-ratios in Table 5 show that these effects are statistically significant in predicting the odds of approval. The last column in the table shows that the odds of approval among whites are 65% percent of those among blacks. In other words, whites are less likely than blacks to approve of a woman having a child outside marriage. In addition, non-marital childbearing is least likely to be sanctioned by people aged 50 to 59. The model indicates that the odds of approval of non-marital childbearing exceed those of currently married persons by 1.69 for people who have been previously married and 2.10 for those who have never married. Having never been married increases the odds of approval more than race does. Also, examining gross (1.066) and net (0.744) marital status coefficients in Table 5 reveals that adding the control variables significantly reduced the differences in attitudes between the currently married and the never married. Likewise, the same is true for the observed differences by income.

Gross (.673) and net (.438) comparisons in Table 5 show that the racial differences are reduced by about one-third when other variables are accounted for in the model. About two-thirds of the much smaller sex difference is attributable to composition on the other variables. The relatively low gender coefficient (.051) reflects that men's and women's attitudes on this issue are not significantly different from one another.

Cohabitation

Table 6 shows no differences by race in approval of cohabitation except that, among previously married women, 59 percent of whites but only 42 percent of blacks approve of cohabitation. More men (54%) than women (46%) approve of cohabitation. Some researchers (Schoen and Owens 1992) suggest that cohabitation minimizes long-term responsibilities for men and requires fewer expectations for them to share economic and household responsibilities equally. Furthermore,

Table 6. Percent Agreeing that "It Is all Right for an Unmarried Couple to Live Together": NSFH 1987–1988

	Men		*Women*					
	Black	*White*	*Black*	*White*	*Black*	*White*	*Men*	*Women*
Total	53%	54%	46%	46%	49%	50%	54%	46%
Education								
<12	48	45	45	42	46	43	46	43
12	50	53	43	44	46	48	53	44
13-15	61	57	46	49	52	53	58	48
16+	60	56	56	49	58	53	56	50
Age in Years								
20-29	60	63	54	54	57	59	62	54
30-39	53	61	47	50	50	56	60	50
40-49	55	48	42	40	48	44	49	41
50-59	36	33	30	31	33	32	33	31
Marital Status								
Never married	62	65	56	55	59	61	64	55
Currently married	44	49	39	41	41	45	49	41
Previously married	60	63	42	59	48	60	62	55
Income/Poverty Line								
<.75	56	52	47	46	49	48	52	47
.75-1.49	43	48	47	48	46	48	48	48
1.5-2.49	48	54	36	44	42	49	53	43
2.5x	52	55	45	43	48	49	55	43

in these relationships as well as in marriage men often have more economic resources than women, and thus tend to have more power. Perhaps the combination of autonomy, economic independence, and lack of commitment makes cohabitation more attractive for men than for women.

Older and less educated persons are least likely to approve of cohabitation. While Bumpass and Sweet (1989, p. 624) observe that "cohabitation is most likely among those who did not complete high school," clearly a majority of respondents in this category does not approve of cohabitation. Table 6 shows that 46 percent of blacks and 43 percent of whites with less than a high school education approve of cohabitation. The prevalence of cohabitation might be a consequence of economic necessity rather than a preferred lifestyle for the least educated.

Of the models specified in Appendix 3, the baseline model (Model A) fits the data best. The t-ratios in Table 7 of the selected model (Model A) show that race, gender, age, and marital status are statistically significant. The odds of approval for women are 71 percent of those for men. The odds of approval are 29 percent greater for whites than for blacks.

The effect of marital status on the odds of approval is also substantial. Compared to those who are currently married, having never been married increases the odds by 1.78 and having been previously married increases the odds by 1.87. Apparently, those who have been previously married are more tolerant of unmarried persons living together than those who are currently married. Bumpass, Sweet, and Cherlin (1991) note that cohabitation is more common among separated and divorced persons than among never-married persons. Between 1980 and 1987, 60 percent of remarried couples had cohabited with a partner before marriage (Bumpass et al. 1991, p. 219).

Gross and net comparisons in Table 7 show that when race, marital status, and gender are included in the model, most of the coefficients increase. While the never-married coefficient decreased by .08 the previously married coefficient almost doubled, which suggests that prior experience of marriage increases the odds of approval of cohabitation. Still being "single" has a profound impact on the odds of approval. Respondents in the oldest age categories (–.945) are

less likely than younger respondents to approve of cohabitation. Comparisons of the gross and net effects show that the race coefficients more than double when other variables are considered. Racial differences in attitudes do not occur because of the composition of factors included in the model. Whites are more likely to approve of cohabitation than blacks. Similarly, the gender coefficient did increase but not considerably. Women are still less likely than men to agree that it is all right for an unmarried couple to live together. Table 7 confirms that the simple bivariate analysis fails to account for the racial differences in age and marital status. The logit increases by 26 percent for previously married people. The estimates from the bivariate analysis are misleading in that they predict that the odds of approval would be higher for the never-married than the previously married. Therefore, the positive coefficients in the logit model indicate that previously married people are

Table 7. Parameter Estimates for the Log Odds of Approval of Cohabitation (Model A)

Parameter	*Gross*	*Net*	*S.E.*	*T-ratio*	*Odds*
Constant		0.008	0.172	0.05	1.01
(Black)					
White	0.115	0.258	0.100	2.58	1.29
(Men)					
Women	–0.279	–0.340	0.082	–4.15	0.71
(Income <.75)					
.75-1.49	–0.050	–0.087	0.154	–0.56	0.92
1.5-2.49	–0.031	–0.026	0.140	–0.19	0.97
2.5x	0.024	0.133	0.121	1.10	1.14
(Age 20-29)					
Age 30-39	–0.100	–0.098	0.103	–0.95	0.91
Age 40-49	–0.625	–0.677	0.121	–5.60	0.51
Age 50-59	–0.889	–0.945	0.134	–7.05	0.39
(<High school)					
High school	0.146	0.043	0.122	0.35	1.04
Some college	0.259	0.018	0.134	0.13	1.02
College +	0.312	0.085	0.142	0.60	1.09
(Currently married)					
Previously married	0.353	0.617	0.103	5.99	1.85
Never married	0.653	0.579	0.121	4.79	1.78

more likely to approve of cohabitation at about the same rate as those who have never married. Furthermore, the educational differences mentioned earlier become smaller once other factors are considered. For example, the "college +" coefficient declines from .312 to .085. Blacks are not more tolerant of cohabitation despite their well-documented patterns of delayed marriage patterns and lower remarriage. Whites are even more likely than blacks to approve of cohabitation when background characteristics are considered.

Marriage

Wilson and Neckerman's (1986, p. 256) theory of marriage hinges on the lack of employment of black men. They suggest that lack of stable employment for black men has had a profound impact on the increasing delay in first marriage, the low rate of remarriage, and the increasing difficulty among black women in finding a suitable spouse. Surprisingly, Table 8 suggests that a majority of both blacks and

Table 8. Percent Agreeing that "It Is Better for a Person to Get Married Than to Go Through Life Single": NSFH 1987–1988

	Men		*Women*					
	Black	*White*	*Black*	*White*	*Black*	*White*	*Men*	*Women*
Total	47%	50%	35%	34%	41%	42%	49%	34%
Education								
<12	57	57	51	45	54	51	57	47
12	41	48	35	35	37	41	47	35
13-15	44	47	26	30	34	38	46	29
16+	43	53	25	31	33	43	52	31
Age in Years								
20-29	39	44	30	32	34	38	44	31
30-39	40	43	34	30	36	37	43	31
40-49	55	54	41	37	48	45	54	37
50-59	67	67	46	42	55	53	67	42
Marital Status								
Never married	39	40	27	27	32	35	39	27
Currently married	53	54	40	37	47	45	54	38
Previously married	47	44	40	28	42	33	44	30
Income/Poverty Line								
<.75	52	62	39	38	42	47	60	38
.75-1.49	60	48	39	35	46	41	50	36
1.5-2.49	60	47	37	30	48	37	49	31
2.5x	41	51	32	36	37	43	51	35

whites are accepting of singleness as an adult status. Wilson and Neckerman's theory of marriage behavior for black men falls short of addressing why only 41 percent of black men in the high income and education categories do not agree that it is better to be married. Highly educated black men (43%) are less likely than highly educated white men (53%) to agree that it is better to be married. Evidently, those most "suitable" for marriage are most tolerant of those who choose to remain single.

In addition, Wilson and Neckerman (1986) do not consider the possibility that women might prefer to remain single rather than get married. Men (51%) are less willing to accept singleness as a legitimate status than women (66%). These results are consistent with the earlier finding by Bumpass and Sweet (1989) that men are more likely than women to agree that it is better to be married. Most white men (62%) and 52 percent of black men with a needs adjusted income less than 75 percent of the poverty line approve of marriage over singleness.

In contrast, most women in all categories do not agree that it is better to be married than to be single. Compared to men, women at all education levels are more tolerant of singleness as a preferred status. Furthermore, 38 percent of currently married women agree compared to 54 percent for currently married men. While one might speculate that these women are probably not happy in their current marriage, one only has to look at the similar patterns of agreement by all women including those who have been previously married and never married. Although men are financially

Table 9. Parameter Estimates for the Log Odds of Agreeing that "It Is Better for a Person to Get Married Than to Go Through Life Single" (Model K)

Parameter	*Gross*	*Net*	*S.E.*	*T-ratio*	*Odds*
Constant		0.882	0.262	3.37	2.42
(Black)					
White	–0.120	–0.222	0.243	–0.91	0.80
(Men)					
Women	–0.562	–0.442	0.459	–0.96	0.64
White women	–0.193	–0.277	0.207	–1.34	0.76
(<High school)					
High school	–0.620	–0.432	0.269	–1.61	0.65
Some college	–1.132	–0.857	0.302	–2.84	0.42
College +	–0.983	–0.839	0.344	–2.44	0.43
Women high school	0.108	0.011	0.244	0.55	1.01
Women some college	–0.199	–0.306	0.267	–1.15	0.74
Women college +	–0.375	–0.384	0.275	–1.40	0.68
White high school	0.061	0.087	0.255	0.34	1.09
White some college	0.576	0.616	0.294	2.10	1.85
White college +	0.725	0.879	0.344	2.56	2.41
(Income <.75)					
.75-1.49	0.113	–0.134	0.157	–0.85	0.88
1.5-2.49	0.266	–0.287	0.145	–1.98	0.75
2.5x	0.190	–0.363	0.124	–2.93	0.70
(Age 20-29)					
Age 30-39	0.161	–0.210	0.108	–1.94	0.81
Age 40-49	–0.287	0.207	0.123	1.68	1.23
Age 50-59	–0.651	0.523	0.133	3.93	1.69
(Currently married)					
Previously married	0.298	–0.448	0.107	–4.19	0.64
Never married	0.447	–0.538	0.126	–4.27	0.58

better off being single, the value of household services provided by their wives cannot be measured in strict economic terms. Perhaps men recognize the tradeoff between earnings and household services to be more favorable in their behalf.

Moreover, education appears to be a distinguishing factor in the levels of tolerance of singleness. Those with less education are more likely to see marriage as a preferred state. Whereas 54 percent of less educated blacks and 51 percent of less educated whites prefer marriage, both blacks and whites with higher levels of education are more tolerant of non marriage. White college educated men (53%) proved to be the exception to the rule. These men are most likely married. Table 8. shows that married men are less tolerant of singleness.

While the higher tolerance of singleness among women might shed some light on the delayed marriage phenomenon, it does not fully explain the divergent racial patterns. To the extent that marriage requires a consensus between a man and woman and marital homogamy still exists, finding a "suitable" spouse requires finding one who is both suitable and willing to get married. Although most of the research on marriage assumes that it is women in the market place looking for a husband, the alternative perspective should be considered.

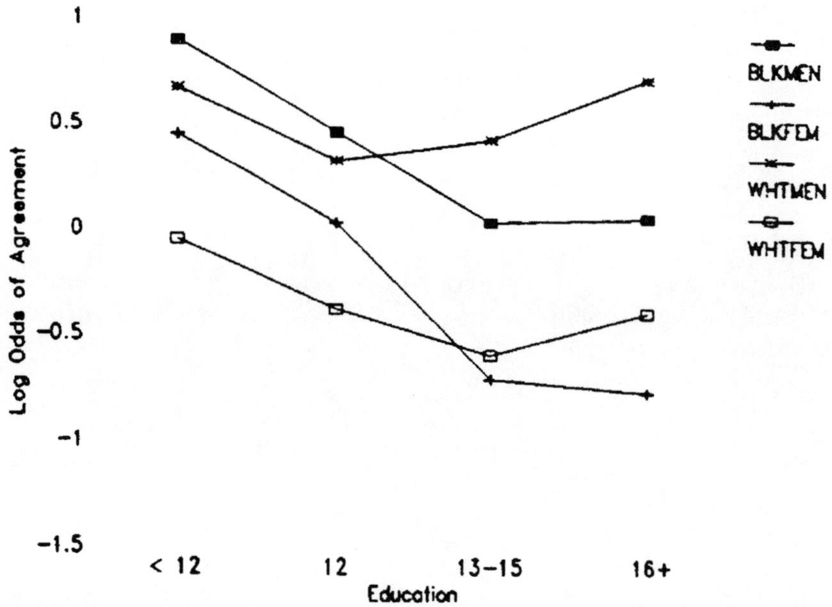

Figure 3. Gross Effects on the Log Odds of Agreement: "It is Better to Get Married Than to Remain Single" by Education, Gender, and Race (Based on Model K)

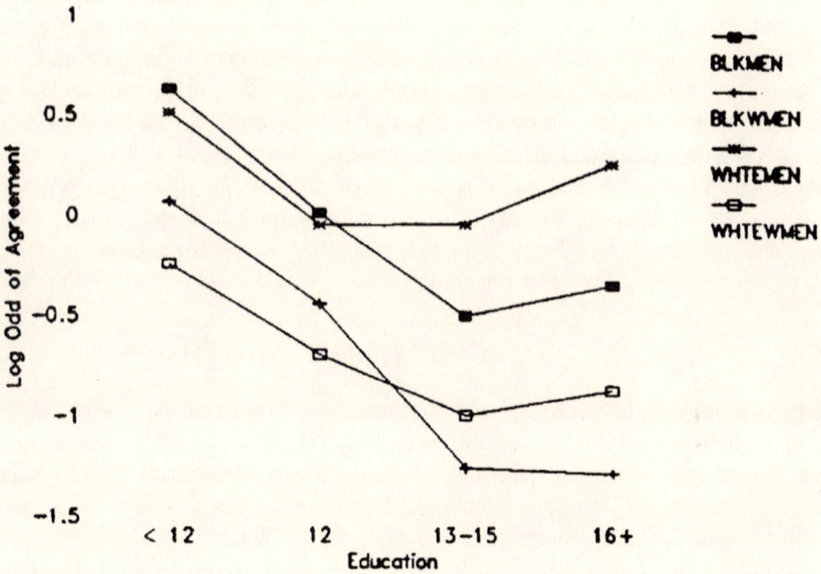

Figure 4. Net Effects on the Log Odds of Agreement "It Is Better to Get Married Than to Remain Single" by Education, Gender, and Race

After examining several models in Appendix 4, Model K, the multiplicative model with three two-way interactions among race, sex, and education, was chosen. This model requires one to compare the odds of approval of marriage for all groups at each education level. Table 9 and Figure 3 confirm the original bivariate findings.

The gross and net estimates in Table 9 require a somewhat more complex interpretation than the three earlier models. Figures 3 and 4 help to make the interpretations of the gross and net effects more intuitive. The omitted categories of the variables are currently married young adults aged 20 to 29 with incomes less than 75 percent of the poverty level. The U-shape patterns of agreement for men in both graphs are quite similar; however, their distinctions are quite subtle. Both graphs support the earlier finding that among the least educated, both black and white men have higher probabilities of agreement than women that it is better to be married. At the other extreme of the education continuum, white men tend to favor marriage more than singleness. In particular, men at all educational levels are less tolerant of singleness than are women. Figure 4 shows the net effects of the other variables of income, age, and marital status included in the model. The differences between men and women are larger than the differences between blacks and whites. Tolerance for singleness as a legitimate state increases with education. However, a substantial discrepancy exists between persons with college education and those without. Persons who have not gone to college are more likely to view marriage as a preferred state. The differences in opinions between blacks and whites and men and women increase with education as well. Although Bennett, Bloom, and Craig (1989) find that education is negatively correlated with marriage for white women, the graph shows that as education increases white women are more likely to agree that it is better to be married than to remain single. In spite of this finding, the graphs also exhibit a substantial difference in opinions regarding marriage between white men and women. The similarity between Figures 3 and 4 suggests that the other variables in the model do not account for the race, sex, and educational differences.

CONCLUSION

When the attitudes of blacks and whites are analyzed comparatively in a format that controls for demographic compositional factors, it is found that race is not the strongest determinant of attitudes about marriage and family behavior. Gender, age, and marital status have stronger effects than race in predicting attitudes. Thus, it is unlikely that the racial differences in behavior are attributable to racial differences in attitudes. For example, whereas blacks are more likely to cohabit, whites are more likely to approve of cohabitation. Similarly, whereas blacks have lower marriage rates than whites, blacks are more likely to agree that it is better to be married than to remain single.

Despite the small racial differences, one finds substantial differences by gender. Men's and women's opinions on marital and family behavior vary significantly. Men are more likely to agree that it is better to be married than to remain single. Perhaps as the division of labor in the household becomes more equitable the differences between men's and women's opinions about marriage might converge.

The major factor in determining differences in attitudes appears to be marital status. Marital status has considerable effect on approval of premarital sex, cohabitation, non-marital childbearing, and marriage. Those who were either previously married or never married were more likely to approve of premarital sex, cohabitation, non-marital childbearing, and marriage.

In addition, we were unable to substantiate the assertion in the literature of the "social acceptance" of non-marital childbearing. We find that while race is a significant factor in the approval of non-marital childbearing, a majority of both blacks and whites do not approve of women who have children without getting married. Whereas blacks were more approving of nonmarital childbearing, the age structure and the marital status distribution of the black and white populations account for some of the important racial differences in attitudes toward this issue.

In sum, the racial differences in behavior have less to do with differences in cultural norms than differences in the way society is structured. Moreover, although the link between attitudes and behavior was not fully addressed in this research, these findings shed some light on the social acceptance of these behaviors.

Appendix 1. Logistic Model Specifications Predicting the Odds of Approval of Premarital Sex

Model		*G2*	*D.F.*
	Baseline (no regressors)	903.70	503
A	RACE+SEX+INCOME+AGE+EDUC+STAT	625.1	490
B	A+RACE*STAT	625.03	488
C	A+SEX*EDUC	624.71	487
D	A+RACE*EDUC	623.46	487
E	A+RACE*EDUC+SEX*EDUC	623.15	484
F	A+RACE*AGE	617.80	487
G	A+RACE*SEX	617.26	489
H	A+RACE*SEX+SEX*EDUC	616.2	486
I	A+RACE*SEX+RACE*EDUC	615.21	486
J	A+RACE*SEX+RACE*EDUC+SEX*EDUC	614.18	483
K	A+RACE*SEX+RACE*AGE	608.64	486
L	J+RACE*SEX*EDUC	605.79	480
M	K+RACE*SEX*AGE	603.95	480

Appendix 2. Logistic Model Specifications Predicting the Odds of Approval of Non-marital Childbearing

Model		*G2*	*D.F.*
	Baseline (no regressors)	686.18	501
A	RACE+SEX+INCOME+AGE+EDUC+STAT	556.01	488
B	A+RACE*SEX	555.95	487
C	A+SEX*AGE	555.62	485
D	B+RACE*AGE	555.55	484
E	A+SEX*EDUC	552.74	485
F	B+SEX*EDUC	552.59	484
G	A+RACE*STAT	551.35	486
H	A+RACE*EDUC	550.16	485
I	B+RACE*EDUC	550.15	484
J	D+RACE*SEX*AGE	548.90	478
K	A+RACE*EDUC+SEX*EDUC	546.33	482
L	A+RACE*SEX+RACE*EDUC+SEX*EDUC	546.24	481
M	L+RACE*SEX*EDUC	546.14	478

Appendix 3. Logistic Model Specifications Predicting the Odds of Approval of Cohabitation

Model		*G2*	*D.F.*
	Baseline (no regressors)	760.82	497
A	RACE+SEX+INCOME+AGE+EDUC+STAT	611.21	484
B	A+RACE*SEX	610.61	481
C	A+SEX*EDUC	608.67	480
D	A+RACE*SEX+RACE*EDUC	608.24	481
E	A+SEX*EDUC	608.19	480
F	A+RACE*SEX+SEX*EDUC	607.79	481
G	A+SEX*INCOME	606.95	478
H	A+RACE*EDUC+SEX*EDUC	605.87	477
I	A+RACE*SEX+RACE*EDUC+SEX*EDUC	605.61	478
J	A+EDUC*INCOME+RACE*INCOME	605.25	484
K	A+EDUC*INCOME	604.64	478
L	A+SEX*INCOME+SEX*EDUC	603.70	481
M	A+RACE*AGE	602.45	480
N	A+RACE*SEX+RACE*AGE	601.97	480
0	H+RACE*SEX*EDUC	600.66	474
P	H+RACE*SEX*AGE	599.14	474
Q	A+EDUC*RACE*INCOME	598.52	460
R	H+SEX*EDUC*INCOME	591.94	460

Appendix 4. Logistic Model Specifications Predicting the Odds of Agreeing that Marriage is Better than Being Single

Model		*G2*	*D.F.*
	Baseline (no regressors)	835.69	497
A	RACE+SEX+INCOME+AGE+EDUC+STAT	626.38	487
B	A+RACE*AGE	624.50	484
C	A+RACE*STAT	622.80	485
D	A+RACE*SEX	622.59	486
E	A+RACE*SEX+RACE*AGE	620.60	483
F	A+SEX*EDUC	620.49	484
G	E+RACE*SEX*AGE	616.55	477
H	A+RACE*SEX+SEX*EDUC	617.46	483
I	A+RACE*EDUC	613.57	484
L	A+RACE*SEX+RACE*EDUC	611.09	483
J	A+RACE*EDUC+SEX*EDUC	608.36	481
K	A+RACE*SEX+RACE*EDUC+SEX*EDUC	606.56	480
L	K+RACE*SEX*EDUC	600.95	477

ACKNOWLEDGMENT

A previous version of this paper was presented at the Annual Meeting of the American Sociological Association, August 1994, Los Angeles, California. The National Survey of Families and Households was funded by Grant HD21009 from the Center for Population Research of the National Institute of Child Health and Human Development, and the analysis was supported under Grant HD22433, using facilities provided under Grant HD05876. This study was prepared while the author was supported by a research assistantship under Larry Bumpass at the Center for Demography and Ecology, University of Wisconsin-Madison.

REFERENCES

Agresti, A. 1990. *Categorical Data Analysis*. New York: Wiley.

Bachrach, C. 1987. "Cohabitation and Reproductive Behavior in the United States." *Demography* 24: 623-637.

Becker, G. S. 1973. "A Theory of Marriage: Part I." *Journal of Political Economy* 81: 813-847.

______. 1981. *A Treatise on the Family*. Cambridge, MA: Harvard University Press.

Becker, G. S., E. Landes, and R. Michael. 1977. "An Economic Analysis of Marital Instability." *Journal of Political Economy* 85: 114-1187.

Bennett, N. G., _. Bloom, and _. Craig. 1989. "The Divergence of Black and White Marriage Patterns." *American Journal of Sociolog* 95: 692-722.

Bernard, J. 1966. *Marriage and Family among Negroes*. Englewood Cliffs, NJ: Prentice-Hall.

Brown, S. V. 1985. "Premarital Sexual Permissiveness among Black Adolescent Females." *Social Psychology Quarterly* 48: 381-387.

Bulcroft, R. A., and K. Bulcroft. 1993. "Race Differences in Attitudinal and Motivational Factors in the Decision to Marry." *Journal of Marriage and the Family* 55: 338-355.

Bumpass, L. L. 1979. "The Changing Linkage of Nuptiality and Fertility in the United States." Pp. 195-209 in *Nuptiality and Fertility*, edited by L. T. Ruzicka. Liege, Belgium: IUSSP.

______. 1990. "What's Happening to the Family? Interactions Between Demographic and Institutional Change." *Demography* 27: 483-498.

Bumpass, L. L., and S. McLanahans. 1989. "Unmarried Motherhood: Recent Trends, Composition, and Black-Whites Differences." *Demography* 26: 279-286.

Bumpass, L L., and J. A. Sweet. 1989. "National Estimates of Cohabitation." *Demography* 26: 615-625.

Bumpass, L. L., J. A. Sweet, and A. Cherlin. 1991. "The Role of Cohabitation in Declining Rates of Marriage." *Journal of Marriage and the Family* 53: 913-927.

Caplow, T. 1991. "Trends in Sex." Pp.144-156 in *American Social Trends*. Fort Worth, TX: Harcourt Brace Jovanovich.

Cuber, J. F. 1972. "How New Ideas About Sex Are Changing Our Lives." In *Intimate Lifestyles*, edited by J.S. Delora and J.R. Delora. Pacific Palisades, CA: Goodyear.

Cutright, P., and H. L. Smith. 1988. "Intermediate Determinants of Racial Differences in 1980 U.S. Non-marital Fertility Rates." *Family Planning Perspectives* 20: 119-123.

Duberman, L. 1974. *Marriage and Other Alternatives*. New York: Praeger.

Eggebeen, D. J., and D. T. Lichter. 1991. "Race, Family Structure, and Changing Poverty Among American Children." *American Sociological Review* 56: 801-817.

Elirenreich, B. 1983. *Hearts of Men: American Dreams and Flight from Commitment*. Garden City, NY: Doubleday.

Ellwood, D T., and M. J. Bane. 1984. "The Impact of AFDC on Family Structure and Living Arrangements." *Report to DHHS Grant No. 92-A-82*. Cambridge, MA: Harvard University.

Espenshade, T. J. 1985. "Marriage Trends in American: Estimates, Implications, and Underlying Causes." *Population and Development Review* 11: 193-245.

Farley, R., and S. M. Bianchi. 1987. "The Growing Racial Difference in Marriage and Family Patterns." University of Michigan, *Population Studies Center, Research Report* No.87-107.

Feinberg, S. E. 1978. *The Analysis of Cross-Classified Categorical Data*. Cambridge, MA: MIT Press.

Furstenberg, F. F., Jr. 1976. *Unplanned Parenthood*. New York: Free Press.

Glick, P. C., and A. J. Norton. 1977. "Marrying, Divorcing and Living Together in the U.S.A. Today." In *Population Bulletin* 32(5): 1-39. Washington, DC: Population Reference Bureau.

Goldscheider, F. K., and L. J. Waite. 1986. "Sex Differences in the Entry into Marriage." *American Journal of Sociology* 92: 91-109.

Gove, W. R., and M. Hughes. 1979. "Possible Causes of the Apparent Sex Differences and Physical Health." *American Sociological Review* 44: 126-146.

Hamilton, L. C. 1992. Pp.217-249 in *Regression with Graphics: A Second Course in Applied Statistics*. Brooks/Cole.

Hanushek, E. A., and J. E. Jackson. 1977. *Statistical Methods for Social Scientist*. Orlando, FL: Academic Press.

Hoffman, S. D., G. Duncan, and R. Mincy. 1991. "Marriage and Welfare Use Among Young Women: Do Labor Market, Welfare and Neighborhood Factors Account for Declining Rates of Marriage Among Black and White Women." Paper presented at University of Wisconsin Conference on Poverty, June 1991.

Hogan, D. P., and E. M. Kitagawa. 1985. "The Impact of Social Status, Family Structure, and Neighborhood on the Fertility of Black Adolescents." *American Journal of Sociology* 90: 825-855.

Kiecolt, K. J. 1988. "Recent Developments in Attitudes and Social Structure." *Annual Review of Sociology* 14: 381-403.

Kluegel, J. R., and E. R. Smith. 1986. *Beliefs about Inequality: America's Views of What Is and What Ought to Be*. New York: Aldine De Gruyter.

Lichter, D. T., F. LeClere, and D. McLaughlin. 1989. "Local Marriage Market Conditions and the Marital Behavior of Black and White Women." Population Issues Research Center, Working Paper Series No.1989-11.

Lichter, D. T., D. McLaughlin, G. Kephart, and D. Landry. 1992. "Race and the Retreat from Marriage: A Shortage of Marriageable Men?" *American Sociological Review* 57: 781-799.

London, K. A. 1991. "Cohabitation, Marriage, Marital Dissolution, and Remarriage: United States, 1988." *Advance Data* 194: 1-8.

Mare, R. D., and C. Winship. 1991. "Socioeconomic Change and the Decline of Marriage for Blacks and Whites." Pp.175-202 in *The Urban Underclass*, edited by C. Jencks and P. E. Petersen. Washington, DC: The Brookings Institution.

Moore, K. A., and T. M. Stief. 1991. "Changes in Marriage and Fertility Behavior: Behavior Versus Attitudes of Young Adults." *Youth and Society* 22: 362-386.

O'Connell, M., and K. Moore. 1980. "The Legitimacy Status of First Births to U.S. Women Age 15-24: 1939-1978." *Family Planning Perspectives* 12:15-25.

Oppenheimer, V. K. 1992. "American Marriage Formation in the Eighties: How Important Was Women's Economic Independence?" Presented at Annual Meeting of the Population Association in Denver Colorado, April 30-May 2.

Parnell, A. M. 1989. "Non-marital Pregnancy and Marriage in the United States." Paper presented at Annual Meeting of Population Association of America.

Pagnini., D. L., and R. R. Rindfuss. 1993. "The Divorce of Marriage and Childbearing: Changing Attitudes and Behavior in the United States." *Population and Development Review* 19: 331-347.

Preston, S. H., S. Lim, and S. P. Morgan. 1992. "African-American Marriage in 1910: Beneath the Surface of Census Data." *Demography* 29(1): 1-15.

Rindfuss, R. R., and J. A. Jones. 1991. "One Parent or Two? The Intertwining of American Marriage and Fertility Patterns." *Sociological Forum* 6(32): 311-326.

Rindfuss, R. R., and A. Parnell. 1989. "The Varying Connection Between Marital Status and Childbearing in the United States." *Population and Development Review* 15: 447-470.

Rindfuss, R. R., and A. Vandenfleuvel. 1990. "Cohabitation: A Precursor to Marriage or an Alternative to Being Single?" *Population and Development Review* 16: 703-726.

Robertson, I. 1987. "Sexuality and Society." Pp.225-233 in *Sociology.* New York: Worth.

Schoen, R. 1985. "Marriage and Divorce in Twentieth Century American Cohorts." *Demography* 22: 101-114.

Schoen, R., and J. R. Kluegel. 1988. "The Widening Gap in Black and White Marriage Rates: The Impact of Population Composition and Differential Marriage Propensities." *American Sociological Review* 53 :895-907.

Schoen, R., and D. Owens. 1992. "A Further Look at First Unions and First Marriages." Pp. 109-117 in *The Changing American* Family, edited by S. J. South and S. E. Tolnay. Boulder: Westview Press.

Schuman, H., and M. P. Johnson. 1976. "Attitudes and Behavior." *Annual Review of Sociology* 2: 161-207.

Singh, B. K. 1980. "Trends in Attitudes Toward Premarital Sexual Relations." *Journal of Marriage and the Family* 42: 387-393.

Smith, J. P. 1988. "Poverty and the Family." Pp.141-172 in *Divided Opportunities*, edited by G. D. Sandefur and M. Tienda. New York: Plenum.

Smith, T. W. 1992. "Attitudes Towards Sexual Permissiveness: Trends and Correlates." National Opinion Research Center, University of Chicago. General Social Survey Change Report No. 35.

South, S. J. 1991. "Sociodemographic Differential in Mate Selection Preferences." *Journal of Marriage and the Family* 53: 928-940.

______. "Racial and Ethnic Differences in the Desire to Marry." *Journal of Marriage and the Family* 55: 357-370.

Spanier, G. B., and P. C. Glick. 1980. "Mate Selection Differentials between Whites and Blacks in the United States." *Social Forces* 58: 707-727.

Staples, R. 1985. "Changes in Black Family Structure: The Conflict Between Family Ideology and Structural Conditions." *Journal of Marriage and the Family* 47(4): 1005-1013.

Sweet, J. A., and L. L. Bumpass. 1990. "Young Adults' Views of Marriage, Cohabitation, and Family." NSFH Working Paper No.33, Center for Demography and Ecology, University of Wisconsin-Madison.

Sweet, J. A., L. L. Bumpass, and V. R.A. Call. 1988. "The Design and Content of the National Survey of Families and Households." NSFH Working Paper No.1, Center for Demography and Ecology, University of Wisconsin-Madison.

Tanfer, K. 1987. "Patterns of Premarital Cohabitation Among Never Married Women in the United States." *Journal of Marriage and Family* 49: 483-498.

Tanfer, K., and M. C. Horn. 1985. "Non-marital Cohabitation Among Young Women: Findings from a National Survey." Paper presented at the Annual Meetings of the Population Association of America, Boston, March 18.

Thornton, A. 1988. "Cohabitation and Marriage in the 1980's." *Demography* 25: 497-509.

______. 1989. "Changing Attitudes Toward Family Issues in the United States." *Journal of Marriage and the Family* 51: 873-893.

Thornton, A., and D. S. Freedman. 1982. "Changing Attitudes Toward Marriage and Single Life." *Family Planning Perspectives* 14: 297-303.

Waite, L. J., F. K. Goldscheider, and C. Witsberger. 1986. "Nonfamily Living and the Erosion of Traditional Family Orientations Among Young Adults." *American Sociological Review* 51: 541-554.

Walsh, R. H. 1989. "Premarital Sex Among Teenagers and Young Adults." Pp.162-186 in *Human Sexuality: The Societal and Interpersonal Context*, edited by K. McKinney and S. Sprecher. Norwood, NJ: Ablex.

Wilson, W. J., and K. M. Neckerman. 1986. "Poverty and Family Structure: The Widening Gap between Evidence and Public Policy Issues." Pp.232-259 in *Fighting Poverty: What Works and What Doesn't*, edited by S. Danziger and D. H. Weinberg. Cambridge, MA.: Harvard University Press.

Zelnick, M., and J. Kanter. 1980. "Sexual Activity, Contraceptive Use and Pregnancy among Metropolitan-Area Teenagers: 1971–1979." *Famity Planning Perspectives* 12: 230-237.

Zelnick, M., and T. Shah. 1983. "First Intercourse among Young Americans." *Family Planning Perspectives* 15: 67-70.

MARRIAGES, INFORMAL UNIONS, AND GENDER IDEOLOGIES IN REVOLUTIONARY RURAL NICARAGUA

Suzanne M. Baker

INTRODUCTION

The Nicaraguan revolution of 1979 touched off a series of far-reaching changes in the lives of the Nicaraguan people. In addition to social, political, and economic changes, Nicaraguans experienced profound ideological transformations designed to change the way they viewed themselves and the world around them. One of the central issues in this ideological transformation was an intent to modify gender roles and identities. By 1988, this attempt was well under way, but with mixed results. One arena in which the mixed results were evident was that of female-male relationships, with more and more Nicaraguan women questioning traditional aspects of those relationships, such as women's financial dependence upon men, extramarital freedom for men, and ways of forming unions, including marriage.

This is not to overstate, however, the impact the ideological transformation had in ten short years. The seeds of discontent with traditional patriarchal family relations existed in many women prior to revolution, and in part had prompted some to actively participate in the revolution. But, by providing a new way of evaluating gender roles and identities, the ideological transformation gave women the tools to critically examine their choices regarding marriage and other types of unions. Furthermore, new economic opportunities for women provided by the revolution gave many the hope of supporting themselves and their children, an option that also had an impact on union choices. Because the employment opportunities necessitated the belief that women had both the right and ability to work outside the home, the economic changes were therefore both part of, and supportive of, the ideological changes.

It was not so much that the new ideology and its concomitant socioeconomic changes furnished new ways of forming marriages or other types of unions, but that they provided new ways of evaluating, discussing, and justifying union choices. For example, informal unions continued to grow in popularity among many Nicaraguans, particularly women—an outcome the Catholic-influenced revolutionary government[1] probably did not anticipate. Socioeconomic changes also altered Nicaraguans' lives in such a way that marriage and other types of unions were affected. As women's economic options broadened, for example, and men did not have to migrate in search of work (due to the forming of cooperatives), the balance of relations between men and women began to change.

Lastly, the coexistence of both the traditional and new gender ideologies provided Nicaraguans with a variety of beliefs and norms on which to draw when making decisions or engaging in social interaction. The diversity in available ideological beliefs acted as tools that women and men could use to make their own adjustments and negotiations to the personal, social, political, and economic changes taking place in their lives. By picking and choosing among various ideological components, Nicaraguans were able to mobilize certain beliefs and norms to support their actions, including their preferences for forming, not forming, or leaving unions, such as marriage.

NATIONAL CONTEXT

In July 1979, the Sandinista National Liberation Front (FSLN) took power in Nicaragua after more than a decade of armed struggle against the dictatorial government headed by Anastasio Somoza Debayle. Upon assuming control, the FSLN initiated a series of revolutionary changes in Nicaragua designed to reactivate the economy, redistribute resources to rural and urban poor, and empower the poorest sectors of society, including peasants, workers, and women.

In one of their more ambitious strategies for change, the FSLN attempted to transform the traditional Nicaraguan ideology of gender, with its emphasis on *machismo* and patriarchy,[2] into an ideology predicated on gender equality. Building upon Peggy Sanday's (1990, pp. 4-5) definition, a gender ideology is a culturally constructed system of beliefs about the way in which members of the two sexes are "perceived, evaluated and expected to behave." Gender ideologies legitimate gender roles (assigned on the basis of one's sex) and guide the behavior of women and men *as women and men*, and thus, act as a guide for social action and interaction.

Table 1. Traditional Nicaraguan Gendered Identities

	Women	*Men*
Descriptions of who they are	• Highly moral/religious • Passive • Self-sacrificing • Inferior in logic • Inferior in strength • Submissive to men's will • Dependent on men	• Sexually voracious • Aggressive • Active • Superior in logic • Superior in strength • Locus of familial authority • Sole provider in family
Descriptions of what they do	• Defined by their role as mothers • Dedicate themselves to sacrificing for their children and are responsible for child care • Obey men's decisions • Act primarily in the domestic domain • Rely upon men for financial support	• Require more than one woman • Demonstrate virility through the number of their offspring • Make decisions • Act in the public domain • Engage in economic activities

The stated aim of the FSLN's ideological transformation was to eradicate those beliefs and practices that relegated Nicaraguan women to a subordinate status vis-à-vis men, and to allow them to exit the perceived isolation of the domestic realm, participate in productive and revolutionary tasks, and realize their full potential. At the same time, the new government promoted changes in education, health care, legal statutes, access to resources, and employment opportunities that both required and created a foundation for this ideological transformation.

Table 1 lists the features of the gendered identities embodied by the traditional gender ideology. Gendered identities consist of descriptions both of who men and women *are* (often termed "innate qualities") and what women and men *do* (often described as "custom" or "tradition") based solely on their respective sexes.

Although no single, concise definition was given of the male and female gendered identities the FSLN hoped to promote, many official proclamations were given by the revolutionary government to demonstrate the ideological changes the FSLN hoped would occur. In 1982, for example, Tomas Borge (1982, p. 13), Minister of the Interior and member of the Sandinista Joint National Directorate, gave a public speech in which he declared that:

> We have to struggle against the habits, customs and prejudices of men and women. We have to embark upon a difficult and prolonged ideological struggle—a struggle that equally benefits men and women. Men must overcome a multitude of prejudices.
>
> How can we not guarantee that a woman can be both a mother and a worker, both a mother and a student, both a mother and an artist, both a mother and a political leader, both fulfill all the tasks the revolution demands of her and at the same time fulfill the beautiful work of a self-sacrificing, capable, and loving mother?

As the revolution continued the ideological message became more defined. For example, the FSLN's 1987 Proclamation (National Directorate of the FSLN 1987, p. 28) declared that Nicaraguans had to work against "discriminatory laws and policies, the subordinate position of women in society and in the family, paternal irresponsibility, physical and moral abuse, and machismo, all of these obstacles that must be overcome by men and women."

During the period from 1988 to 1990, Nicaraguans continued to receive a plethora of information about the ongoing changes in gender ideology. This information was transmitted in a largely explicit manner, and was debated, elaborated, and repeated. There were two primary sources for the transmission of these messages: the mass media (particularly radio and newspaper) and the conferences, seminars, and workshops held by nongovernmental organizations linked to various political streams.

In their attempt to transform the gender ideology in revolutionary Nicaragua, the FSLN primarily emphasized expanding the definition of what women *do* (such as adding work outside the home to domestic chores), with a fairly strong, secondary focus on changing the ideological description of who women *are* (such as intellectually capable of running income-generating projects). However, the FSLN placed less importance on re-conceptualizing who *men* are and what they do. Owing in part to this unbalanced approach, the old, traditional ideology and the new, egalitarian ideology continued to coexist uneasily into the 1990s. Nicaraguan women and men, therefore, had to grapple with the complexities and contradictions of two gender ideologies in the events of their daily lives, including making arrangements for entering and exiting marriages or other types of unions.

The revolutionary changes did, however, provide women and men with an ideological environment in which they could use the new conceptions regarding women and men to create compromises and syntheses of ideological beliefs about gender pertinent to their own particular, current socioeconomic situations. These syntheses and compromises were vital for people's *personal* adjustments to the social and economic changes taking place, and the new critical gender ideology provided women and men with tools they could use precisely at the time adjustments and negotiations were necessary.

On the other hand, the unbalanced change in gender ideology did not result in an ideology of gender equality. Because their notions of who women and men *are* remained essentially the same, Nicaraguans' schema for acting and interacting in the world also remained virtually the same. Women and men continued to be situated in personal relationships primarily predicated on the traditional ideology's notion of who women and men are. Because of this, people were not always able to change their own social identity or roles, nor did they always have the desire to do so. Most Nicaraguans' engagement with the gender ideologies remained at the level of trying to accommodate or repudiate the changes, creating compromises or syntheses, or making sense of the *new* definitions of what women *do* within a schema of mainly *old* definitions of who women and men *are* in terms of social identities.

Choosing to form a union, whether to marry, how to marry, and whether to leave a marriage or a union, therefore, were decisions taken within a confused, contradictory, and ever-changing ideological environment. Such decisions had always been at least partly based on a wide variety of economic, social, and even political, factors. With the new socioeconomic opportunities afforded women by the revolution, however, their options were now at least a little broader and their horizons a bit wider. How this would impact marriage practices of the future remained to be seen at the end of 1990, but through an examination of preferences (and some practices) of older women in 1988 to 1990 in one rural hamlet, as well as a review of the attitudes and intentions of 14- and 15-year old women, some changes in attitudes, if not practices, become apparent.

LOCAL CONTEXT

This paper examines the types of unions between men and women existing and being formed in the community of Isiquí during the period of 1988 through 1990. Located 11 miles to the east of Estelí in the northwestern part of Nicaragua, Isiquí lies in a valley, with mountains to the south and north. The road from Estelí, a city of 60,000 people, is dirt and rock, and travel between Estelí and Isiquí takes about one and one-half hours. Isiquí is a completely rural community of 46 houses. Although many of the inhabitants had lived in the area for some time, the actual community of Isiquí was fairly new, having been established along with an agricultural cooperative in the very early 1980s. Prior to the revolution, most of what was considered as contemporary Isiquí did not exist, and the families that did live in the area worked the land of the local large landowner, either as sharecroppers or as wage laborers.

In addition to houses, Isiquí had, in the late 1980s, a two-room school that provided classes for grades one through four, a small health center, and an unfurnished chapel where a visiting priest form Estelí occasionally celebrated Mass. There were also several buildings that made up a chicken farm collective and the agricultural cooperative.

There were 309 inhabitants, in 46 households in Isiquí at the time of research. When Isiqueños discussed the issue of economic survival, it was in terms of what was necessary for the survival of the household (those who lived together in one house, generally pooling income and engaging in consumption activities as a whole), rather than for an individual or for the family. The family, to Isiquenos, was a unit composed of a mother and her children, to which might or might not be added a father. A man was considered necessary for the optimal survival of a household in economic terms, but a man was not necessary for the existence of a family unit per se. Of these 309 inhabitants, 47 percent were children under the age of 15 years. By 15, most Isiqueños[3] were establishing families (either as a married couple or as a single mother), working full-time as members of the cooperative, or working as individual farmers. Of the 165 adults, 51.5 percent were women.

Owing to FSLN's Literacy Crusade of 1980, most Isiqueños had some ability to read, and several were quite literate. Newspapers purchased frequently in Estelí were one crucial medium through which many Isiqueños became acquainted with the new ideas regarding men and women. Isiqueños also learned about the new socioeconomic and ideological changes by listening to the radio and by attending seminars and conferences held for members of the National Union of Farmers and Ranchers (UNAG), the FSLN, and the Association of Nicaraguan Women "Luisa Amanda Espinoza" (AMNLAE).

The focus of economic and social life in Isiquí was the agricultural cooperative, which consisted of 36 members, all of them men. In addition, some men were individual farmers or day laborers. Most women of Isiquí engaged in informal economic activities, such as selling chickens or food items, while young daughters of the poorest families often worked as general house servants for other Isiqueño families. The majority of the adults in the community classified themselves as Sandinista supporters, and the men of the cooperative and the chicken collective were members of UNAG. In addition, many women, particularly those belonging to a new women's collective, were members of AMNLAE, which promoted the new gender ideology.

THE WOMEN'S COLLECTIVE

The women's income-generating project was, for the most part, in the planning and construction phases during the time of field research, and was not yet operational. However, the very idea of an income-generating project, as well as the hands-on experience gained by the women and teenage girls who made up the women's collective during this phase, had already had an effect on women's decisions and thoughts regarding marriage arrangements.

In 1988, UNAG had offered a cream-separating machine to the agricultural cooperative in Isiquí to encourage its leadership to include women as members. When the cooperative declined to purchase it, several Isiqueña women became interested in purchasing the machine themselves in order to establish a collective of women, and seven women formed the Collective Esperanza Flores. The majority of the women in the collective described themselves as Sandinista supporters, and several were quite active in the party.

After many false starts and much delay, the Embassy of the Netherlands provided the women's collective with partial funding in July 1990 and, in late August 1990, the Embassy of Canada agreed to provide the remaining funds for the full execution of the project. The months of August through December 1990 were devoted to the construction of a building to house the project, installation of machinery, and purchase and construction of various implements. Although the collective, which then had 20 members, began production of cheese and cream in October, it was a short-lived attempt and, by late December, full-time production still had not begun.

Apart from participating in the collective, most of the women of Isiquí described themselves as "not working." According to the women of the collective, 80 of the 85 women in Isiquí were not working, meaning that the women did not work for income outside the home. The five who were considered to work were two women who worked as cooks for the chicken farm, a woman who made harnesses, a local nurse, and a woman who sold coffee and cigarettes from her home.

Women also engaged in myriad of activities designed to bring in a small amount of cash. These activities included raising small-size farm animals, and making breads, pastries, or a batch of the local alcohol to sell. Women also purchased various items, such as fruit, beer, or bread, in Estelí to be resold in Isiquí. The problem with all of these activities, according to the women of Isiquí, was that they were sporadic and unpredictable. It is for this reason that women decided to establish an income-generating collective.

When discussing their reasons for initiating the project, the women of the collective named three primary goals they hoped to achieve, all of them financial in nature. First, the women wanted to earn a stable income that would allow them to make certain purchases for their families. Second, they wanted to earn an income that would be separate from that of their husbands/*compañeros,*[4] often because of their drinking, gambling, or womanizing. Third, the women wanted to be able to support themselves and their children in case *something* would happen in the future. Although some women said that their concern was that their husbands/*compañeros* would fall ill or otherwise be unable to work, the majority were concerned about the questionable longevity of the relationship with their husband or *compañero*. The project, by contributing to financial stability, was therefore primarily seen by Isiqueña women as a way to deal with the special problems presented by men (either by their spending habits or by the eventual break-up of relationships).

FORMING UNIONS

By their late teens, most Isiqueños had begun to establish families, either formally through marriage (through civil or religious ceremonies, or both), or informally. Informal unions in Nicaragua consisted of two types in particular: *uniones de hecho* (literally, unions of fact, or common-law arrangements) and *estados de acompañado* (informal unions of "companionship," or informal unions). Although defining *marriage* in such a way that it is applicable in all instances across time and space is difficult,[5] Kingsley Davis's (1985, p. 5) description of five types of sexual unions (marriage, common-law, consensual, cohabitation, and liaison) provides a basis for differentiation. According to Davis, marriage and common-law arrangements differ in that for the former there is official and unofficial public recognition, whereas for the latter there is not, although both are assumed to imply permanence. In the case of common-law unions, the official recognition may or may not be given over time. Common-law and consensual unions differ in that the latter does not necessarily imply permanence.

In Isiquí, however, the popular assumption was that most unions would be impermanent, and no one spoke of permanent, informal unions (common-law arrangements). Therefore, the term *informal unions*, covering both Davis's common-law and consensual unions, is used here to highlight the fact that Nicaraguans do not necessarily consider permanence to be a certainty of any union, and particularly not of those lacking official sanction. Davis (1985, p. 7) reports that in some parts of Latin America, the consensual union "constitutes a majority of all unions." Other research indicates that in Latin America, consensual unions may "significantly outnumber" formal unions in the lower classes (Ingoldsby 1995b, p. 336).

Whereas most Isiqueña women described themselves as *casadas* (married), many were actually involved in informal unions. According to the data presented in Tables 2, of the 60 Isiqueña women involved in a union type, 24 (40%) were in a formal union (church or civil marriage), while 36 (60%) were in informal arrangements. This pattern was also reflected in national statistics of that time, in which less than 50 percent of all couples were married, with informal arrangements being the more frequent type of union (Stead 1991, p. 62). Official marriages could be contracted through civil or religious ceremonies.

Informal Unions

There were several ways in which informal arrangements were established in Isiqui. For example, when a young woman became pregnant, she and her *novio* (boyfriend) might decide to live together without undergoing any formal marriage ceremony. In other cases, a couple would first decide to live together after a period of dating and then to begin a family, again without any formal marriage ceremony. In both situations, community members described these couples as *married,* particularly after the birth of their first child. In theory, these couples had the option

Table 2. Formal Versus Informal Unions for All Isiqueña Women Age 15 and Over in Unions, 1989–1990 (n = 60)

Type of Union	*Number of Women*	*Percentage of Total*
Formal (church and civil)	24	40%
Informal unions	36	60%

to formalize the union at a later date and receive official recognition, although few took this next step.

According to most of the women in Isiquí, informal arrangements (or less so, civil marriages, which are described later) were their preferred ways of establishing unions or families, or both. This preference, however, was not shared by the men of Isiquí, nor by Nicaraguan men in general. According to Roger Lancaster (1992, p. 308), although Nicaraguan men were less concerned with divorce than were women, they were more interested in formal marriage than were women. The reason for this, according to Lancaster, was that "a married woman would be much less likely to leave her husband." In addition, despite the revolutionary changes, Nicaraguan women continued to feel that formal marriages benefited only men, and that consequently informal unions were more advantageous for women.

Lancaster (1992, p. 46) links this difference in preferences and benefits to the relative difference in power between a man and a woman. A husband could refuse to support his wife or he could beat her, but the only real alternative a woman had was to threaten to leave her husband, an action that presented more difficulty with a legal marriage than with an informal arrangement. A man maintained his relative sexual freedom within a legal marriage, because both the traditional and new gender ideologies accepted the belief that men required more than one woman. At the same time, a formal marriage was a union that also helped to ensure that his wife would not leave him. A woman, on the other hand, was more constrained by a legal marriage than was a man, because both gender ideologies condemned any adulterous affairs by women. For her, formal marriage, with its legal ties, was also a union that lessened the possibility that she might leave.

For women, informal unions were a way of maintaining a bit of freedom and keeping open their options as a partial response to the constraints imposed by the traditional gender ideology. Lancaster (1992, p. 47) notes that one of his informants (an older woman) hypothesized that "with the new laws and with some men trying to be more responsible and less *machista*" a legal marriage might become more feasible for women. Some Nicaraguans, therefore, believed that with the transformation of the gender ideology, a practical change in establishing unions would occur. As Lancaster (1992, p. 45) points out, the "informal union as an institution occurs in the context of the culture of *machismo*," rather than creating the "irresponsible sexual behavior" of men that is called for, and supported by, the traditional ideology.

However, in Isiquí women were not making a change to formal unions, and those teenaged women who were establishing unions were still doing it through informal arrangements. In a breakdown by age groups (Table 3), 100 percent of all women aged 14 to 19 who were in a union were in informal arrangements. For women aged 23 to 30, the percentage was 56 percent, while for women over 30, the it was 48 percent. The preference for informal arrangements did not necessary decrease with changes in the ideology of gender, and may have actually increased.

Table 3. Distribution Of Isiqueña Women in Informal Unions, Formal Unions, and No Unions, by Age Group, 1989–1990 ($n = 85$)

Type of Marriage	*Age Groups*		
	14–19	*20–30*	*Over 30*
Church marriage	0(0%)	3(11%)	4(11%)
Civil marriage	0(0%)	5(18%)	12(33%)
Informal unions	11(52%)	10(36%)	15(42%)
Single mother/No union	4(19%)	7(25%)	4(11%)
No children/No union	6(29%)	3(11%)	1(3%)
Totals	21	28	36

Young Nicaraguan women (both rural and urban) were the most vocal of all ages in declaring no desire to be married or to form a union of any kind with a man, stating that they would rather work and spend their money as they saw fit. On one occasion, a 16-year-old member of the collective (and supporter of AMNLAE and the FSLN), Berna, declared to her mother, Ana, that she was "planning to work forever in the women's collective." She had "no plans to feed anyone" but herself. This firm statement was made in response to Ana's criticism of Berna's half-hearted attempts to learn to prepare acceptable rice and tortillas.

Alicia, another young member of the women's collective who was also a supporter of AMNLAE and the FSLN, was 14 years old. Despite being desired by most men in the community, both unmarried and married, Alicia had no *novio*. On the contrary, Alicia declared more than once that she just wanted "to work in the women's collective, because men are such trouble. They will tell me what to do and what to wear, and I don't like that." This negative attitude toward unions, held by both Berna and Alicia, contradicts Bron Ingoldsby's prediction (1995b, p. 337) that as "more women in Latin America enter the labor force, it may be that marriages will shift from traditional [based on parental choice and approval] to more companionate [based more on affection and self-choice]. . . ." For young Isiquena women, entering the (paid) labor force meant a hoped-for freedom from the perceived restrictions implicit in any union with a man.

The greater economic and educational opportunities that were now open to women were calling into question the traditional assumption that a woman must have a man to support her. With women now able to envision a future in which they could support themselves, more and more were voicing a preference for informal arrangements or no unions at all. In Isiquí, single mothers who were not in any union at all made up 19 percent of the women aged 14 to 19, and 25 percent of the women aged 20 to 30. As the woman in Lancaster's example conjectured, formal unions *might* have become more attractive if men were "more responsible and less *machista*." However, because this particular aspect of the transformation was occurring very slowly due to a lack of emphasis on *men's* roles, informal unions and single motherhood remained a preference for many young women in Isiquí.

Abductions: One Way to Establish an Informal Union

According to the women of Isiquí, many informal unions also began as an abduction (*rapto*)[6] of a girl by a young man.[7] While occasionally such an abduction was done without the consent of the girl, most occurred with her connivance. Therefore, these arrangements could be viewed as elopements, although they resulted more frequently in informal unions rather than formal marriages.[8] According to Bron Ingoldsby (1995a, p. 144), this type of abduction has as its roots the capture of women that took place "in patriarchal societies in which women were often considered property."

The primary reason a couple in Isiquí would choose the route of abduction to begin an informal union, rather than simply to live together, was the opposition of the girl's parents (particularly that of the father), to the relationship. For example, Bianca (age 15) was abducted when she was 14 years old by her *novio* (who was 24 at the time). Because Bianca's parents were not in favor of her marrying the man (because they thought she was too young), she and her *novio* arranged to meet in secret one night after dark, and to run away together to Estelí.

They hid overnight in his brother's house while Bianca's father searched Isiquí, gun in hand. In the morning, Bianca and her boyfriend took the public transport to Estelí. After residing in Estelí for a month with the man's relatives, the couple returned to Isiquí. By this time, Bianca's parents had resigned themselves to the situation, because the union was now a fact. Bianca and her *compañero* continued to live in his parents' house in Isiquí, and their relationship with her father and mother remained chilly, despite the fact they lived a few feet away.

According to Isiqueño informants of all ages, abduction due to parental opposition had long been a common way, especially in the rural areas, to establish a union. Historian E. Bradford Burns (1991, pp. 66-67), in his book on Nicaragua of the 1800s, begins his discussion of patriarchy with the story of the trouble caused in 1851 when two young adults ran away to marry without the permission of the young man's father. The incident sparked a furious national debate on patriarchal authority. The reason for this furor, according to Burns, was that the patriarchal family had been the glue holding together Nicaraguan society since the 1600s, and marriage was the "very foundation stone of patriarchal society." It was not the marriage of the couple that had caused the furor, but the fact that patriarchal authority, a cornerstone of the traditional gender ideology, had been subverted.

Most importantly, in Isiquí these cases were always described as abductions, despite the fact that in the majority of the cases, both past and present, the girl played just as active a role as her *novio*. While Isiqueños admitted that girls were actively choosing to defy parental opposition, the way in which they described such cases highlighted their view of girls as passive participants: young men kidnap girls. However, the seeds for a gender ideology that would give public recognition to the active role of women in society, the economy, and the family, clearly existed prior to the Nicaraguan revolution of 1979. Girls had always played active roles in abduction dramas, but because of the framework provided by the traditional gender ideology, they continued to be portrayed as passive participants.

FORMAL UNIONS: CIVIL CEREMONIES AND WEDDING CEREMONIES

In addition to informal living arrangements, Isiqueños also established families by means of formalized unions, or marriages. According to the people of Isiquí, there were at that time two ways of being married in Nicaragua, a civil marriage ceremony and a church wedding. In order to have a church wedding, they reported, a civil ceremony had to take place first, but there was no legal requirement that a civil ceremony had to be followed by a church ceremony. This was the way in which all Isiqueños described the legal aspects of formal unions, and this is indeed the way it had been prior to the revolution. According to the 1974 Nicaraguan Constitution (Morgan 1984, p. 486), only the civil marriage was recognized by law, while all informal unions or common-law arrangements were not recognized.

However, changes in the legal code since 1979 nullified the requirement of a civil ceremony prior to a church wedding, and laws enacted in 1981, as well as the Constitution of 1987, stated that "marriage and *stable common-law unions* were protected by the State" (Lancaster 1992, p. 307).[9] The fact that no one in Isiquí was aware that a civil ceremony was no longer required indicated that a significant number of Nicaraguans, especially in the rural areas, were not aware of many of the structural changes that had occurred since 1979. In part, this lack of knowledge was due to the logistical problems faced by government officials when they attempted to disseminate information, particularly in remote, rural areas. However, the Isiqueños' lack of knowledge about such legal changes also rested upon the fact that, for most of them, informal arrangements were the preferred form of establishing a union, and therefore most inhabitants had no reason to seek information on marriage laws.

In Isiquí, 71 percent (17 of 24) of the women who had formally married had done so with just a civil ceremony (see Table 4), in which a judge was contracted to perform the marriage vows and write up the marriage certificate.

According to Isiqueña women, there were several reasons for choosing to have just a civil ceremony, without the following church ceremony. Margarita, age 23 and originally from Leon,[10] had celebrated a church wedding in the mid-1980s. According to Margarita, a Sandinista *militante*[11] and member of AMNLAE, "not

Table 4. Distribution of All Isiqueña Women Aged 15 and Over in Informal Unions, Formal Unions or No Unions, 1989–1990 ($n = 85$)

Type	*Number of Women*	*Percentage of Total Women*
Church marriage	7	8
Civil marriage only	17	20
Informal unions	36	42%
Single mother/No union	15	18
No children/No union	10	12
Totals	85	100

all people are married in a church, unless they are very religious, since you have to pay the priest, wear a white and long dress, with a veil, white gloves and white shoes." For this reason, a church ceremony, with the expected—and nearly mandatory—fiesta afterward, was considered by most Isiqueños to be too great an expense. In a community where most women reported not having purchased new clothing in over five years, the cost of a church wedding was prohibitive.

A civil ceremony, on the other hand, was less expensive, and required no new nor formal clothes, and there was no priest to pay. There was, however, payment for the judge and a fiesta afterward, so a civil ceremony was not without its costs. A church ceremony, because of its higher price tag, demonstrated not only one's religious devotion, but also one's economic level. For Margarita, who had her teaching degree, a church wedding was an important mechanism for demonstrating to her new family members and neighbors the economic resources of her family back in Leon.

Sara, also 23 years old and a member of the FSLN and AMNLAE, had only a civil ceremony in the early 1980s. Although she did not have a church ceremony, Sara was careful to point out that "these special clothes [wedding dress, etc.] are not really necessary, especially if the couple has lived together before the church wedding. Then you can wear just any clothes." Still, many women indicated a preference for arrangements other than a church wedding, even though they considered themselves religious. Some other factor must have mitigated against a woman's desire for a church wedding, and this factor is the availability of divorce.

ENDING UNIONS

Prior to 1979, divorce was very difficult for Nicaraguan women to obtain, but it was an option open to and utilized by many men. With unilateral divorce at the request of either spouse guaranteed by the 1987 Constitution, both women and men could petition for divorce, and if neither contested it the papers were quickly signed at little or no cost. According to the women of Isiquí (who were quite knowledgeable about the legal changes regarding divorce, in contrast to their lack of knowledge about the changes in the legalities surrounding marriage), these changes made it easier for both women and men to get a divorce.

In reality, however, if either spouse refused to attend the paper signing, the one who solicited the divorce was charged a higher fee. The new law, therefore, was still disadvantageous for women because they were less likely to have the necessary financial resources to initiate a divorce against the wishes of their husbands. However, the law did increase women's right to petition for a divorce.

Although divorce became available in Nicaragua under the new law, only the State, and not the Catholic church, recognized divorce. One Isiqueña woman described the situation this way: "One can never be divorced with a church wedding, or really, one can get a civil divorce, but the church won't recognize it. The church won't recognize any remarriage either. And you can't get remarried in a church either." According to Sara and Margarita, if one were to divorce and remarry, "that person would have two spouses, one recognized by the church and the other one recognized in civil law." Both women found this hypothetical situation offensive, but agreed that it was sometimes necessary.

Although men continued to have greater access to divorce they expressed less interest in this option than women did. As is described later in this paper, Ana (Sara's mother, age 42 and an FSLN supporter) was divorced by her husband several years prior to the revolution, so that he could immediately marry another woman. But men in Isiquí rarely availed themselves of this solution, because as one older woman (45 years old) said, in reference to her ever-womanizing husband, "why should he divorce me when he can have it all? He has us both! I do all the work, and he has the fun. What can I do—I have to put up with it." Many women, especially the older ones expressed similar sentiments: why should men divorce their wives when they can enjoy other relationships on the side anyway?

The establishment of cooperatives in many communities had an unintended consequence that affected men's extramarital and extra-union affairs. Because most men no longer had to migrate in search of work, they were conducting their secondary relationships with women in their own, small communities, rather than establishing other families in other towns or regions. This meant that these affairs were often very public, acknowledged by the whole community, and thus, an inescapable fact for the women with whom the men had unions.

For example, in response to Sara's anger at her husband Manuel's womanizing, her mother gave what Sara believed was a less than sympathetic explanation. According to Ana, the only problem with Manuel's behavior was that he had chosen another woman who lived in the same community, rather than in some other locale. This was one of the most frequent complaints women had regarding the womanizing of their husbands. For many women, it was impossible to completely resign themselves to the situation when it was ever-present. When the offensive behavior was out of sight, however, resignation was easier to achieve. With the establishment of agricultural cooperatives, there was less migration, which made it easier for many men to remain with their families year-round. However, there was no concomitant emphasis on male monogamy by the new gender ideology, and relationships between men and women became even more problematic.

On the other hand, for Isiqueña women, and in accordance with the traditional gender ideology, living with one man, while having a relationship with another man was simply not feasible. Such action in the past could certainly have brought legal accusations of adultery, and even with the revolutionary changes could have brought a severe beating, as well as community assessment of the woman as *vaga*[12] or worse, a *puta.*[13] Even so, there were a few women in Isiquí who had fleeting relationships with men or maintained relationships with men who had wives or *compañeras*, and so were considered *vaga* or *putas*, but even these women were not living with other men while doing so. For men, however, such behavior was completely acceptable, even expected, and divorce was generally not necessary.

If the revolution provided greater ease when ending unions, it also provided new ways in which to remain in unions when problems arose. For example, as a result of the installation of Isiqui's community clinic in the late 1980s, a more liberal governmental policy toward birth control, and the promotion of the view that men and women should participate equally in decisions regarding the family, the women of Isiqui were able to purchase contraceptives at low cost without their husbands' permission. According to the local nurse, many women of the community used contraceptives, but nearly all kept it secret from their husbands.

While most women are hesitant to discuss the topic, Sara said that one way in which she would use the eventual salary from the women's collective was to purchase birth control pills. Sara and her husband, Manuel (age 27) had agreed to not have any more children until their economic situation improved. Sara also reported that she keeps her use of the pills strictly hidden from Manuel, fearing that he would be angry. Sara, like other women in Isiquí who are members of AMNLAE, the FSLN, and the women's collective, recognized their newly won right to control family size, but like their mothers and grandmothers before them, realized that their men might resist and leave them for women with whom to have more children. The compromise of these women was to practice hidden birth control, a solution facilitated by both the policies and attitudes of the revolutionary government.

Men like Manuel, on the other hand, were ambivalent about limiting family size, and often expressed a desire for more and more children. While the overt reasons given included both the joy children brought and the help they could be in the house and in the field, men also indicated that they felt *más hombre* (more manly) with more children. Having numerous children was still an overt way to demonstrate to others (especially other men) that a particular man embodied all the *innate* characteristics of men, such as sexual voraciousness, aggressiveness, and activeness, according to the traditional gender ideology.

Manuel, a member of the FSLN, UNAG, and the agricultural cooperative, represented the position many revolutionary men found themselves in, caught in ambivalence between the contradictory views regarding the nature of men in the traditional and new ideologies. On the one hand, Manuel agreed with the need to limit family size, and stated that "we are going to wait some time for the next one." On the other hand, Sara felt that Manuel's attitude showed that it was necessary to hide her use of birth control pills. The ambivalence Manuel felt was shown even more clearly when his mistress had a baby boy in late 1989. Manuel clearly enjoyed his status as a new father and strutted about the community, receiving the congratulations and jokes of other men.

Manuel, in having two women, was able to reconcile at least part of the requirements of both gender ideologies, and remain in a marriage that was not providing him with enough children to maintain his *más hombre* image. With Sara, he advocated responsible paternity (ostensibly a characteristic of the new gender ideology), while with the other woman he was able to fulfill his *innate* role as a man in the traditional ideology—defined by the number of his women and children. Manuel is literally a man in transition, with a foot in both ideological camps.

Thus, Manuel and Sara illustrate the confusion felt by many Isiqueños when trying to live as a couple within the dictates of both the traditional and new gender ideologies. They were young adults in their 20s who had lived half of their lives prior to the revolution, yet who also had been significantly exposed to the new gender ideology through their organizational memberships (as well as through media) during the formative years of late adolescence and early adulthood. In dealing with issues of family planning and maintaining a marriage, each found a compromise while neither entirely refuted the traditional ideology, nor totally embraced the new gender ideology.

NO UNIONS

Not all women is Isiquí with children established a union and a household with a man. Of the 15 single mothers in Isiquí, 12 still lived with or had returned to their parents, and three lived alone (see Tables 2 and 4). Four of the single mothers were older, and either abandoned, divorced (for example, Ana), or widowed.[14] The younger single mothers had either been recently abandoned, or had never initially settled down with the fathers of their children. While these latter fathers lived in Isiquí, and visited the children on occasion, there was no attempt to constitute a family unit that included the father. This type of "union" would best be described as a liaison in Kingsley Davis's typology (1985, p.5) since there was no cohabitation or assumed permanence to the union, nor any of the features Davis highlights as aspects of other types of unions.

For the majority of the young single mothers who had never left their natal home, life continued much as it had previously, except now they had to care for children as well. In the case of the other single women, such as Ana, most had returned to their natal home, after being abandoned, divorced, or widowed by their husbands. They returned home, they say, because they could not survive on their own without a man to support them. According to Ana, "I had no choice but to come home, so that my parents could help me," and many, like Ana, remained in their parents' home, even as their own children became parents themselves.

The situation of single mothers living along with their children was an anomaly in Isiquí, and there were only three cases. Two of the women were sisters,[15] with five children between them, and they lived together in a house that was merely ten steps away from that of their mother (where yet another sister with children lived). The two households were closely linked economically, and these two single mothers had never really left their natal home. The third single mother[16] was a complete anomaly, in that she lived completely alone, having not only kicked out her husband, but also forced him to take their three young children with him. This husband had returned with the three children to his own mother's house.

While a woman living along was uncommon in Isiquí, it was even more uncommon for men to live alone. In Isiquí there were unattached sons still living in their parents' homes, young men living with their aunts, and single fathers living with their parents. But no man, with or without children, lived alone. Men and women alike explained this situation in question form: "Who would cook for him?" "Who would take care of the house?" "Who would do his laundry?" To the people of Isiquí, men needed women around to take care of the domestic chores, and if there was no wife, there was always a mother, aunt, or sister who would be there to provide these services.

For both men and women, family members, whether spouses, parents, or other relatives, were necessary for survival, but the reasons differed. For women, it was economic necessity—they need someone to help support them. Most Isiqueña women believed that women, especially as single mothers, would still be unable to support themselves financially. For the women of the collective, however, there was the hope that in the future they would have more economic options. Men, on the other hand, needed someone to provide the services that men say they cannot and will not do—women's tasks (as set forth in the traditional gender ideology, but not negated by the new gender ideology), such as cooking and cleaning.

With the new ideology and the revolutionary socioeconomic changes, women gained greater access to education, economic prospects, and other opportunities once reserved mostly for men. In reality, however, this meant that women's responsibilities increased because they were still responsible for all the activities considered to be women's work. Even under the new ideology, there was little emphasis on prompting men to take over some of these responsibilities, and this effectively limited men's freedom to seek other lifestyle arrangements, such as running their own households.

A SPECIAL TYPE OF SINGLE MOTHER

A special category of single mothers in Isiquí included women designated as *putas* (literally whores, but in the sense of very "loose" women, rather than prostitutes) by the other women of the community. In Isiquí four women, all sisters in their late teens and 20s, were considered by most women of the community to be *putas*. These four women lived with their 11 children in two houses in the center of Isiquí, along with their mother and five other siblings. To support their households, two of the sisters worked as cooks at the local chicken collective and their mother made items for sale in Estelí. The younger women also worked their own fields (loaned to them by the local agricultural cooperative) and did most of the chores generally considered to be men's work, such as gathering firewood. But the women barely made ends meet, and the households were two of the poorest in Isiquí, with very dilapidated homes, and under-dressed and barefoot children. The children, however, were always very active and cheerful, and most attended the local school.

According to Isiqueño opinion, what marked these women as *putas*, rather than simply as single mothers who had not formed unions with the fathers of their children, was the fact that their 11 children were fathered by various men in the community, most of whom were living with other women, either in married or informal unions. According to the four sisters, however, they were simply victims of circumstance. One sister said she had a husband, although the man in question was known to have a wife and home in Estelí. The other sisters said that the wives/*compañeras* of their men were to blame for the situation. As one sister described it, "she [the *compañera*] won't let him leave. My man wants to be with me, but he feels obligated to stay with his other woman." None of the sisters appeared openly ashamed of their situation but, through their explanations, strove to cast themselves as "good" women in unions with men who just happened to live elsewhere. Although their lifestyle was different from that of other women in the community, the sisters viewed themselves as ordinary mothers, albeit disadvantaged ones.

But it was their lifestyle that caused the other women of Isiquí to view them as *putas*. None of the sisters had ever lived with any father of her children, and each had children by several different men. It was because of this that the women of Isiquí considered the four sisters to be *putas* (loose women) and not prostitutes. Whores and prostitutes, to the women of Isiquí, were not the same at all. Prostitution was seen by the women of Isiquí as the last road a mother would take to ensure the survival of her children. It was understood by all that some mothers, to support their children, had to turn to prostitution. To the women of Isiquí, prostitution, while reviled, was at least understandable.

On the other hand, *putas*, according to the women of Isiquí, enticed and stole husbands away, and least understandable of all, received little or nothing in return for their sexual services. Implicit in the women's discussion of these "other women" was the suggestion that the sisters were either doing it merely for sexual pleasure, or (expressed more explicitly) were exceedingly stupid because they were gaining neither financially supportive unions, husbands, nor payment in return, only more and more children to feed through their own labor. This assessment, well-known by the sisters themselves, did not seem to bother them in the least. Furthermore, they declared that their husbands/*compañeros* did give them a variety of assistance, from milk allotments to firewood to cash. The truth of this assertion was difficult to ascertain.

FORMING UNIONS: PREFERENCES AND REASONS

Preferences for Civil, Religious, or Informal Unions

As mentioned previously, according to Isiqueña women a simple civil marriage caused fewer problems in the long run for women than did church marriages. For example, there were fewer initial financial challenges to be met with regard to the ceremony. If a marriage was then so problematic and ceased to be an economic asset for a woman, a civil ceremony also afforded her more choice about the longevity of the union than did a church wedding. A civil marriage provided more freedom to leave by means of a divorce, and an informal union afforded even more ease in severing the union. Studies of divorce and separation rates in Latin America support Isiqueña women's belief that it is easier to exit some unions than others. For example, Goldman's (1981) research in Colombia, Peru, and Panama has demonstrated that consensual unions are three to seven times more likely to end in separation than are marriages (quoted in McKenry and Price 1995, p. 199).

This is precisely the argument made by Kingsley Davis (1985, p. 8) in his discussion of divorce, although he is focusing solely on the need in contemporary society to bind men, not women, to marriage. According to Kingsley, ". . .the long-run commitment of the male . . . is reinforced in ritual and in the involvement of his relatives and allies . . . not only does the public and ceremonial recognition of the new union instill a sense of permanence, but it also provides a basis for enforcement later on."

It is not surprising, therefore, that informal unions, followed by civil ceremonies, were considered by women in Isiquí to be the preferred ways of forming unions. This is not to say that whatever preference a woman had for the type of union she wished to enter would be the one to which her *novio* would agree. However, many women expressed a strong preference for informal unions or civil ceremonies, and the overall state of their unions did reflect their preferences. Future research should examine the way in which men and women in Nicaragua reach decisions about what type of union to undertake, focusing on the relative contributions of each to the process as well as the differential power each holds.

CHOOSING WHETHER TO FORM A UNION

Financial Reasons

Across all age groups and backgrounds, financial support was the most frequently given reason for forming a union, whether formal or informal, by Isiqueña women. According to Ana, "a woman needs a man to support her, to feed and clothe the children. Without a man, what can a woman do?" Women openly admitted that they settled down with a particular man because they felt he would support them—love and companionship were rarely mentioned at all. Researchers elsewhere have also registered the absence of love and affection in descriptions of mate selection preferences. For example, Ingoldsby (1995a, p. 143) noted that "free mate choice based on love which is common to the modern West has not been the way people in most other societies have selected their mates."

Economic support was also the reason given for staying in a union that was otherwise unsatisfactory (because of mistresses, beatings, etc.) for the woman. When Bianca (the young woman who had been abducted) became pregnant, Ana and Sara both pointed out the overt womanizing of Bianca's husband, saying "she will have to stay, she should have left before. She can't leave now, who will support her and the baby?"

When Sara was having difficulties with her husband, Manuel, also due to "another woman," Ana told her, "You have to think of the children. Who will support you? Let him have his woman. He gives you all the money, he supports you." It is not surprising, therefore, that one of the reasons given most frequently by the women in the collective for having an income-generating project was their desire to support their families without having to rely on a man. According to these women, the project would enable them to gain the freedom to make better choices about forming unions and staying in relationships.

Lack of financial support, physical abuse, and general unreliability (owing to gambling, alcoholism, laziness, and womanizing) were reasons given most frequently by Isiquena women for having left, or more commonly, wanting to leave, their husbands/*compañeros*. Recent research on divorce around the world demonstrates similar reasoning by other women. According to one study, three primary reasons women seek divorce around the world are incompatibility, financial failings, and physical violence. The same study, however, demonstrated that men seek divorce on slightly different grounds: incompatibility, reproductive failure, and infidelity (Pasternak et al. 1997, p. 196).

Nicaraguan men also held the belief that women in unions needed men to provide financial support for them and their children. For example, several Isiqueño men once decided to take matters into their own hands in order to force a man to financially support his wife. In 1990, a member of the *pollera*[17] left for another community in order to live with his "other woman." In doing so, he had taken everything in the house, "the food, children's clothes and dishes," leaving his wife and five young children with nothing. When his wife went to the *pollera* to complain about his actions, they quickly took up a collection and supplied her with food and other necessities. When the man returned several days later (having found a less than welcoming reception from his "other woman"), the leaders of the *pollera* (who were also the local leaders of the FSLN) agreed to take him back as a member, with the understanding that the *pollera* would pay his wages directly to his wife, in order to "force him to support her."

This is in stark contrast to Ana's experience in the early 1970s. When Ana's husband abandoned her in order to marry his mistress, and took all the household possessions with him, Ana appealed to his employer for help. His employer, a Somocista deputy to the National assembly, refused to intercede on Ana's behalf, saying that it was her husband's "right to leave her, and to take the things with him." Ana said that "he dismissed me with a wave, telling me to get out of his office."

Supernatural Reasons

There also existed in Isiquí a different incentive for older girls to marry, or at least settle down. This inducement was a belief in the *duende,*[18] a little boy all dressed in red who appeared only to unmarried girls, particularly after dark. Described as a "mean spirit," the *duende* "gives flowers and other pretty things" to unmarried girls, and woos them "just like a boyfriend." The *duende* was considered "mean" because it enticed girls to do "silly things," such as accept his gifts of flowers and wander away at night to meet him, and kept them from looking for *novios*.

According to the women of Isiquí, the best way to prevent the unwanted attention of the *duende* was to get married, or a least have a *novio*, because that would cause the *duende* to become very jealous and angry and he would eventually leave in search of another unmarried girl. Many older women volunteered that they had seen the *duende* before getting married, and that it was quite frightening. Doña Isabel (age 63) described her encounter with the goblin as a "shock—I could not move, I just looked at him in his red suit. He was on the road, outside my house. Then someone called my name, and he left." It was with good reason that unmarried girls often scurried home at dusk, saying they did not want "to meet the *duende*."[19] While the *duende* approached unmarried girls, it was not necessary to be married; it was enough to simply have a *novio*. The *duende* was very accommodating about the loose and flexible nature of relationships in Nicaragua.

Religious Pressures

The visiting priest from Esteli, on the other hand, was not as accommodating as the *duende.* Although he came infrequently to Isiquí, the Esteliano priest never missed an opportunity to point out the Catholic Church's desire that everyone get married in a religious ceremony, and to remind the Isiqueños of the repercussions of not doing so. In 1990, for example, during a mass to commemorate the death of five Isiqueños in an ambush five years earlier, the young priest spoke at length about the need for children to evangelize, particularly to their wayward parents. Children, he said, should ask their parents if they were married by the Church because a civil ceremony was not enough.

If their parents were not married by the Church, he continued, children should tell them that they do not want to live in sin and beg their parents to get married. Those Isiqueños that sinned by living together or sinned without confessing, he said, would not be allowed to take communion. Finishing his sermon, he fumed, "I am tired of being here for all the baptisms but never for the marriages." Notably, only a handful of children, one elderly woman, and two teenage girls took communion that day.

Therefore, while the choice to live together in informal unions or civil marriages provided certain benefits to women (such as greater freedom to leave abusive situations), it was not without repercussions for them. In Isiquí it was the women who attended church and who considered themselves at least moderately religious, not the men. Sanctions imposed by the church, such as a ban on communion, therefore fell more heavily upon women than men. Yet the style of union preferred by the women clearly prohibited them from participating in key rituals of their church. In order to maintain their freedom to deal with problematic or abusive unions, Isiqueña women engaged in unions that contradicted their self-perceived religiosity, as well as their gendered identity as highly moral and religious people under the traditional gender ideology.

Children from a Previous Union

Several women expressed reluctance to marry, or even live with, another man because of children from a previous relationship. Ana, whose husband left her prior to 1979, chose not to remarry despite a few proposals. As Ana explained it, she never took another husband "because stepfathers never love the children who are not their own. Maybe one stepfather in three hundred would love another man's children. I did not marry again because I didn't want my children to resent me."

This type of rejection by stepfathers is not limited to Nicaragua. In writing about street children in Colombia, Bron Ingoldsby (1995b, p. 345) points to rejection by stepfathers as one of the causes of so many children, particularly boys, living on the streets. He states that stepfathers are generally "not interested in the fruits of a previous union," and directly or indirectly force the children onto the streets.

Although some Isiqueña women did form unions with other men despite children from previous relationships, this idea of inherent conflict between stepfathers and children was commonly held, and the reason given by women for these unions was nearly always financial. In some cases, women had to choose between their new union and their children. When one Isiqueña woman married a man from a neighboring hamlet, for example, he refused to allow her five children from a previous relationship to live with them. The five children, all under 15 years old, were then parceled out among the woman's two sisters and her mother, but it is not a common practice for women to leave their children in this manner and begin a new family elsewhere.

CONCLUSION: RELATIONSHIPS AND POSSIBLE IMPACT OF THE REVOLUTION

Women's union-style preferences in Isiquí before, during, and after the revolution of 1979 were, for the most part, based on sound foresight about what the future might hold for them, rather than on purely short-run financial considerations, romantic love, or matters of convenience. The types of unions preferred by the majority of women, both young and old, demonstrated much about the way they viewed marriage, relationships between men and women, and the longevity and benefits of both. Often, there was a trade-off that had to be made between the type of union they were in (or not in) and other parts of their lives.

For example, Isiqueña women gave up their right to confession and communion in exchange for the possibility of completely extricating themselves from relationships that were no longer beneficial (in whatever manner), or to be completely and legally rid of a man who has left them anyway (either with or without a divorce). Clara Murguialday (1990, p. 25), in her study of feminism and revolution in Nicaragua, makes a similar point when she says that although most Nicaraguan women would prefer a monogamous and stable union, informal unions are chosen by more than one-half of all women in relationships with men because of "the conviction that little possibility exists that relations will be long-lasting." Another example of this type of trade-off occurred when women decided not to form a union with a man because of the existence of children from a previous marriage.

In addition to weighing the pros and cons of various types of unions and non-unions, women—particularly younger women—were also engaging in precaution-taking behaviors. Taking precautions (such as safeguarding one's own income and establishing informal unions) against the inevitable problems caused by men, while maintaining an attitude of resignation, was how most women of Isiquí dealt with their relationships with men. Younger women focused more and more upon precautions—such as striving to be self-sufficient or avoiding unions altogether—actions made possible by the changes instituted by the FSLN to increase the legal, educational, and economic options for women.

Rather than engaging in precaution-taking behaviors, most older women in Isiquí utilized resignation as a coping mechanism, and they attempted to encourage this resignation in the younger generation. Ana, for example, told her daughter, Sara, that she had to resign herself to Manuel's womanizing, because that was "what men do." Other older women (in general, those over 35 years old), whose husbands/*compañeros* were engaging in extramarital affairs, expressed similar feelings of resignation : "what can I do, I have these children. . . ."

Weighing the financial, social, and emotional pros and cons of union types, assessing trade-off, taking precautions, and using resignation as a coping mechanism are strategies women have undoubtedly used for centuries in Nicaragua and elsewhere in the world. What is crucial about this particular time period, however, is the way in which the new gender ideology and expanded economic opportunities were expected, by women, to give them greater freedom when making these decisions. Furthermore, the framework provided by the new gender ideology gave both men and women new ways of discussing these issues, brought new conflicts to light, and furnished new ways of (or reasons for) staying together or leaving the unions. With the change in government in 1990,[20] it remained to be seen whether these new frameworks would survive.

ACKNOWLEDGMENTS

The work upon which this article is based was supported by grants from the National Science Foundation, Sigma XI, and the Boston University Women's Guild, as well as special assistance from the Inter-American Foundation and Dr. Allan Hoben. I am grateful to Hilda Baker for her careful editing and thoughtful suggestions, and to Robin Miller and Sandra Browning for inviting me to contribute to this anthology. With the most profound respect and admiration, I want to thank the women and men of Isiquí, who are given pseudonyms to preserve their privacy. Finally, as is fitting in an anthology about cross-cultural perspectives on marriage, I would like to warmly thank my husband, Gerhard Krabbe, and my daughters, Mei-Ling and Marloes, for providing me the loving space and environment in which to write this article.

NOTES

1. *Sandinismo,* the guiding principle of the revolution, combined elements of Catholicism (in particular, liberation theology) and patriotism with socialism to create a revolutionary doctrine that was specific to Nicaragua. There was, for example, no emphasis on atheism and there was continued support for the family as the basic unit of society. For a full discussion of *Sandinismo*'s unique blend of principles, see Giulio Girardi (1986).

2. *Machismo* refers to a set of beliefs predicated on a series of oppositions about men and women. In particular, *machismo* emphasizes male virility and all things considered "masculine." Although *patriarchy* is often used interchangeably or in conjunction with *machismo*, *patriarchy* specifically refers to familial relations while *machismo* refers to the beliefs about men and women regardless of the domain being discussed (public, private, familial, workplace, etc.). In its essence, *patriarchy* refers to an idealized family system in which the husband/father is the sole provider and the locus of authority.

3. *Isiqueños* refers to either the men form Isiqui, or both the men and women from Isiquí, depending on the context. *Isiqueñas* refers to the women of Isiquí.

4. *Compañero/a* refers to a man or woman in an informal union; it can also refer to a friend, coworker or comrade. Isiqueñas would use *compañero*, *esposo* (spouse), and *marido* (husband) interchangeably to describe the man with whom they were currently united.

5. See Pasternak, Ember, and Ember (1997, pp. 82-85) for a discussion of the difficulty in defining marriage.

6. It is informative to note that the verb *raptar* is used to indicate the abduction or kidnapping of a girl; it is also used to mean *to ravish* or *to rape*. *Secuestrar* means to kidnap in general, and was never used by Isiquenos to indicate abductions. With *secuestrar*, the person taken is called a *rehen* (hostage). With *raptar*, no such designation was ever made, perhaps indicating that the abducted girl is not considered a hostage against her will.

7. Because the people of Isiquí described abduction as the kidnapping of a *girl* rather than a *young woman*, and because most of the kidnapped girls were in their early to mid-teens, I also use the term *girl* here, while recognizing that by their activities and responsibilities they were young women.

8. Niurka Peréz Rojas (1986, p. 49), in her study of a rural Nicaraguan woman living in the department of Matagalpa, also mentions the prevalence of abduction in the countryside.

9. Chapter 4, Article 72 of the 1987 Nicaraguan Constitution.

10. Leon is a city toward the western coast of Nicaragua.

11. *Militante*: a full member of the FSLN.

12. *Vaga*: unreliable, lazy, vague; implies moral looseness.

13. *Puta*: whore, but more in the sense of a very "loose" woman, rather than a prostitute.

14. While census-takers often list these women as "divorced" or "abandoned," the women describe their state as *madre soltera*, or "single mother."

15. These are the sisters considered *putas* by the other Isiqueña women, discussed later in this section.

16. The woman is considered abnormal by the other inhabitants of Isiquí because of her odd public behavior and thus her actions, particularly that of throwing out her young children, cannot be regarded as typical of Nicaraguan women as a whole.

17. *Pollera*: chicken farm—the name given to the poultry collective.

18. *Duende*: a goblin all dressed in red that preys upon and woos unmarried girls.

19. The Nicaraguan anthropologist Milagros Palma (1988, pp. 213-219) discusses the regional variants of the *duende* in Nicaragua, who in some parts also harasses unbaptized children.

20. Violeta Chamorro, a conservative candidate, won the presidential election in February 1990, and took power in April 1990, along with a substantially transformed legislature in which the FSLN was now the opposition.

REFERENCES

Borge, T. 1982. *Women and the Nicaraguan Revolution*. Managua: Ministerio del Interior.

Burns, E. B. 1991. *Patriarch and Folk: The Emergence of Nicaragua, 1798-1858*. Cambridge: Harvard University Press.

Davis, K. 1985. "Introduction: The Meaning and Significance of Marriage in Contemporary Society." Pp. 1-21 in *Contemporary Marriage: Comparative Perspectives on a Changing Institution*. New York: Sage Foundation.

Girardi, G. 1986. Volumen I: La confluencia *Sandinismo,marxismo, cristianismo en la Nueva Nicaragua* (*Sandinismo, Marxism, Christianity in the New Nicaragua*, Volume I: The Confluence). Mexico, D.F.: Ediciones Nuevomar.

Goldman, N. 1981. "Dissolution of first unions in Colombia, Panama, and Peru." *Demography* 18: 659-679.

Ingoldsby, B. 1995a. "Mate Selection and Marriage." Pp. 143-160 in *Families in Multicultural Perspective*, edited by B. Ingoldsby and S. Smith. New York: Guilford Press.

______. 1995b. "Poverty and Patriarchy in Latin America." Pp. 335-351 in *Families in Multicultural Perspective*, edited by B. Ingoldsby and S. Smith. New York: Guilford Press.

Lancaster, R. 1992. *Life is Hard: Machismo, Danger, and the Intimacy of Power in Nicaragua*. Berkeley: University of California Press.

McKenry, P., and S. Price. 1995. "Divorce: A Comparative Perspective." Pp. 187-212 in *Families in Multicultural Perspective*, edited by B. Ingoldsby and S. Smith. New York: Guilford Press.

Morgan, R. 1984. *Sisterhood is Global: The International Women's Movement Anthology.* Garden City, NY: Anchor Books.

Murguialday, C. 1990. *Nicaragua, revolución y feminismo (1977-89)* (*Nicaragua, Revolution and Feminism, 1977-89*). Madrid: Editorial Revolución.

National Directorate of the Sandinista National Liberation Front (FSLN). 1987. *Women and the Sandinista Revolution.* Managua: Editorial Vanguardia.

Palma, M. 1988. *Nicaragua: Once mil vírgenes: imaginario, mítico-religioso del pensamiento Mestizo Nicaragüense* (*Nicaragua: Eleven Thousand Virgins: Imaginary, Mythic-Religious Thinking of Nicaraguan Mestizos*). Bogotá: Tercer Mundo Editores.

Pasternak, B., C. Ember, and M. Ember. 1997. *Sex, Gender, and Kinship: A Cross-Cultural Perspective.* Englewood Cliffs, NJ: Prentice Hall.

Peréz Rojas, N. 1986. *El hogar de Ana: Un estudio sobre la mujer rural Nicaraguense* (*Ana's Home: A Study of Rural Nicaraguan Women*). Havana: Editorial de Ciencias Sociales.

Sanday, P. 1990. "Introduction." Pp. 1-19 in *Beyond the Second Sex: New Directions in the Anthropology of Gender*, edited by P. Sanday and R. Goodenough. Philadelphia: University of Pennsylvania Press.

Stead, M. 1991. "Women, War, and Underdevelopment in Nicaragua." Pp. 53-87 in *Women, Development, and Survival in the Third World*, edited by Haleh Afshar. London: Longman Press.

COHABITATION, NON-MARITAL BIRTHS, AND TRADITIONALISM

MARRIAGES IN RURAL SIBERIA FROM A LIFE CYCLE PERSPECTIVE

Anna Mikheyeva

INTRODUCTION

The phenomenon of off-modal forms of marriage such as cohabitations, consensual unions, and step marriages have been noticed by researchers in nearly all advanced nations (Lesthaege 1992; Popenoe 1993, etc.). The degree of public tolerance for them as a form of family depends on whether the transition from "traditional" to "modern" type of family has or has not been completed. In societies where, over several generations, men and women were free in choice of their marriage partners, number of births, and time of childbearing, and where family was of an egalitarian and nuclear type, cohabitation is widely spread and accepted by the public with tolerance. The similarity of these processes in many countries was a reason for specialists to conclude that a new stage in the world demographic development had begun (i.e., the second demographic transition). The succession in which these changes regularly proceed also has been noted: first, a drop in births, then a rise in divorces, after that a rise in the number of cohabitations, and, lastly, higher rates of out-of-wedlock births (Lesthaege 1992).

Of course, mating, family, and reproduction patterns in Siberia are still far behind those in such regions as France, Sweden, and Germany; but, similarity of demographic trends may indicate the presence of deep-seated grounds for new types of behavior. Among the present changes in the family sphere, the most noticeable are those that take place in sexual behavior (Kontula and Haavio-Mannila 1995; Maddok 1995). These patterns are based on social and historical phenomena and concurrently changing broader cultural models. Sexual behavior proves to be regulated by different types of social norms, that is, law, custom, morality, and stereotypes. Sexual morality is characterized by the rigidity of its norms (if we take society as a whole and not vanguard or marginal groups) and continues to be reproduced by cultural codes of other spheres: "religion, language, education, and culture (both elite- and mass-related) set up a certain conception of women and men and accentuate it. In the result, established are certain moral norms based on gender differences" (Lorber 1994). Marriage stereotypes include, first, gender (sex-role) stereotypes determined by men's and women's family roles (husband-wife, mother-father-children, etc.); second, marriage stereotypes as such, for example, the opposition of the married (legal) to non-married (cohabiting) family, also includes, therefore the stereotype of the non-married family. Traditional gender stereotypes within the family provide the basis for a functionalist approach: quite different and invariable roles of men and women (woman keeps the home and raises children, man provides family with livelihood). Under such a role arrangement "each marriage is two marriages—'his' and 'hers' that often do not coincide" (Maddock 1995). Stereotypes of legal families in contrast to illegal ones are conventionally based on the opinion that only a family based on marriage can best satisfy men's and women's needs for intimacy and domesticity, ensure the future by assuring partners fidelity and guarantee normal socialization of children. So legal marriage is not only moral, but also good in itself. The social stereotypes of marriage are, therefore, based mostly on its value in a given cultural system. Along with this (or just owing to this), in many societies, both previous and present, being married and belonging to a married family were essential for social ascent. And, in turn the negative stereotype of cohabiting couples may have resulted from the belief that this family pattern is peculiar to deprived, marginal social groups (Lewis 1975), and that in a one-parent family a child cannot be properly socialized into the "right" gender role.

It is not quite clear, however, if the gender role taken by a child through socialization will be adequate to that reality in which he or she will have to shape family life; rather, it may be impossible to reproduce those gender stereotypes and patterns that were common for the parents' generation. The stability of stereotypes depends on how often the groups that set a stereotype are in contact with those who adopt it. Therefore, the spread into all walks of life of alternative family patterns erodes and equates stereotypes, making it difficult to speak about particular family patterns as a necessary and regular feature of a particular social group. The deterministic statement that married family is "good" and non-married family is "bad" in every respect disappears. This leads to relativization of legally married unions; they no longer can improve social status or be of intrinsic value.

Here one can pose some questions. How far has the process of liberalization and egalitarianism in sexual morality, and similarity and interpenetration of stereotypes, in the sphere of marriage gone in Siberia (Russia)? What are the specific elements of culture in Russian society and what cultural transformations are behind these processes? What is the current state of sexual morality, gender relations, and stereotypes? It is possible to make some assumptions. First, liberal norms of sexual morality in Russia as a whole were caused by specific historical circumstances—the Soviet propaganda of gender equality, and industrialization, with almost 100-percent participation of women in social production due to the post-war disproportion of women to men. Second, it is a reflection of a large "poor" segment of the population, whose subcultural element, according to Lewis (1966), is defiance of generally accepted moral norms, early sexual contacts, free unions, and high frequency of abortions. The third factor seems to be the current crisis situation of Russian society in which further erosion of moral norms and stereotypes, as well as expansion of marginal layers and groups "free" from restrictions in sexual and family behavior, takes place.

This paper analyzes current trends in the formation of marital unions, legal and illegal, and age characteristics of brides and grooms in rural Siberia. The analysis chiefly draws on the data from the 1970, 1979, and 1989 Russian population censuses, and the microcensus of 1994. The obtained regularities are explained in terms of weaker patriarchal family norms and stereotypes. First let us turn to the derivation of traditional family behavior.

SIBERIAN PEASANTS' TRADITIONAL FAMILY IDEOLOGY

Traditional norms of marital and family behavior of rural residents were established over the course of a long history of communal-tribal peasant life in Russia. The most important factors in the origin and existence of ethical norms in the sphere of family and marriage were high mortality, lack of communal or governmental support for the aged, and, most important, family farming as the only possible form of peasant farming (Chayanov 1925; Thomas and Znaniecki 1971; Shanin 1987). The land was in the possession of the commune which, from time to time, redistributed fields and meadows among households in proportion to the number of married men in the family; neither women, nor single men were taken into account. Accordingly, marriage was the chief condition for access to land use. The unmarried were not full members of the family; bachelors and maidens (i.e. peasants without family and land) were seen as unlucky, ill-fated people. The main principle regulating the behavior of peasants (members of the rural commune) in every sphere of life was subordination of individuality, in favor of absorption by, and fusion with the commune. The commune exercised strict social control and censorship of morals, and there was no other sphere where peasants were as cruel as in that of family life (Mironov 1977; Gromyko 1991). Therefore, the only possibility for both men and women to obtain full-rights social status was to get married. A similar peasant ideology was also characteristic of Poland, a western neighbor of Russia (Thomas and Znaniecki 1971) and of China, an eastern neighbor where, as was marked by Fei Hsiao Tung (1971),

> the traditional ideology suppresses individualism in favor of familism. The meaning, or value, of the individual's existence is defined by its being a link in the chain of social continuity which is concretely conceived in terms of descent. The most important task of a man is to continue the family line. Of the three traditional charges against an undutiful son, failure in giving offspring comes first.

Another cultural-historical factor that strengthened the Russian peasants' desire for marriage and family was the deep religiosity of rural dwellers and the enormous role played by the church as originator of criteria of good and evil, honor and disgrace. The events of birth, marriage, and death were sacral, accompanied by religious rites with participation of all members of the family. Wedding ritual incarnating marital union with the union of Christ and Church was one of the principal Orthodox church rituals. Family was for peasants a refuge where all misfortunes could be diverted.

The peasants' acknowledgment of the high role of family for an individual's material and moral welfare was reflected in numerous proverbs and sayings widely spread in Russia, as for example: a single man is a half-man; a family pot always boils; life without a husband is an unsavory mess. So it is no surprise that at the turn of nineteenth and twentieth centuries, the proportion of the married population in Russia was 95 percent among women and 96 percent among men. These figures were among the highest for European nations (Hajnal 1965). The relevant indicators for rural dwellers who then made up 80 percent of the population in Russia were even one percentage point higher. Marriage was universal and early—economic and moral need made peasants marry at the first opportunity at hand. The act confirming the acknowledgment by the commune of a marital union was the church ritual of marriage ceremony and wedding party with participation of almost the whole village.

The view of the family as the most important and indispensable factor in the life of every peasant was accompanied by strong condemnation of premarital affairs and conjugal infidelity. The wife's infidelity was condemned more severely than a husband's. She was publicly soiled with soot and for a year she was scorned and prohibited from visiting her neighbors, or even entering somebody's house (Gromyko 1991). A distinctive feature of the peasant ethical norms was that they were unchanged over several centuries, and the countryside lived according to the "forefathers' principle." Instead of a critical view of the ancestors' experience, the peasants remained undeviating to the old customs and traditions. Therefore, a very deep traditionalism was typical of the whole order of peasant life in Russia and, especially, in its European part.

Life in Siberia, however, has its specifics determined by historically recent (since the end of the seventeenth century) settlement of these lands. It attracted active, energetic, freedom-loving people from central Russia who arrived hoping not to pay taxes and dues. "The old" and "the new" were mixed. New generations of Siberian people honored the norms of their forefathers, but did not shun novelties they could take from the indigenous cultures, customs, and practices. After all, considerations of common sense prevailed over conservative traditions. In Siberian towns and villages, family life was very different from the stringent ancient regulations. This could be seen in the woman's higher standing, both in family and communal life. Some historical evidence shows that women in Siberia also had a certain freedom in the choice of a husband. There were known incidents of women hiding from undesirable marriages, and they seem to have aroused sympathy instead of resistance and condemnation (Liutsidarskaya 1992).

Later on, with the introduction of constraints imposed by rural communes and strict feudal regimentation, Siberia gradually lost its specificity of location under colonization and its freedoms became more subordinated to the state legal norms. But life in central Russia in the mid-nineteenth century was changing, and new phenomena affected, among other things, the marriage and family sphere. With the penetration into the countryside of commodity-money relations and the development of migrant trades and other similar events at the turn of the twentieth century, many traditional values and customs began to lose their significance and were replaced by new ones. Thus, the average marriage age for brides and grooms began to increase as early marriage had lost its former importance for people who already had left the land and its rural communes (Vishnevski 1977). In addition, with gradually increasing citizen consciousness, even in peasants who had come in contact with urban culture, the character of family relations began to change, and the value of marriage as a voluntary partnership between a man and a woman was increasing.

CONTEMPORARY TRENDS IN THE FORMATION OF MARITAL COUPLES

First of all, it should be noted that Western Siberian rural population, as is shown by the long-term dynamics of its reproduction rates, evolved to the current model of family as late as two or three decades ago. And a distinctive feature of its demographic situation (in comparison to the national rural average) is slightly retarded development of the population: parameters of many demographic processes have been unstable, structure of the population is younger, and, in general, demographic transition to the current mode of reproduction postponed. Also, in many aspects of marital and family life, the behavior of Siberian rural people, even in comparison to their counterparts in the European part of Russia, is of a more traditional character.

The former pattern of early marriage and desire for "ritual" entry into it is still present in Siberia. It is shown essentially in the data on marital status in the modern censuses of 1959, 1970, 1979, 1989, and the 1994 microcensus of the Russian population. An important feature of the census data, in terms of spread of illegal marriages, is that all categories of marital status (married, divorced, widowed, never married) are recorded according to people's answers, without any documentary support.[1]

On the whole, the dynamics and level of marital data for the Western Siberian rural population in censuses characterize its marriage rate as stable and unchanging from generation to generation. As in the rest of Russia, its postwar stabilization was completed by the end of the 1970s. By this time the generations of males affected by World War II were no longer of reproductive age. But even when postwar disproportions in the number of men and women were still felt (1959, 1970), the marriage rate of rural Western Siberia was notably above the national figure. At present it is also higher: in 1989 it was 682 per 1,000 women 16 years of age and older, as compared to the national figure of 616. The relevant figures for men are 744 and 730, respectively. The high percentage of married people recorded in censuses provides a "cross-sectional view" of the marital lot of different generations of rural people: actually, all men and women in each generation get married at least once. But, as was repeatedly noted by sociologists and demographers, matrimonial behaviors and the relevant cultural norms regulating them (or, more exactly, their strictness) may depend on, and often are really determined by the situation in the marriage market (Henry 1966; Festy 1971). The behavior of males often depends on the availability of marriageable women and vice versa. Thus, though the previously mentioned figures reflect a fairly high marriage rate in rural Western Siberia, the lower rates among men in the 1970s and 1980s were caused by a "brides deficit" that emerged in the second half of the 1970s and is still evident in many villages.

Bearing in mind the specific features of the census data (i.e. self-specification of marital status), these statistics can be compared to data on the registration of marriages and divorces in the rural side. As can be seen, the number of new marriages is decreasing, and that of dissolutions increasing. The dynamics of the marriage rate here shows inconsistency: A rise according to respondent self-specification, and a drop according to registration record. The explanation can be the spread of *unregistered* marital unions. Indeed, according to the 1994 microcensus data, where free unions (cohabitatations) were first recorded, they made up 12.3 percent of all unions in the Western Siberian rural population (the relevant indicator for urban population was 8.3%). The percentage of out-of-wedlock births was 18.5 percent of all births. In spite of this, one can hardly suppose that the value (and attractiveness) of marriage lessened among all social groups in Russia, and in Siberia, in particular. It is the attitude toward legal marriage that had changed. And though increasing numbers of Russians are refusing to enter legal marriages, this does not mean that they renounce family life as such, which is shown in the increased number of out-of-wedlock births.

COHABITATION AS A UNION PRECEDING THE FIRST MARRIAGE

The proliferation of cohabitation reflects complicated events in the life and minds of rural people. On the one hand, after "church disestablishment" and other Soviet transformations, the social controls that demanded rituals and customs regulating marital, sexual, and family life were weakened so new generations of rural dwellers sometimes even repudiate these norms. Taking into account these "new" social features, the spread of cohabitation can be interpreted as the result of individualistic trends, insistence on recognition of reality, and value of personal choice among forms of marital unions.

The desire for marriage and family remains, however, a fully traditional behavior when people mature and seek the status of an adult member in the rural community. The fact that present-day rural grooms and brides are little different in age from their counterparts at the end of the last century can be seen from the results of demographic analysis of the appropriate indicators. Table 1 shows the dynamics of the calculated (singulated) average age of first marriage.[2]

This indicator, calculated from the data of population censuses, has an advantage in that it characterizes the marital status of people by their self-specification, does not depend on the age structure, and can be taken as showing just the marital behavior of this population in the given time.

By typology of first marriage entry ages among women, these indicators are low. Researchers of historical types of marriage in Russia (Vishnevski and Toltz 1988), note that in the nineteenth and early twentieth century, the Siberian population was similar in this process to European Russia. And at the turn of the twentieth century, the calculated average age of entering first marriage in Siberia was 20.8 for females, and 24.2 for males; that is, they were fairly close to the current characteristics of the rural population (see Table 2).

Table 1. Singulated Average Age of Entry into First Marriage, Rural Population, in Years

	Western Siberia		*Russia*	
Years	*Men*	*Women*	*Men*	*Women*
1979	23.6	19.7	23.4	18.6
1989	24.1	20.2	23.9	20.6

Table 2. Age Characteristics of First Marriage, Entrants in Western Siberia, Rural Populations

	Median Age of First Marriage (years)		*Percent First Marriage Entrants (among all marriage entrants)*			
			20 Years of Age		*25 Years of Age*	
Years	*Men*	*Women*	*Men*	*Women*	*Men*	*Women*
1969/70	22.5	20.2	8.7	31.7	64.6	74.5
1978/79	21.4	19.3	6.6	38.2	77.6	84.5
1988/89	22.5	19.4	6.0	42.4	67.0	82.6
1993/94	21.7	18.3	12.4	53.4	76.8	89.6

But the general recent trend for rural national and Siberian populations was an increased calculated average age at the first marriage. This tendency in the period from 1979 through 1989 is not, in principle, contradicted by the dynamics of median age of entrants into legal marriage according to the current data (Table 2). The difference between grooms' and brides' ages in 1970 and 1979 was about two years; at the end of the 1980s, three years. In the mid-1990s, however, the average age at marriage registration among rural grooms and brides remarkably decreased (see Table 2). So the essential tendency in the process of entry into first marriage, according to the 1970, 1979, 1989 data, is that marriage rates are stably high both for men and women in the rural Western Siberia.

It is possible to assume that marriage rate dynamics simply reflect the features of the demographic situation in rural Siberia, or, more exactly, the concrete situation in the "marriage market." The brides deficit that emerged in the 1970s determined the marital behavior of rural men; namely, entry into marriage at a fairly young age, often immediately after returning from military service, to very young brides. And in the next decade, despite the normalization of the sex ratio, this model of marital behavior was rather common for rural males: in 1989, 15 percent of women in the 16- to 19-year-old age group had, by age of 20 years, already entered their first marriage (in 1979 this was 12 %; in 1970, 8%). The contemporary juvenilization of first marriage in the Western Siberian countryside is mainly due to the increased percentage of the youngest (18 to 20 years of age) grooms and brides. Among older age groups, on the contrary, the proportion of never-married men increases, but the proportion of women who report being single decreases (the 1994 microcensus data; see Table 3).

Female rates seem to be more revealing for behavioral aspects. They also permit one to suppose that the emerging trend of postponing registration of marriage may spread in the rural population of Western Siberia. Even more revealing (for behavioral aspects) are female coefficients that also indicate the increased tendency of postponing registration of marriage in the rural population of Western Siberia.

Table 3. Percentage of the Never-married, Western Siberia, Rural Population

	Percent Among Men		*Percent Among Wormen*	
Year	*25–29 Years*	*30–34 Years*	*25–29 Years*	*30–34 Years*
1979	13.5	6.4	5.8	3.1
1989	17.8	9.1	5.8	3.3
1994	18.2	10.1	4.3	2.8

The consistency of the dynamics of indicators estimated from census data (through self-specification of marital status) with the characteristics of marriage registration provides grounds for assumption that cohabitation as an alternative to first marriage hardly can be described as widespread in Siberian rural places. But the presence of a great number of out-of-wedlock births, including first births, registered according to the application of both parents, does show the existence of marriages in the youngest age groups (see Table 4). It seems that cohabitation preceding the legalization of the first marriage is a preliminary process to entering marriage, dependent upon tolerance of the rural community to sexual connections outside of legal marriage. In the villages, this process is often complicated by childbirth, as the spread of premarital sex in the youngest age groups occurs in a culture with limited access to contraception or pregnancy interruption.

In the Western Siberian countryside, the youngest mothers (15 to 19 years of age) made up one-fourth or one-fifth of the total of women who gave birth to out-of-wedlock children. In the period of 1969 through 1979, this share rose from 18.3 to 27 percent, and in the 1980s it decreased, dropping to 21.2 percent in 1989.

The conclusion about widespread out-of-wedlock births among the youngest rural women can be made also on the basis of the percentage of births by unmarried women in total births. These data also show that the most frequent out-of-

Table 4. Distribution of Mothers by Their Ages at First Out-of-wedlock Births, Western Siberia, Rural Population

	Percentage of Births			
Mother's Age (years)	*1969/70*	*1978/79*	*1988/89*	*1993/94*
15–19	46.5	47.5	45.6	38.2
20–24	36.3	38.8	35.9	37.1
25–29	6.1	10.1	12.1	14.4
30–34	6.5	2.6	4.5	6.3
35–39	2.8	0.8	1.6	3.3
40–49	1.8	0.4	0.3	0.7
Average (Me) mother's age (years)	20.5	20.3	20.4	20.7

wedlock births were among the youngest women (15 to 19 years of age): nearly one out of three births by mothers of this age in Siberian villages was out-of-wedlock; the same proportion two decades ago as at the end of the 1980s and at present. No substantial shifts in the percentage of natural children among the youngest women during the period in question occurred. Changes did occur in the figures for natural births in the 20- to 29-year-old age group: they decreased perceptibly from 1969 to 1989, but the current level is still above the national rural level.

Very high figures of natural births among young women indicate a fairly high occurrence of premarital and extramarital connections in rural Siberia. However, if such connections result in births they generally lead to first, very early, marriages. Most cohabitants expect (hope) to get married sometime, and the birth of a child expedites the matter. The aforementioned trend in marriage postponement is the result of a more pronounced sexual culture and increased contraception opportunity. Legal marriages can be postponed, but not natural. Therefore, the spread of cohabitations in the countryside does not imply permanent abstinence from marriage. The assumption that mothers of natural children do not stay single is also supported by data on the very high, almost 100 percent, marital status of women in the region.

Public tolerance of premarital sexual connections also was supported by data from a mail survey of rural inhabitants of the Altai Krai in Siberia (Russia) that I conducted in the winter of 1996.[3] The opinion that men's premarital affairs are "always" or "almost always blamable" was shared by only 18 percent of respondents. Women's affairs were condemned by 36 percent. But most respondents thought this to be "normal" and "sometimes bad" for women (17%). Rural people were less lenient toward premarital connections among very young people (under 16 years of age). They were frowned upon by 57 percent of respondents, and those among young girls were disapproved of by 75 percent.

In general, one can conclude that in rural Siberia, people are rather tolerant of premarital sexual connections. This opinion seems to have been established not recently but some time ago, perhaps during the 1940s and 1950s. In that time, there was a leniency towards postwar ("compensatory") marital behavior under conditions of a very high disproportion between the number of men and women. Later, this liberalism gradually took root in the marital-family patterns of the rural population.

COHABITATION AS STEPMARRIAGE

The principal factor in matrimonial patterns of rural people (i.e. desire for early marriage) is in contradiction with modern patterns of other social behaviors—education, skills, economic independence—and is the cause of a large number of divorces at early stages of marriage. Almost two-thirds of all divorces in the Western Siberian countryside, and in Russian as a whole, occur in the first ten years of married life. Half of the divorces take place after six years of marriage. In the late 1980s, 17 women and 15 men per 1,000 persons in the 25- to 29-year-old age group were divorced.

Another, and it seems substantial, reason for dissolution of even young marriages in the countryside is widowing. By my estimate, after only seven to eight years of marriage, the predominant cause of marriage termination in the Siberian countryside is widowing (Mikheyeva 1993).

As can be seen, both causes for marriage dissolution are rather substantial in the rural population. Despite this, percentages of widowed and divorced men and women recorded in the 1979 and 1989 censuses and the 1994 microcensus are very small. This is obviously a result of stepmarriage, which is widespread in the countryside, be it legal marriage (see Table 5) or free cohabitation recorded in the census as a marital union (by specification of the spouses).

A high level of stepmarriages also shows an adherence of rural dwellers to the traditional model of family behavior. As was noted by Mironov (1977), norms of this behavior were such that the second marriage of widowed people was not condemned since widowhood, according to the peasants, was "God's penalty, great misfortune." However, third marriage of a widowed person was strongly condemned. In prerevolutionary Russia, the percentage of stepmarriages was 20 percentage (Vishnevski and Tolz 1988).

On the other hand, the current trend of more frequent stepmarriages and increased number of divorces can be a result of the normal desire for marriage but on the modern, quite new grounds of social regulation of marital family behavior; that is, the consideration of each person as an individual; recognition of the right to personal freedom; to intimate, private life; and the possibility of alternative norms of family behavior. Furthermore, as mentioned earlier, such a type of matrimonial behavior among rural dwellers is fostered by the real demographic situation in Western Siberia and in Russian villages in general (i.e. "deficit" of brides).

The greatest occurrence of stepmarriages is observed in younger age groups of rural populations: among those less than 30 years of age, nearly 35 percent of widowed and divorced men create stepfamilies; among young women, this proportion is about 25 percent (1988/89). Average (Me) age of entrants into stepmarriage among men is 34.5 years; among women, 32.4 years. These data make it possible

Table 5. Dynamics of Stepmarriages in 1959-1994, Western Siberia, Rural Population

	Share (%) of Stepmarriages to Total Marriages Arranged				
	1958/59	*1969/70*	*1978/79*	*1988/89*	*1993/94*
Men	8.2	9.6	10.6	19.6	18.4
Women	7.7	9.4	10.6	22.9	19.9

Table 6. The Structure of Out-of-wedlock Births by Their Number, Western Siberia, Rural Population

	Share (%) in Total Births				
Years	*First Births*	*Second Births*	*Third Births*	*Fourth (etc.) Births*	*Average Birth Number*
1969/70	35.8	16.1	16.3	31.8	2.92
1978/79	53.0	16.3	13.7	16.9	2.03
1988/89	42.4	23.9	17.2	16.6	2.33
1993/94	64.2	19.7	9.0	8.0	1.66

to conclude that matrimonial behavior of young people who were once married can be even more "free" (with regard to legal registration) than behavior of those who enter marriage for the first time. For people who already have had a marital experience, maybe a bitter one, cohabitation can be an alternative to legal marriage. Living together (illegally) may have an advantage of allowing cohabitants to retain their social relief and other benefits.

Divorced people may cohabit in order to avoid the legal and social troubles associated with a new marriage, as well as because of doubt about the new (step) partner. In Siberian villages, the spread of cohabitation is caused, to a certain extent, by the social phenomenon of migrant labor—arrival of construction and transportation teams of men and women who, for different reasons, avoid legalization of their marriages. This is supported very convincingly by data on extramarital second, third, fourth (etc.) births (Table 6).

Remarkable is the high proportion of out-of-wedlock, fourth or more births in all years under consideration (% of legal fourth, etc., births decreased in the respective years: 23.2, 10.6, 12, and 4.4%). The distribution of extramarital births also determines a fairly high average birth rate, the dynamics and level of which is little different from the corresponding figures for marital births. These figures are obviously affected by the registration practice. The record of a fourth birth by a woman out-of-wedlock, for example, must not be taken as a sign that her three preceding births were also out-of-wedlock. The increased percentage of late, out-of-wedlock births conforms to the spread of stepmarriages among widowed and divorced women, as some of these cohabitations are stimulated by the out-of-wedlock births. The number of this birth can be fairly high with regard to the woman but not to the given unit or to the child's father. It is very probable that second, third, fourth, and so on out-of-wedlock births are the consequence of the dissolution (because of divorce or widowing) of the first marriage and the woman's desire to remarry. This is confirmed by the data of the percentage of step to total marriages.

Another corroboration is a rather long (six years in 1989) interval between first (marital and extramarital) births, and extramarital second births. Such a long interval quite obviously shows that after the first birth these women had decided to stop

Table 7. Age Characteristics of Mothers at Marital and Non-marital Births, Western Siberia, Rural Population

	Share (%) in Total Births			
Years	*First Birth*	*Second Birth*	*Third Birth*	*Fourth (etc.) Birth*
Marital Births				
1969/70	21.9	24.6	30.3	34.2
1978/79	21.9	24.1	27.5	33.3
1988/89	21.2	25.2	28.9	31.9
1993/94	21.5	266.2	29.5	32.5
Non-marital Births				
1969/70	20.5	28.2	30.3	35.3
1978/79	20.3	26.2	28.0	35.4
1988/89	20.4	26.6	30.3	33.4
1993/94	22.6	27.0	30.3	33.5

childbearing. At the same time, the data in Table 7 show that the pauses between extramarital births are little different from intervals between marital births. This seems to be accounted for by the same factors as the high average number of extramarital births; namely, by changes in marital status of rural women during their lives.

The previously mentioned data on extramarital births may lead to an assumption of a high percentage of one-parent families with children born and reared outside of legal wedlock. The census data on the number and composition of families with children, however, give quite the opposite picture. Nuclear families consisting of a married couple with children under 18 years of age are an absolute majority of all families with children; and during the intercensus decade, they increased 10 percent, totalling 78 percent in 1989. At the same time, the percentage of one-parent families tended to decline and was only 8 percent of all families with minor children. The data obtained from the 1994 microcensus recorded no retreat from this trend. Therefore, in spite of rather high figures of extramarital fertility in Western Siberia over the past 25 to 30 years, no increase of out-of-wedlock (one-parent) families as a social phenomenon has occurred. Some percent of nuclear families with children are based on stepmarriages (legal or illegal) of the mother or father, or on legalized, out-of-wedlock births. Such families, as was shown by special studies, may have social-psychological, socialization, or other problems, but this can also be the case in first marriage families.

The spread of cohabitation and extramarital births has been almost indifferent to the time characteristics of processes of marital union formation and sibships in rural families. Average values of "central" points in the marital life of rural men in Siberia are as follows (1994): entry into the first marriage—21 to 22 years of age, divorce—30 to 31, entry into a step-marriage—34 to 35 years of age. For women the respective ages are 18 to 19, 30 to 31, and 32 to 33 years. In the childbearing cycle there is also a trend toward age stability: at present, as 20 and 10 years ago, 60 to 70 percent of married couples give birth to two children within six to seven years after marriage; another 15 to 20 percent, after three or five years give birth to the third child. Therefore, 90 percent of rural couples in Siberia finish their childbearing period by the wife's thirtieth birthday.

CONCLUSION

The current spread of off-modal forms of marriage such as cohabitations and free unions (primary or step) in Siberia, and in Russia as a whole, cause an increase in extramarital fertility. This phenomenon seems to be associated with liberalization trends, including more rights obtained by men and women, and the increased value they attach to their individualism. It may also be a kind of anomie, embarrassment experienced by the Russian public. So far, however, there is not much evidence to state, a priori, that the mother, as a result of her out-of-wedlock birth, will stay unmarried for the rest of her life. In this case, the family may assume a form different from the traditional and may incur disapproval. But often it is in these forms that the quite traditional or even ritual men's and women's drives to be married, to form a complete family, can be realized. The public should help to make this process as painless as possible for them.

ACKNOWLEDGMENT

This paper was translated from the Russian by D. I. Shtirmer. The paper uses data and results of investigations supported by the Russian Humanitarian Scientific Foundation, Grant No. 96-03-04473 and the John D. and Cathrine T. MacArthur Foundation, GA 96-41124A-FSU.

NOTES

1. Unfortunately, none of the previously mentioned categories of marital status had not been taken into account until 1979, 1989, and 1994. The category "illegal marriage" was first introduced in the 1994 microcensus, and has already been included in the program of the forthcoming 1999 census of Russian population.
2. The calculation method developed by Hajnal (1953).
3. This survey was supported by the Research Support Scheme of the Higher Education Support Programme (G. Soros Foundation System), Grant No. 228/1995. The rate of return was 252 questionnaire forms (31%); sex-age-specific characteristics of the array of respondents were representative of the rural population of Western Siberia.

REFERENCES

Chayanov, A. V. 1925. *Organizatsiya krest'yanskogo khozyaictva*. Translated as *The Theory of Peasant Economy,* edited by D.Thorner, R. E. F. Smith, and B. Kerblay. 1966. Burr Ridge, IL: Irwin.

Fei Hsiao Tung. 1987. "Peasantry as a Way of Living." Pp. 57-59 *in Peasants and Peasant Societies. Selected Readings*, 2nd ed., edited by T. Shanin. New York: Basil Blackwell.

Festy, P. 1971. "Evolution de la Nuptialit en Europe Occidentale, Depus la Guerre." *Population* 2.

Gromyko, M. M. 1991. *Mir Rossiiskoi derevni(The World of the Russian Countryside)*. Moscow: Molodaya Gvardia.

Hajnal, J. 1953. "Age at Marriage and Proportions Marring." *Population Studies* 2: 125-156.

{3em}. 1965. "European Marriage Patterns in Perspective." Pp. 98-154 in *Population in History,* edited by D. V. Glass and D. E. Eversley. London.

Henry, L. 1966. "Perturbations de la nuptialite resultant de la guerre 1914–1918." *Population* 2: 273-315.

Kontula, O., and E. Haavio-Mannila. 1995. *Sexual Pleasure. Enhancement of Sex Life in Finland, 1971-1992*. Andershot, Brookfield: Dartmouth.

Lesthaege, R. 1992. *The Second Demographic Transition in Western Countries: An Interpretation.* Brussels: IPD Working Paper.

Lewis, O. 1966. *La Vida*. New York: Random House.

{3em}. 1975. *Five Families: Mexican Case Studies in the Culture of Poverty.* London: Souvenir Press.

Liutsidarskaya, A. A. 1992. *Starojily Sibiri. Istoriko-Etnograficheskiye Ocherki, XVII–nachalo XVIII v.* (*Old Residents of Siberia. Historical-Ethnographic Essays, XVII–Beg. XVIII*). Novosibirsk: Nauka.

Lorber, J. 1994. "Gender kak socialnaya kategoriya" ("Gender as a Social Category"). *Tezis*(*Thesis*) 6: 127-136.

Maddock, J. U. 1995. "Semeinaya cshizn I seksualnost" (Familiy Life and Sexuality"). Pp. 86-109 in *Sem'ya na Poroge Tret'yego Tysyazcheletiya* (*Family at the Turn of the Third Millennium*) edited by A. A.Antonov, M. S.Matskovski, J. U. Maddock, and M. J. Hogan. Moscow: Institute of Sociology, RAN.

Mikheyeva, A. R. 1993. *Selskaya sem'ya v Sibiri: Cshiznennyi Tsikl I blagosostoyaniye* (*Rural Family in Siberia: Life Cycle and Well-being*). Novosibirsk: IEIE SB RAN.

Mironov, B. N. 1977. "Traditsionnoye demodraficheskoye povedeniye krest'yan v XIX–nachale XX v" ("The Traditional Demographic Behavior of Peasants in the XIXth–Beg. XXth Century"). Pp. 83-104 in *Brachnost, Rocshdayemost, Smertnost v Rossii I v SSSR* (*Marriage, Birth, Death Rates in Russia and in the USSR*), edited by A.Vishnevski. Moscow: Statistika.

Peasants and Peasant Societies. Selected Readings, 2nd ed. Edited by T. Shanin. New York: Basil Blackwell.

Popenoe, D. 1993. *Sociology* (9th ed.). Englewood Cliffs, NJ: Prentice Hall.

Shanin, T. 1987. "A Peasant Household: Russia at the Turn of the Century." Pp. 21-27 in *Peasants and Peasant Societies. Selected Readings*, 2nd ed., edited by T. Shanin. New York: Basil Blakwell.

Thomas, W. I., and F. Znaniecki. 1971. "A Polish Peasant Family." Pp. 16-20 in *Peasants and Peasant Societies. Selected Readings*, 2nd ed., edited by T. Shanin. New York: Basil Blakwell.

Vishnevski, A. G. 1977. "Ranniye Etapy Stanovleniya Novogo Tipa Rocshdayemosti v Rossii" ("Early Stages in the Establishment of a New Type of Natality in Russia"). Pp. 105-134 in *Brachnost, Rocshdayemost, Smertnost v Rossii I v SSSR* (*Marriage, Birth, Death Rates in Russia and in the USSR*), edited by A. Vishnevski. Moscow: Statistika.

Vishnevski, A. G., and M. S. Toltz. 1988. "Evolutsiya Brachnosti I Rocshdaemosti v Sovetski Period" ("Evolution of Marriage and Birth Rates in the Soviet Period"). Pp 75-114 in *Naseleniye SSSR za 70 let* (*The Population of the USSR over 70 years*), edited by L. Rybakovski et al. Moscow: Nauka.

HOMOSEXUAL RELATIONSHIPS IN THE NETHERLANDS

Paul Vlaardingerbroek

INTRODUCTION

In the last half of the twentieth century, family law and civil rights in Western countries have undergone considerable change as a consequence of societal changes. One of the most important changes concerns the legal recognition of equal rights between (married) women and men. Also noteworthy is the decrease of difference in legal treatment between married and cohabiting unmarried couples. Implementation of social security benefits follows the same general legal rule.

Equal legal treatment between married and unmarried couples does not imply that all differences have disappeared (for example, in matters of old age pension and adoption). Successive governments are tending, however, to wipe out differences in treatment on the basis of sex or the legal status of the cohabitation arrangement. There is a preference for individualization of family law, which must give tailor-made discretionary solutions that suit the individual and the individual families. Nowadays one of the greatest challenges in family law policy is how to achieve suitable solutions and regulations for de facto marriages. This is especially difficult for homosexual couples with children. In this paper, I provide information about the demographic situation in the Netherlands and deal with the new proposals in the field of family law for couples living outside marriage.

DEMOGRAPHIC CHANGES IN THE NETHERLANDS

The Netherlands is a very small country (41,548 sq km) that lies on the edge of the northern European plains. It is a flat country; only in the most southern regions does one find some hills (the highest one is 321 meters). The Netherlands consists of a delta of three main rivers and it is one of the most densely populated countries in the world. More than 15 million people live on those 42,000 square kilometers (thus more than 350 people per sq km). In 1998, 86,956 marriages and 32,459 divorces were inscribed at the registrar's office. To give some more figures, in 1998, 199,408 children were born alive; 41,439 were born outside marriage. The total number of people living together outside marriage was 1,012,000; of these 924,000 were living with (their) children while 196,000 were living in an arrangement without children. In the past two decades, a large number of immigrants have arrived mainly from former Dutch colonies (Surinams and Antillians), and from the Mediterranean area (particularly from Turkey and Morocco). The past few years have also seen a growing number of refugees (from Zambia, Ruanda, and Yugoslavia) who have settled in the Netherlands.

CONCEPT OF FAMILY LAW

In the Netherlands, "family" and "family life" are not limited to traditional concepts of the family. Apart from the traditional family of a man and a woman who are married and have children, the notion of a "family" is also used to describe *other primary living units in which the care and upbringing of children takes place;* for example, single-parent families (predominately mothers with children), or unmarried persons living together, of the opposite sex or of the same sex. Regardless the composition of the unit, as soon as there are children to be reared, in the Netherlands it is called "a family."

In the Netherlands, there are already more than 1 million single-parent families and there is a widespread tolerance for unmarried couples. As for people of the same sex who live together, more than 20,000 homosexual and lesbian couples are living together on a durable basis, also caring for children. However, while in practice these gay families are fully accepted, in the Dutch regulations they are still discriminated against as compared to heterosexual couples.

In the opinion of the Netherlands government, family life is considered to be an *absolutely private affair.* So, the government does not interfere with the way in which people shape their way of life or their family life, *unless there is a very good reason like the interest of the children* (e.g., if parents maltreat or abuse their child[ren]).

In Dutch jurisprudence, especially with regard to Article 8 of the European Convention on Human Rights, there is a growing tendency to avoid discrimination against alternative living arrangements. In many cases the legal consequences of marriage (or divorce) have been used for non-marital relationships.

A HOMOSEXUAL MARRIAGE?

According to Dutch law a man can be married to only one woman at a time; this monogamy rule also applies for women. Whereas many people do not wish to marry, there are also people who would like to but for whom this option is barred by law. Here I refer to people of the same sex. Article 1:33 BW (BW is the civil code) concerns the sex of the potential partners: "a man can at the same time be married only with a woman, a woman only with a man." Various attempts have been made to convince the judge that, on the basis of the European Convention for the Protection of Human Rights, homosexuals, too, are entitled to contract a civil marriage with each other.

Relying on Article 12 read in conjunction with 8 and 14 ECHR, two women requested the judge to oblige the county clerk to perform a marriage ceremony (Hoge Raad, October 19, 1990, NJ 1992, 129). The Hoge Raad decided that the basic assumption of the Dutch statutory system would be undermined if a marriage between two persons of the same sex were allowed. Pointing to the protected rights of Article 8 ECHR, the Hoge Raad underlined the possibility that any man or woman might marry. Furthermore, it held that Article 12 and also Article 23 ICCPR are intended to protect the traditional marriage between two persons of the opposite sexes. "One of the characteristics of a marriage is that the partners are physically capable of having sexual intercourse in such a way that it could bring about sexual reproduction." In its conclusions, the Hoge Raad pointed to the judgment of the European Court of Human Rights (October 17, 1986, NJ 1987, 945) in which Article 12 ECHR is cited as being concerned only with the traditional marriage of two persons of different sexes. The Hoge Raad opened a door to possible new developments in future, by suggesting that it could constitute an inadequate justification simply to link a specific legal consequence to marriage and not to two persons of the same sex, living together. The declining of these applications was justified, I think, because the Convention is based on traditional matrimony joining a man and a woman. The judge does not have any say in this matter. It is up to the legislature to decide upon the legalization of marriage for homosexuals.

Since this decision in 1990 of the Dutch Supreme Court, some local authorities have given the opportunity to gay and lesbian couples to register their cohabitation in a so-called marriage record. Although this record does not give any legal status to their domestic arrangements (or attach any legal consequences), homosexual partners sign up because of the statement this makes to the outside world. Although such a registration fulfills the couple's desire for acknowledgment of their relationship, for most homosexual couples this registration does not go far enough. Actually, although this registration also has a function in some regulations (renting a house, pension rights, tax law of succession), there still is discrimination toward homosexual couples as compared to heterosexual couples who have the opportunity to marry with many legal consequences (rights and duties) for the man and wife.

Same-sex marriages are not permitted. During the discussions in the Dutch Parliament on the new bill regarding the registration of couples, most members of Parliament pleaded for the opening of marriage to homosexual couples. The government said that this would cause too many international legal problems (such as the recognition of this "marriage" by other countries, problems with regard to the affiliation of the children born within the marriage, and adoption). To meet the wish of the majority (the Christian parties are against same-sex marriages), Parliament created a commission of experts for advice. The report of the commission was sent to the government in 1997. So until the legislature has the political will to enable homosexuals to marry, gay families have to obtain their rights by way of other regulations, such as by contractual agreement, registration of their partnership, modern law on parental authority, and so forth. In the following pages I discuss these possibilities, including the new act regarding the registration of partnership.

REGISTERED PARTNERSHIP

In the early summer of 1997 the Dutch Parliament accepted the bill with regard to registered partnership. This act was to take effect on January 1, 1998 (Staatsblad 1997, 325). According to this legislation, two (adult) persons will e able to enter into partnership through a registration at the registrar's office. The registrar must investigate whether all requirements are satisfied (such as that none of the partners is married to, or has a relationship with another person; that both partners are capable to perform legal acts, etc.). The rules regarding investigation in the case of marriage are almost the same as in the case of a registration (except of course the heterosexual criterion). Registration will not only be possible for Dutch nationals, but also for European nationals who lawfully reside in the Netherlands (or who have been lawfully resident for at least one year immediately preceding the application).

The legal consequences of this registration include most of the provisions concerning marriage, such as a mutual duty to support each other, the same system of matrimonial property, mutual rights of inheritance, and the ability of the partners to assume the other's family name. The same applies to rules concerning pension rights, safeguards for widows and widowers, and other social benefits. With regard to these social benefits, a cohabiting couple generally has the same rights as a married couple. The same applies to tax law. However, some legal consequences of marriage are excluded in the registered partnership. In case of a registration of the partnership, the law of affiliation and the law of adoption are excluded. So, when a heterosexual couple has been registered and the woman gives birth to a child, in order to establish a legal filiation link with the child, the father has to recognize the child. Another consequence of this exception will be that a homosexual or lesbian couple still cannot adopt a child, although one of the partners can adopt a child alone (the child would then, of course, be educated by both). And if two women register themselves as partners and one of gives birth to a child,[2] the girlfriend of the mother will not be the legal parent of the child (second mother or "step-mother"). If the "second mother" wants to have some rights with regard to the child, there are few possibilities (except access rights in case there has been "family life" between the applicant—the second mother—and the child).

JOINT PARENTAL AUTHORITY

From January 1, 1998 on, the parent who has the parental authority of the child and his or her partner[3] can ask the judge to give joint "parental" authority to both. Of course, the parent with parental authority must agree. Unmarried partners who have never held joint custody can already obtain it simply by applying to the civil status registrar and requesting registration as joint parents (Art. 252 Dutch Civil Code). According to the new Act, it will also be possible for homosexual couples to apply for a custodian order.[4] A person, not the legal parent, who is living with and sharing the care for a child, and has a "close personal relationship" with the child can apply for a custodianship order (the so-called *gezamenlijk gezag*). This custodianship order is almost equivalent to the consequences of joint custody. Where there is a legal parent without custody, the request for custodianship will only be granted if the custody order has been in force for at least three years (e.g., the parents were divorced and the father did not obtain parental authority). Before the judge will decide upon the request, the other parent will be invited to give his or her opinion to the judge. Even if the custodianship order has been made, the parent without custody will have the rights of visitation, of information, and consultation about the education of the child. If the relationship ends by separation, both partners can ask the judge for custodianship, even if only one of them is the biological and legal parent of the child. The judge will then have to decide which of the partners will obtain custodianship, based upon the best interests of the child, of course. In case the partnership ends by the death of the custodial parent, the other custodian becomes the sole custodian. The noncustodial parent, however, has the right to apply for another custody order to obtain custody himself or herself on the grounds of a change in circumstances. If one or both custodians want to terminate the joint custodianship, they may apply for this. The custodianship will also end if the judicial parent's right of custody ends. This may occur when parental rights have been taken away from the custodian (e.g., in case of maltreatment or sexual abuse of the child), or when custody is given to the other legal parent.

In the case of a child who is educated by a foster parent, the educator may apply for the custodianship *(voogdij)*. If the child is educated by two foster parents, both may apply for a custodianship order (the so-called *gezamenlijke voogdij*). This joint custody order can be given to married or non-married (heterosexual or homosexual) foster parents. Of course, the parent(s) of the child and the child himself or herself (when he or she has reached the age of 12 years) will be heard by the court before the court reaches a decision.

With this new Act, the Dutch government hopes to satisfy persons who plead for more rights for the men and women who are responsible for the daily care and rearing of the child. And, as already mentioned, the Dutch government also wants to take away discriminatory rules, especially because of the impact of Articles 8, 12, and 14 of the European Convention on Human Rights. Of course the Act has already been criticised. Homosexual and lesbian groups say that, although they can apply for a joint custody order, this still is not the same as joint parental authority. In my opinion, however, the difference between "parental authority" and the "custodian order" is very small. They also say that the best thing the Dutch government can do is to allow gay people to marry.

OBJECTIONS TO HOMOSEXUAL MARRIAGES

The question is, of course, why gay marriage is not allowed. It is often said that this is not possible because of legal aspects, or that it would undermine the meaning of marriage: the procreation of kinship. Others say that it is immoral or prohibited by religious traditions. With regard to the legal objections, I think that it is very difficult to continue saying that it is legally impossible to allow gay lovers to marry. Of course, it is often said that this prohibition goes back to the early days of humankind, but at some point in the past, marriage as a heterosexual relationship was coded into law. Western societies also forbid polygamy while other cultures allow it. In the Scandinavian countries and some western European countries, there is a slight move toward a more open minded attitude about homosexuality. However, even in these countries, gay persons still face discrimination.

The objection that marriage is meant for the procreation of children can no longer be supported. Love for each other is the basic element for partnership or marriage, or both, although financial motives can play an important role, too. Fortunately the elderly are permitted to marry, even if it will be impossible for them to procreate. And who would argue that a marriage should be annulled because the man or wife (or both) are infertile? Of course, one may have ethical or religious objections to homosexuality and especially to gay marriage, but this is not a legal argument. In my opinion, the only two arguments that sincerely count are these:

- A Dutch homosexual marriage would not be recognized outside the country. However, this is more a problem for foreign law, and for the couple itself, than for Dutch law. As long as the couple lives in the Netherlands this objection does not apply.

- There are currently legal consequences for children born inside a homosexual marriage. This is a serious problem, because the Dutch affiliation law says that when a child is born within marriage, the legal father of the child is the husband of the mother. So, if this law were changed it would not be a barrier for the gay marriage (e.g., by making a rule that a child can only have one legal mother and one legal father. The same can be said with regard to adoption: a child can only be adopted by a man (and eventually by his female partner) or by a woman (and her male partner).

The solution for the legislation is, of course, not as simple as I have suggested. Firstly, if it would be the wish of the government to enable homosexual and lesbian persons to marry, new rules would have to be made and the entire legislation verified, but this work would not be impossible. Secondly, legislation can only be changed if there is a democratic majority in Parliament, and although a majority of the Second Chamber in Parliament has asked the government for the opening of marriage to homosexual partners, I am not sure whether such a bill would gain the majority in the First Chamber.

The political willingness to carry out such a dramatic modern change of the marriage law is, in my opinion, the only obstruction to gay marriage—not the legal, ethical, or religious objections. The wish of two homosexual partners to be allowed to marry cannot simply be dismissed by an application to culture and tradition. As matrimony loses its distinctive quality compared to other modes of living, it seems necessary to examine the issue of why marriage has to remain forbidden territory for homosexuals. I do not mean to suggest here that Dutch society is now ripe for homosexual marriage. Meanwhile, however, a radical change in the legislation would be preferable. This could be an arrangement such as that introduced in Denmark, where a number of provisions in the legislation on matrimony have been declared applicable to registered homosexual relations, such as the duty of maintenance, the law regarding matrimonial property, and pension provisions. Joint responsibility for children should be another consequence, in my opinion.

ADOPTION FOR HOMOSEXUALS

According to Dutch legislation, only married partners have the right to adopt a child. A request for adoption by an unmarried man or woman cannot be accepted by the courts. This topic, like the previous ones, is the subject of discussions and debates. An important question in this matter is whether the rules on adoption have to be changed in such a way that one man or one woman can adopt a child, or that two unmarried persons (heterosexuals or homosexuals) can adopt a child.

In my opinion, a child can be nurtured in any primary-life form. This need not, however, imply that the relationship between the caregivers and the child should be legalized through adoption. The discussion is, however, far from over.

Three different tendencies are reflected. The first one is that adoption should be open to other than married couples, provided that the adoptive parents are of different sexes. According to a second point of view, adoption should also be possible for cohabiting partners of the same sex. The third tendency opposes the principle of adoption because, through adoption, legal family rights are broken and replaced by those of parents who are not genetically kin. As already noted, the Dutch government will allow singles to adopt a child. So, a part of the problem will be solved.

Nowadays children are cared for and socialized in a variety of settings that are very different from the nuclear family environment. Examples of these settings are: stepfamilies based or not on a marriage bond, children who live with foster-parents, children born or brought up in homosexual or lesbian relationships, children brought up by nonkin single parents, and children procreated or born through modern insemination methods. The expression "social parenthood" has been coined in the Dutch language *(sociaal ouderschap)* to cover these different ways of bringing up children.

Adoption especially concerns persons who, at a later time, establish a relationship with the child, whether this relationship is combined with legal or social parenthood, or stands alone. They have "taken the responsibility to take a child that has already been born." This category of social parenthood is especially relevant for stepfamilies that derive from either marriage or cohabitation. I, myself, do not favor full adoption *(adoption plenière),* because the legal ties between the child and its parent(s) are cut. In my opinion, the child can be raised by others, but they should not become his or her legal parents, because the child already has legal parents. But, in cases where the child does not have legal parents (i.e., a foundling) adoption should be possible. In other cases, the educators must have the same duties and rights as "normal" parents, but they will not become the child's legal parents. For them, the term "social parenthood" could be introduced. In this matter of social parenthood the following questions are relevant:

- Which legal relationships merit consideration in the case of social parenthood, and what should their underlying principle be? Should a legal connection be made between parental authority and an obligation to provide for the child?
- Does the child become a party to a legal relationship, too, through the legal relationships binding the social parent(s)?
- Which legal procedures should be followed?
- Could legal relationships be initiated by the child as well?

The following three recommendations can be made:

1. Whether it is desirable or not for the child whose social relationship to an adult derives from legal rights to also claim rights from the family(ies) of the social parent(s), by either legal or other means, should be investigated.
2. It should be recommended that the legal consequences of a relationship—which has not as yet a legally binding character—between an adult declaring the intention to care for the child and the child to be procreated should be investigated. According to the law now in existence, the legal consequences are: authority, obligation of support, name, nationality, and right of contact with the child. To these are added the so-called prescriptions by extension of legal rights, such as the right to give consent to marriage, and the right to be called by the judge when a request for custodianship is filed by a relative.
3. It is desirable to further examine some of the possible consequences. For example, the concept of "legal kinship" could be replaced by that of "legal parenthood." A "parent" would then be the person who made known his or her determination to take the responsibility for the child upon himself or herself. Legal parenthood would thus be separated from the concepts of biological and legal kinship. Legal rights would also be attached to parenthood.

Adoption with the present legal consequences (the forming of new legal ties with the adoptive parents and the cancellation of the former ties with the biological parents) is unnecessary, because by giving more legal consequences to social parenthood, almost the same results are possible. So, in my view it is not necessary to change the law of adoption only to make it possible for homosexuals to adopt a child. By step-parenthood or the so called "social parenthood," they can obtain almost the same legal consequences (starting January 1, 1998) because of the introduction of the custodianship order.

CONCLUSION

As the foregoing discussion has noted, there is a growing tendency in Dutch society to give more rights to homosexuals. In this paper I dealt only with the rights of homosexuals in family life, especially with regard to the right to marry and to build a family. In my opinion, the wish of two homosexual or lesbian partners to marry cannot simply be dismissed by an application to culture and tradition. I believe that the introduction of the registration of homosexual partnerships will only be an interim solution and that, slowly but surely, we are on our way to eliminating discriminatory rules with regard to marriage. Meanwhile, some discrimination can be removed by declaring a number of provisions in the legislation on matrimony applicable to registered homosexual relations, such as the duty of maintenance, the law regarding matrimonial property and pension provisions, the law of succession, and joint custodianship in cases where children are educated by two men or women.

After having finished my paper, the government sent two bills to Parliament. In July 1999 a bill regarding the opening of marriage for homosexual couples (TK26672) and a bill regarding adoption by homosexual couples (TK26683) were sent to Parliament. Unfortunately these bills could not be commented on in this paper.

NOTES

1. This nondiscriminatory approach is completely in harmony with Rec. No R (88) 3, adopted within the Council of Europe. The purpose of this recommendation is to ensure that contracts regulating matters relating to the property of unmarried couples are not considered to be invalid solely because the persons are living together as an unmarried couple.

2. Married couples and cohabitants are offered artificial insemination in hospitals, whereas single or lesbian women may obtain insemination by a donor in some hospitals or private clinics, too.

3. Including the married partner: the stepparent.

4. A heterosexual partner (not being the legal parent) of the parent can also ask for a custodian order.

INDEX